Mission Analysis and Design of Lighter-than-Air Flying Vehicles

Mission Analysis and Design of Lighter-than-Air Flying Vehicles

Guest Editors

Alberto Rolando
Carlo E. D. Riboldi

MDPI

Basel • Beijing • Wuhan • Barcelona • Belgrade • Novi Sad • Cluj • Manchester

Guest Editors

Alberto Rolando
Department of Aerospace
Science and Technology
Politecnico di Milano
Milano
Italy

Carlo E. D. Riboldi
Department of Aerospace
Science and Technology
Politecnico di Milano
Milano
Italy

Editorial Office

MDPI AG
Grosspeteranlage 5
4052 Basel, Switzerland

This is a reprint of the Special Issue, published open access by the journal *Aerospace* (ISSN 2226-4310), freely accessible at: https://www.mdpi.com/journal/aerospace/special_issues/lighter_than_air_flying_vehicles.

For citation purposes, cite each article independently as indicated on the article page online and as indicated below:

Lastname, A.A.; Lastname, B.B. Article Title. *Journal Name* **Year**, *Volume Number*, Page Range.

ISBN 978-3-7258-4675-7 (Hbk)
ISBN 978-3-7258-4676-4 (PDF)
https://doi.org/10.3390/books978-3-7258-4676-4

Cover image courtesy of Carlo E. D. Riboldi

Contents

About the Editors

Alberto Rolando

Alberto Rolando is a researcher and educator in the field of aeronautical engineering, with a career spanning over two decades at the Department of Aerospace Science and Technology, Politecnico di Milano. His expertise encompasses flight testing, avionics systems, hybrid-electric aircraft technologies and the development of innovative teaching tools for aerospace education. He is the inventor and designer of Mnemosine, an advanced modular flight test instrumentation system that has been deployed in over 300 test flights across seven different aircraft models. This system has been an integral part of the hands-on training in the Flight Testing course of Politecnico di Milano, providing students with the opportunity to engage with real-world flight data acquisition and analyses. He has been actively involved in several high-impact EU-funded research initiatives aimed at advancing sustainable aviation technologies. His portfolio includes journal and conference papers as well as patents. Combining technical expertise with a passion for education, Alberto Rolando serves as a vital link between academia and industry, fostering the next generation of aerospace engineers.

Carlo E. D. Riboldi

Carlo E. D. Riboldi holds a PhD in Aerospace Engineering from Politecnico di Milano (2012), where he is currently a professor within the Department of Aerospace Science and Technology. Since 2014, he has been a principal lecturer in undergraduate and graduate courses on flight dynamics and aircraft design. His recent research activities include design methodologies for aircraft featuring novel propulsion systems, accurate dynamic simulations, and practical automatic flight control designs for diverse flying craft, including missiles and airships. He has authored and co-authored approximately 100 journal and conference papers in his research areas, as well as two books on aircraft dynamics. He has participated in a number of funded research projects in the field of fixed-wing, rotorcraft, and lighter-than-air aviation. A lifelong aviation enthusiast, with a special interest in the history of aeronautics, he got a pilot's license in 2013, and he currently flies in the general aviation category.

Preface

Over the last few years, the field of lighter-than-air (LTA) technology has been the focus of significant revitalization. Researchers and manufacturers have started investing again in airship and balloon concepts, mostly attracted by the naturally better endurance typical of these platforms when compared to other flying machines, like fixed-wing aircraft or rotorcraft. Furthermore, the significant power-to-weight performance of novel power systems based on electric energy has been a motivation for exploring new topologies, featuring, for instance, distributed propulsion or solar cells to resupply on-board energy storage. The adoption of such new design architectures has starkly increased the ability of novel machines to tackle intrinsic controllability and flight management issues typical in lighter-than-air machines, thus fostering the study of radical configurations, for example, without tails or with non-slender envelopes. On the one hand, these technological developments are starting to demonstrate their potential in relation to airship technology, especially through prototypes and early production models, thus causing the research in this field to gain further momentum. On the other hand, however, this quick-paced growth of LTA technology requires more extensive knowledge compared to what was required in the past, especially when it comes to the design, manufacturing, and characterization of a new lighter-than-air machine. For example, starting from the design phase, new components typically associated with the propulsion and energy-management system need to be produced with dedicated procedures. The aforementioned topology of a new airship needs to be negotiated, often mediating contrasting requirements relating to weight, simplicity, and control effectiveness.

In this perspective, the present Special Issue offers the reader a summary of the many and diverse aspects of the research within the field of lighter-than-air technology, allowing them to determine the state of the art at the time of writing and helping them identify promising areas for future research. As the field is currently very active, this Special Issue explores a broad range of topics. Taken together, these works constitute a useful knowledge base on LTA technology, following its evolution in this peculiar time of its evolution.

Alberto Rolando and Carlo E. D. Riboldi

Guest Editors

aerospace

MDPI

rticle

Multidisciplinary Optimization of Thermal Insulation Layer for Stratospheric Airship with a Solar Array

ang Liu, Ziyuan Xu, Huafei Du * and Mingyun Lv

School of Aeronautic Science and Engineering, Beihang University, Beijing 100191, China; liu312@buaa.edu.cn (Y.L.); zb2005006@buaa.edu.cn (Z.X.); lv503@buaa.edu.cn (M.L.)
* Correspondence: duhuafei@buaa.edu.cn

Abstract: Stratospheric airships with a solar array have demonstrated overwhelming superiority in many aspects, such as earth observation, meteorological survey, and communication relay. The solar array supplies sufficient power for the airship to be in flight for months, but excessive heat is also transferred to the airship, causing high overpressure of inner gas. However, the optimal arrangement of the insulation layer on the airship has not been investigated. The theoretical method, including the geometry, thermal, and energy models, is developed and validated. The distribution of the temperatures and power of the solar cells, with different installation angles on the curved surface, is investigated. The thickness of insulation layer has a significant effect on the solar output power and internal pressure of the airship. An optimized configuration of the insulation structure is proposed, in order to improve the total output energy of solar array. The optimized configuration of insulations helps to reduce the structural mass by 24.9% and increase the payload mass by 9%. Moreover, the optimized arrangement improves the output energy of solar array in a year, and the maximum improvement is 8.2% on the winter solstice. The work displays the optimization of the thermal insulation layer for the stratospheric airship with a solar array, in order to improve the everyday energy acquirement during flight in a year.

Keywords: stratospheric airship; solar array; thermal insulation layer; multidisciplinary optimization; long endurance; renewable energy

1. Introduction

A stratospheric airship, also known as a pseudo-satellite [1], is an ideal platform for high-resolution observation, atmospheric monitoring, and uninterrupted communication [2]. Long-term flight in the stratosphere for several months requires adequate energy. The development of flexible, thin-film materials and conversion efficiency makes solar cells an essential part of renewable energy system on this long endurance vehicle [3].

Research progress, regarding the utilization of the photovoltaic (PV) array on the stratospheric airship, has been spectacular during the past one or two decades [4,5]. Garg et al. [6] proposed a method to estimate the incident solar energy on the airship hull profiles. The arrangement and area of solar panel was designed based on required energy per day. Wang et al. [7] studied the variation of the output power of solar cells with the heading direction of airship on the summer and winter solstice. Appropriate flight directions, at different times of the day, were advised when the cell efficiency was low. Alam and Pant [8] developed a method to optimize the layout of the PV array in different seasons. The area of the solar cell array can be minimized on the premise of the energy requirements for specific days. Li et al. [9] presented a numerical model to simulate the output power of the solar array with the thermal effect. The output power of solar cells was overestimated when the excessively high temperatures of them were neglected. Du et al. [10] investigated the curve surface of the airship and optical paths on different solar cells. The output power and total energy were overestimated, without consideration of angular losses.

With the in-depth study on predicting the energy conversion of solar array on the stratospheric airship, scholars are conscious of the excessive rise in internal temperature, caused by the considerable irradiation absorbed by photovoltaic cells [11,12]. Wang and Yang [13] conducted steady and unsteady simulations of two stratospheric airships, developed in Japan. The temperatures of inner gas and upper film for the airship with PV array were higher than those of another airship. Li et al. [14] pointed out that the PV array may the aggravate superheat of the airship at noon because the solar absorptivity of the PV array was higher than that of the airship hull. They also stated that if the emissivity of the PV array was higher than that of the airship hull, the temperature of the upper part of the airship hull would be lower than the bottom part at night, which was called "supercool". The variation scope of the temperature of lifting gas was expanded, so the structural strength of the airship was affected. Sun et al. [15] proposed a mathematical model to describe the solar radiation on the solar array. A three-dimensional temperature profile of airship with PV array was presented. The difference between the maximum and minimum film temperatures exceeded 20 K.

In order to reduce the significant impact of the PV array on airship temperature, a thermal insulation structure between cell units and outer film was adopted. The insulation structures of the solar array on the stratospheric airship refer to the same structure on the satellite [16]. Sun et al. [17] applied the honeycomb sandwich structure and analyzed its equivalent thermal conductivity. The temperature difference between the upper and lower surface was calculated and in good agreement with the experimental results. Li et al. [18] summarized the common layouts of the insulation structure. The temperature differences of insulation structure, with various combinations of parameters, under different irradiation conditions, were compared. Meng et al. [19] researched the optimal structural parameters of the photovoltaic module with the maximum temperature difference and output power. The independent variables were the thickness and interval of the insulation layer.

Up until now, the researches on insulation structure were concentrated on increasing the temperature difference between the flat solar module and airship film. Once the structural parameters were determined, they would be applied to the other hundreds of solar cell modules on the vehicle. The influence of different installation angles of solar cell units in the airship body reference frame was not considered [20]. Moreover, only the output power of a single solar cell, under different irradiation intensities, was studied. The distribution of output power of solar cell units with insulation layers during the flight process has not been evaluated. The output energy should be maximized by reasonable arrangement of the thermal insulation layers.

In this paper, the geometry model of the solar array on the curved surface of airship is firstly introduced. The thermal and power models of the solar powered airship with thermal insulation layer between photovoltaic array and film are developed and validated. The thickness of thermal insulation layer has a notable effect on the output energy of the solar array and maximum differential pressure of airship. The traditional and optimized configurations of the insulation layer are obtained. The results indicate that the optimized layout functions to decrease the structural weight of insulation layer and improve the capacity to accommodate payloads. The distributions of the temperature and output power of the solar cells with different configurations of insulations are analyzed and compared. The optimized arrangement of the insulation layer significantly increases the output energy throughout the whole year, without breaking the structural integrity of airship film. The multidisciplinary optimization study has valuable reference for reducing the structural weight and improving the everyday output energy of solar powered airships during flight in a year.

2. Theoretical Method

2.1. Spatial Geometry Model

Figure 1 shows the description of the spatial position and attitude of airship. The $O_g x_g y_g z_g$ and $O_b x_b y_b z_b$ are earth reference frame and body coordinate frame, respectively [21]. The axis $O_g x_g$ points to the north. The positive directions of axes $O_g y_g$ and $O_g z_g$ are east and downward perpendicular to the ground. As for the body reference system, the origin O_b is located at the airship stern. The $O_b x_b$ axis coincides with the geometrical longitudinal axis of the airship, pointing to the nose of the airship. The $O_b z_b$ axis is in the symmetrical plane and directs downward. The $O_b y_b$ axis is perpendicular to the $x_b O_b z_b$ plane and points to the right. The positive yaw motion moves the nose of the airship to the right. The positive pitch motion raises the nose of the airship and lowers the stern. The positive roll motion lifts the left side of the airship and lowers the right side.

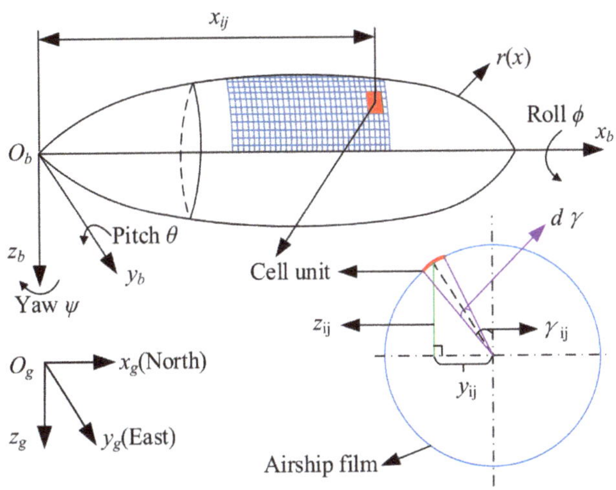

Figure 1. Schematic of airship with solar array.

The surface equation of the airship in the body coordinate frame is:

$$F = y^2 + z^2 - r^2(x) \tag{1}$$

where r is the rotation radius of the hull, according to standardization models, given by the National Physics Laboratory [22], and is written as:

$$r(x_b) = \begin{cases} \frac{b}{a_2}\sqrt{a_2^2 - (x - a_2)^2} & 0 \le x \le a_2 \\ \frac{b}{a_1}\sqrt{a_1^2 - (x - a_2)^2} & a_2 \le x \le a_1 + a_2 \end{cases} \tag{2}$$

where a_1 and a_2 are the lengths of the semi-major axis, and $a_2 = \sqrt{2}a_1$. b is the length of the semi-minor axis.

The upper limit of differential pressure is derived according to the mechanical property of the film of airship hull and can be calculated by:

$$\Delta P_{limit} = \frac{2 \cdot d_f \cdot \sigma_f}{r_{max}} \tag{3}$$

where d_f is the thickness of film, σ_f is the tensile strength of film, and r_{max} is the maximum radius of curvature of film.

The differential pressure of the airship is:

$$\Delta P = P_{he} - P_{atm} \tag{4}$$

where P_{atm} is the pressure of ambient air at the flight altitude [23], and P_{he} is the pressure of the buoyancy gas and can be obtained by:

$$P_{he} = \frac{m_{he} \cdot R \cdot T_{he}}{M_{he} \cdot V_{he}} \tag{5}$$

where m_{he} is the mass of helium, M_{he} is the molar mass of helium, R is the ideal gas constant, T_{he} is the temperature of helium, and V_{he} is the volume of helium.

The installation of the solar cells on top of the airship is described by the central angle γ and x_{ij} coordinate [24]. The other two coordinates of the center point of the cell unit in the body reference system can be derived by:

$$y_{ij} = r(x_{ij}) \cdot \sin(\gamma_{ij}) \tag{6}$$

$$z_{ij} = -r(x_{ij}) \cdot \cos(\gamma_{ij}). \tag{7}$$

The area of the solar cell unit is:

$$A_{ij} = r(x_{ij}) \cdot d\gamma \cdot dx \cdot \sqrt{1 + r'(x_{ij})^2}. \tag{8}$$

The normal vector of airship surface in the body reference frame is defined by:

$$\vec{n}_{b,ij} = \left(\frac{\partial F}{\partial x_{ij}}, \frac{\partial F}{\partial y_{ij}}, \frac{\partial F}{\partial z_{ij}} \right) / \left(\sqrt{\left(\frac{\partial F}{\partial x_{ij}} \right)^2 + \left(\frac{\partial F}{\partial y_{ij}} \right)^2 + \left(\frac{\partial F}{\partial z_{ij}} \right)^2} \right) \tag{9}$$

where $\frac{\partial F}{\partial x_{ij}}$, $\frac{\partial F}{\partial y_{ij}}$ and $\frac{\partial F}{\partial z_{ij}}$ are calculated by:

$$\frac{\partial F}{\partial x_{ij}} = \begin{cases} \frac{2 \cdot b^2 \cdot (x_{ij} - a_2)}{a_2^2} & 0 \le x_b \le a_2 \\ \frac{2 \cdot b^2 \cdot (x_{ij} - a_2)}{a_1^2} & a_2 \le x_b \le a_1 + a_2 \end{cases} \tag{10}$$

$$\frac{\partial F}{\partial y_{ij}} = 2 \cdot y_{ij} \tag{11}$$

$$\frac{\partial F}{\partial z_{ij}} = 2 \cdot z_{ij} \tag{12}$$

The normal vector of the cell unit in the earth coordinate frame is [25]:

$$\vec{n}_{ij} = R_b^g \cdot \vec{n}_{b,ij} \tag{13}$$

where R_b^g is the transformation matrix from the airship body coordinate system to the earth reference frame.

2.2. Thermal Environment

Figure 2 shows the thermal environment of the stratospheric airship with the solar array. The external solar radiation includes direct solar, diffuse solar, and albedo radiation. The external infrared radiation consists of atmosphere and ground infrared radiation [26]. A certain proportion of the solar irradiance can be converted to electrical power through the solar array, but the majority of the solar energy is transformed to heat as a thermal load on the airship. Convective heat transfer also exists between airship and external atmosphere [27]. The internal thermal environment is composed of infrared radiation of the film, inner gas infrared radiation, and convection between the film and internal gases [28].

2.3. Power Model

Figure 3 shows the photoelectric process of solar array on the airship. The total output power of solar array is determined by [29]:

$$P_{PV} = \sum_{i=1}^{m} \sum_{j=1}^{n} P_{ij} \tag{14}$$

where m and n are the numbers of units in the axial and circumferential directions, respectively.

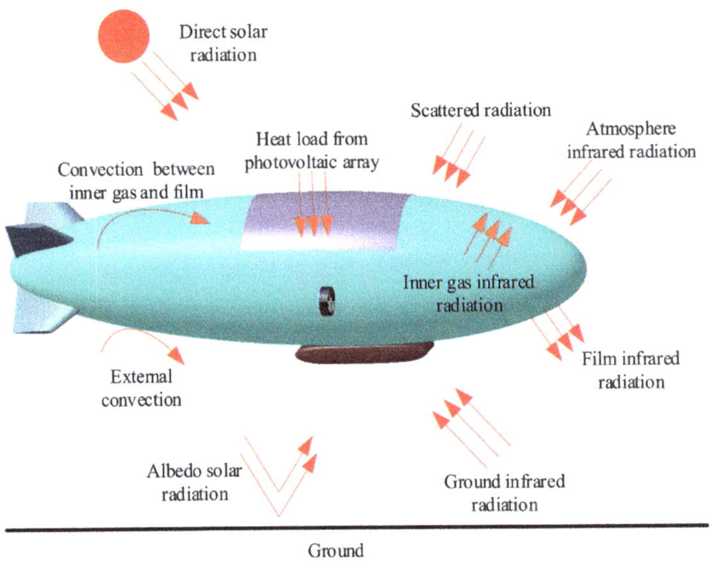

Figure 2. Thermal environment of stratospheric airship with solar array.

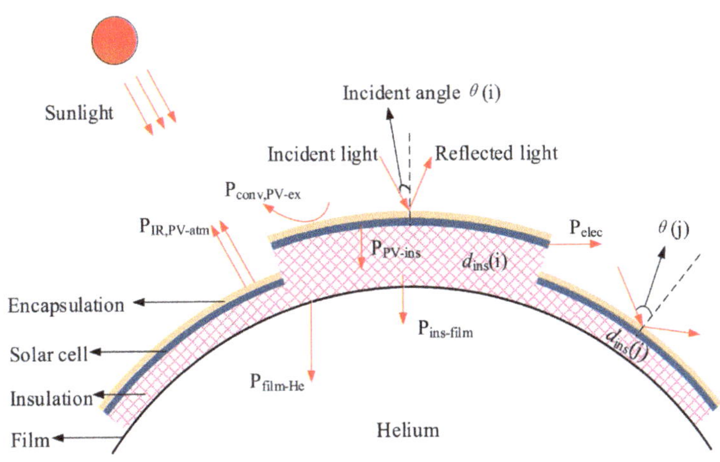

Figure 3. Photoelectric conversion process of solar cell units.

The output power of a single cell unit P_{ij} is:

$$P_{ij} = \eta_{ij} \cdot P_{eff,ij} \tag{15}$$

where η_{ij} is the conversion efficiency of the solar cell, as a function of temperature [30], obtained by:

$$\eta_{ij} = \eta_0 - |c_1| \cdot (T_{PV,ij} - T_0) \tag{16}$$

where η_0 is the conversion efficiency under standard test conditions, T_0 is the standard temperature of 298.15 K, and c_1 is the attenuation coefficient.

The effective incident radiation energy on the cell unit $P_{eff,ij}$ is:

$$P_{eff,ij} = \left(P_{D,ij} + P_{S,ij} + P_{R,ij}\right) \cdot \frac{\tau_{al}(\theta_{ij})}{\tau_{al}(0)} \tag{17}$$

where $P_{D,ij}$, $P_{S,ij}$, and $P_{R,ij}$ are direct, diffuse, and reflected radiation, and $\tau_{al}(\theta_{ij})$ and $\tau_{al}(0)$ are effective transmittance of encapsulation, with incident angles of θ_{ij} and $0°$, respectively. $\tau_{al}(\theta_{ij})$ is calculated by:

$$\tau_{al}(\theta_{ij}) = 1 - \frac{1}{2} \cdot \left(r_s^2(\theta_{ij}) + r_p^2(\theta_{ij})\right) \tag{18}$$

where $r_s(\theta_{ij})$ and $r_p(\theta_{ij})$ are the reflection coefficients of transverse-electric and transverse-magnetic [31].

The govern equation of the solar cell unit temperature $T_{PV,ij}$ is [32]:

$$m_{PV,ij}c_{PV} \cdot \dot{T}_{PV,ij} = P_{D,ij} + P_{S,ij} + P_{R,ij} - P_{ij} - P_{conv,PV-ex,ij} - P_{cond,PV-ins,ij} - P_{IR,PV-atm,ij} \tag{19}$$

where $m_{PV,ij}$ is the mass of photovoltaic array, c_{PV} is the specific heat of the photovoltaic array, $\dot{T}_{PV,ij}$ is the change rates of the photovoltaic array temperature, $P_{conv\_PV\_ex,ij}$ is the external convection heat transfer of the unit, $P_{cond\_PV\_ins,ij}$ is the conduction heat transfer energy between the solar cell unit and insulation layer, and $P_{IR\_atm\_PV,ij}$ is the absorbed infrared radiation from the sky.

The direct solar radiation on the unit is:

$$P_{D,ij} = \mu_{ij} \cdot \alpha_{PV} \cdot I_D \cdot A_{ij} \tag{20}$$

where α_{PV} is the absorptivity of solar cell, A_{ij} is the unit area, and μ_{ij} is the discount factor of direct radiation, defined by:

$$\mu_{ij} = \begin{cases} \left|\vec{n}_{ij} \cdot \vec{n}_S\right| & \vec{n}_{ij} \cdot \vec{n}_S < 0, -\arccos(\frac{R_{Earth}}{R_{Earth}+h}) < \theta_h < \pi + \arccos(\frac{R_{Earth}}{R_{Earth}+h}) \\ 0 & else \end{cases} \tag{21}$$

where R_{Earth} is the radius of the Earth, and $\vec{n}_S$ is the direction vector of sunlight in the earth reference frame and can be expressed by:

$$\vec{n}_S = (\cos\theta_h \cos\theta_{azi}, \cos\theta_h \sin\theta_{azi}, \sin\theta_h) \tag{22}$$

where θ_{azi} is the azimuth angle of sun.

The absorbed diffuse solar radiation from the sky of the cell unit is:

$$P_{S,ij} = \frac{1}{2} \cdot \left(1 - \cos\theta_{z,ij}\right) \cdot \alpha_{PV} \cdot I_S \cdot A_{ij} \tag{23}$$

where $\theta_{z,ij}$ is the included angle between the normal vector and gravity acceleration direction $\vec{n}_z = (0,0,1)$ [33], calculated by:

$$\theta_{z,ij} = \arccos\left(\frac{\vec{n}_{ij} \cdot \vec{n}_z}{\left|\vec{n}_{ij}\right| \cdot \left|\vec{n}_z\right|}\right). \tag{24}$$

The reflected radiation on the unit is:

$$P_{R,ij} = \theta_{R,ij} \cdot \alpha_{PV} \cdot I_R \cdot A_{ij} \tag{25}$$

where $\theta_{R,ij}$ is the self-shadowing coefficient, defined by:

$$\theta_{R,ij} = \begin{cases} 1 & \vec{n}_{ij} \cdot \vec{n}_z \geq 0 \\ 0 & \vec{n}_{ij} \cdot \vec{n}_z < 0 \end{cases}. \tag{26}$$

The conductive heat transfer between PV unit and insulation structure is:

$$P_{cond,PV-ins,ij} = A_{ij} \cdot \left(\frac{T_{ij} - T_{ins,ij}}{d_{PV}/\lambda_{PV} + d_{ins,ij}/\lambda_{ins}}\right) \tag{27}$$

where λ_{PV} and λ_{ins} are the thermal conductivity of solar cell and insulation layer, and $d_{ins,ij}$ is the thickness of insulation layer. The conductive heat transfer between insulation structure and film obeys the same law as:

$$P_{cond,ins-film,ij} = A_{ij} \cdot \left(\frac{T_{ins,ij} - T_{film,ij}}{d_{ins,ij}/\lambda_{ins} + d_{film}/\lambda_{film}} \right). \tag{28}$$

The incident angle θ_{ij} for calculating the $\tau_{al}\left(\theta_{ij}\right)$, considering angular loss, is:

$$\theta_{ij} = \arccos\left(\frac{\vec{n}_{ij} \cdot \vec{n}_S}{\left|\vec{n}_{ij} \cdot \vec{n}_S\right|} \right) \tag{29}$$

2.4. Optimization Method of Insulation Layer Configuration

The incident solar radiation and temperatures of solar cells vary with their positions on the airship film. Therefore, the thicknesses of thermal insulations under each solar cell should be optimized to increase the total output energy of the solar array, especially when the airship flights with insufficient solar irradiance. As shown in Figures 1 and 3, the thicknesses of insulation layers $d_{ins}(\gamma, x)$ are expressed as the function of central angle γ and x coordinate in the body reference system, $O_b x_b y_b z_b$, on the curved surface of the airship.

The objective of the optimization is to obtain maximum output energy of solar array during a day and can be written as:

$$\max: E_{out} = \int_{t=0}^{t=t_{end}} P_{PV}(t)dt. \tag{30}$$

Considering the actual flight of stratospheric airship, four constraints are adopted, as follows:

Constraint I: Considering the carrying capacity of airship, the thicknesses of insulation layers cannot be infinity large, and the upper bound is determined as d_{max}.

$$d_{ins,ij} \leq d_{max} \tag{31}$$

Constraint II: The maximum differential pressure of airship, ΔP_{max}, during a day should not exceed the limit of differential pressure, ΔP_{limit}. The maximum hoop stress of the airship film will be greater than the tensile strength when the ΔP_{max} is larger than ΔP_{limit}.

$$\Delta P_{max} \leq \Delta P_{limit} \tag{32}$$

Constraint III: The center of gravity should be kept in the $O_b x_b z_b$ plane to ensure the lateral static stability of the airship [34]. The thicknesses of the insulation structures, at the symmetrical position, relative to the $O_b x_b z_b$ plane, should be the same.

$$d_{ins}(\gamma, x) = d_{ins}(-\gamma, x) \tag{33}$$

Constraint IV: In order to prevent the array from being overturned, the thicknesses of the insulation structures should be uniform in the direction of flight.

$$d_{ins}(\gamma, x_1) = d_{ins}(\gamma, x_2) \tag{34}$$

Based on the above constraints, the optimization process is simplified to search for the optimal thicknesses of the insulation layers, $d_{ins}(\gamma)$, with different central angles, γ. The particle swarm optimization method [35] is adopted to solve the problem, which uses a set of particles with the properties of location, l, and velocity, v. The location space of particles, L_m, is m dimensional and each dimension corresponds to $d_{ins}(\gamma_1), d_{ins}(\gamma_2), \ldots, d_{ins}(\gamma_m)$. The velocity space of particles, V_m, is m dimensional and each dimension corresponds to the change rate of $d_{ins}(\gamma_1), d_{ins}(\gamma_2), \ldots, d_{ins}(\gamma_m)$. The fitness value of the particle is the output energy, E_{out}, during a day. The update equations of the location and velocity for each particle are written by [36]:

$$x_i^{k+1} = x_i^k + v_i^k \tag{35}$$

$$v_i^{k+1} = \omega \cdot v_i^k + c_1 \cdot r_1\left(p_{best,i}^k - x_i^k\right) + c_2 \cdot r_2\left(g_{best,i}^k - x_i^k\right) \tag{36}$$

where k is the number of iterations in the range of $1, 2, \ldots, k_{end}$, ω is inertia weight, c_1 and c_2 are learning factors, r_1 and r_2 are uniform random numbers in the range of $[0, 1]$, p_{best} is the personal best location, and g_{best} is global best location. The optimization process of the insulation arrangement, using particle swarm optimization, is shown in Figure 4.

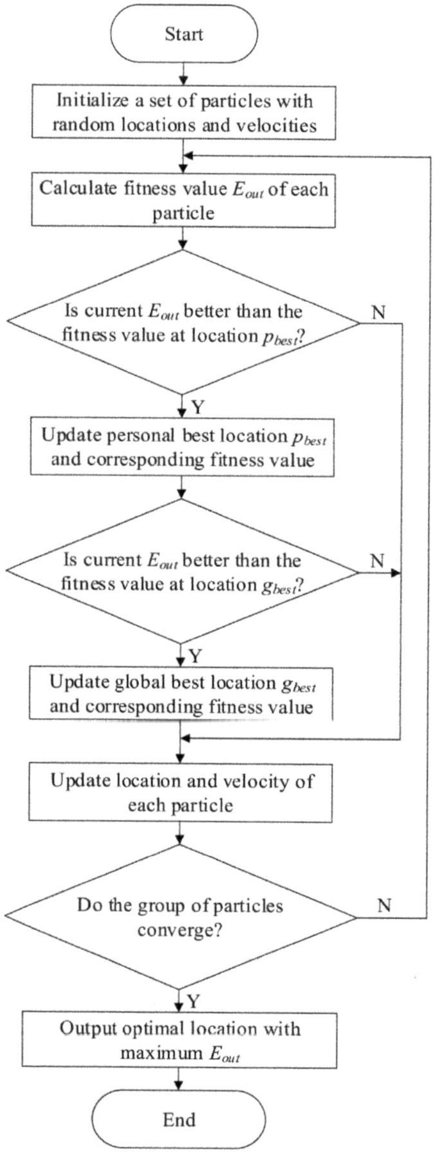

Figure 4. Flow chart of insulation configuration optimization.

3. Results and Discussion

3.1. Model Validation

In order to verify the theoretical method, the experimental result, recorded by Harada et al. [37], is used. The measured points of the maximum temperature of the solar array was from 8:00 am to 16:00 pm in June 2003. The simulated results were obtained under the same conditions. The comparison of the simulated and measured values is presented in Figure 5. The simulated curve coincides with the measured points, and the deviations between them are less than 1.5%, from 10:00 a.m. to 14:00 p.m. The maximum deviation is below 9.7%. Therefore, it is valid to adopt the current method to predict the thermal performance of airship with solar array.

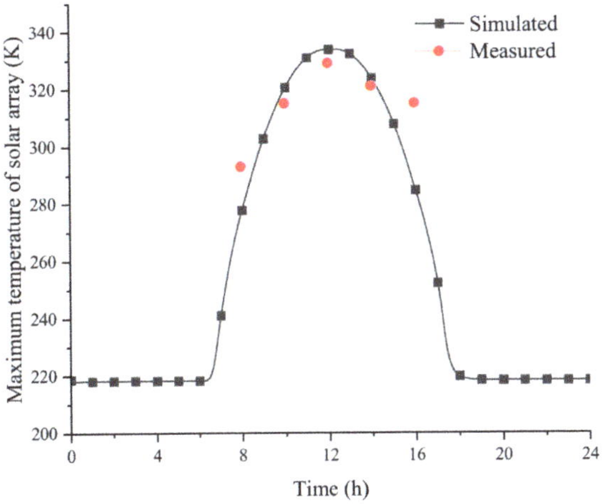

Figure 5. Comparison of simulated and measured results.

3.2. Simulation Condition

The flight requirements and overall design parameters of the solar powered airship are shown in Table 1. At the altitude of 20 km, the west–east component of the wind is greater than the north–south component, based on analysis of the six hourly interval wind data of Jiujiang (30° N, 116° E) in 2020 and 2021 [38]. Therefore, the flight direction of the airship is defined as east–west. The minimum change interval of the insulation thickness for the configuration optimization, with the proposed method, is 0.1 mm.

Table 1. Requirements and overall design parameters of airship.

Requirements	Value	Design Parameters	Value
Flight altitude, km	20	Volume of airship, m^3	71,897
Location	Jiujiang (30° N, 116° E)	Semi-major axis length a_1, m	53.8
Working date	1 January~ 31 December	Semi-major axis length a_2, m	76.2
Flight direction	East–west	Semi-minor axis length b, m	16.3
Airspeed, m/s	15	Total mass, kg	6392
Payload capacity, kg	650	Film mass, kg	2140
Thickness range of insulation layers, mm	0.5~15	Helium mass, kg	882
Central angle range, °	−90~90	Propulsion system mass, kg	957
Allowable stress of skin, MPa	52.5	Storage battery mass, kg	1130
Limit of differential pressure, Pa	1254	Solar array mass, kg	398

3.3. Effect of Insulation Configuration

As for Jiujiang (30° N, 116° E), the strongest solar irradiation occurs on the summer solstice, the 172nd day of 2021, and the airship has the highest overpressure. Figure 6 shows the variations of output energy and maximum differential pressure, ΔP_{max}, and marks up the traditional configuration of the insulation, with a constant thickness of 7.1 mm. The total energy acquirement and maximum differential pressure, ΔP_{max}, both increase when the thickness of insulation layer decreases. The small thickness of the insulation layer is beneficial for the solar array to transfer heat to the film and helium, so decreasing the thickness of the insulation layer functions to increase the average temperature of the inner gas. With the increase of the helium temperature, the differential pressure increases. As for the traditional configuration, the constant thickness of the insulation layer should not be less than 7.1 mm, in order to ensure that the ΔP_{max} does not exceed the limit of differential pressure, ΔP_{limit}. In order to maximize the output energy and minimize the structural mass of the thermal insulation layers, a constant thickness of 7.1 mm is adopted as the traditional configuration for the solar powered airship.

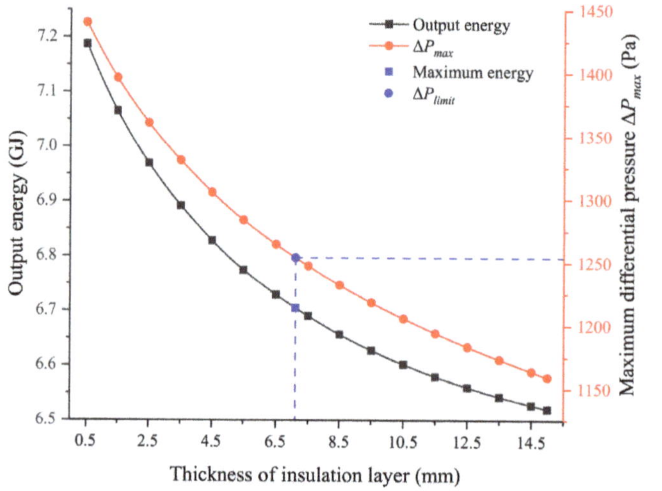

Figure 6. Variations of total energy and ΔP_{max}, with the thickness of the insulation layer.

Figure 7 depicts the configurations of the traditional and optimized thermal insulation layer. As for the traditional layout, the thicknesses of the insulation layers, under different solar cells, are the same as 7.1 mm. For the optimized layout, the thicknesses of the insulation layers, with the central angle, γ, from 0° to ±30°, are thickened to 15 mm, while the others are thinned to 0.5 mm. The parameters of the airships with the above two configurations of thermal insulation are displayed in Table 2. For the airship with the optimized layout, the mass of the insulation structure is decreased by 58.5 kg, and the reduction rate is 24.9%, compared to the traditional arrangement. It is indicated that the payload mass of the airship can be increased by 58.5 kg.

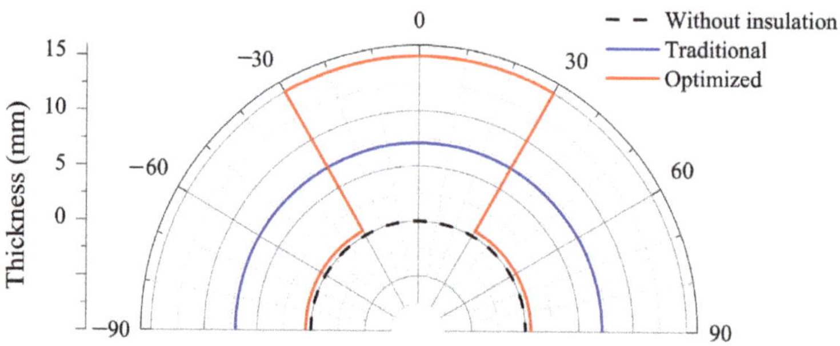

Figure 7. Configurations of traditional and optimized insulation layers.

Table 2. Comparison of airships with traditional and optimized insulation layers.

Parameters	Traditional Layout	Optimized Layout
Insulation layer mass, kg	235	176.5
Limit of differential pressure, Pa	1254	1254
Maximum differential pressure, Pa	1253.9	1253.8

3.4. Comparative Analysis

Figure 8 describes the distribution of the incident solar irradiation intensity at 12:00 on 11 November 2021. The illumination intensity of the solar cells with the central angle, γ, of $-47°$ is the highest. The incident solar irradiance decreases when γ changes from $-47°$ to $\pm 90°$. However, the irradiation intensity of solar cells with γ from $-90°$ to $0°$ is higher than that of solar cells in a symmetrical position with γ from $0°$ to $90°$.

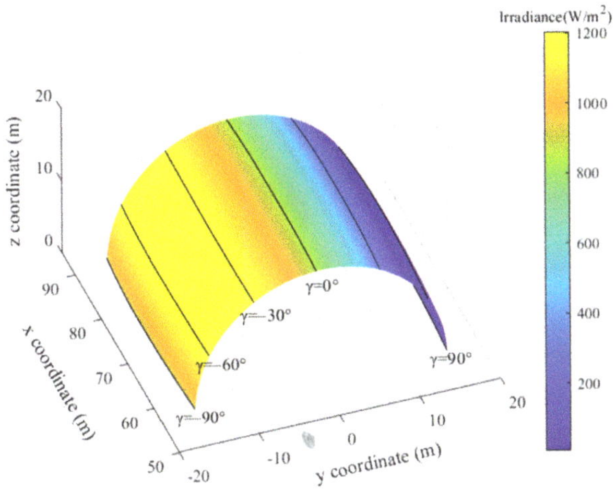

Figure 8. Distribution of irradiation intensity on solar array.

Figures 9 and 10 show the distribution of solar cell temperature with the traditional and optimized configuration of the insulation layer at noon, respectively. The distribution of the cell temperature, with traditional layout insulations, shows the same trend with illumination intensity. Compared to the solar cells with the traditional configuration, the temperatures of the cells with the optimized layout of the thermal insulation layer decrease or increase to different degrees. The temperatures of solar cells with a central angle, γ, of $-90°$ to $-30°$ decrease by 14~26 K. The temperatures of solar cells with γ of $-30°$ to $0°$ increase by 4~9 K. The temperatures of cells with γ of $0°$ to $30°$ have no obvious change. The temperatures of cells with γ of $30°$ to $60°$ have a maximum increase of 7 K. The optimized configuration effectively influences the heat exchange process and, therefore, changes the distribution of the solar cell temperature.

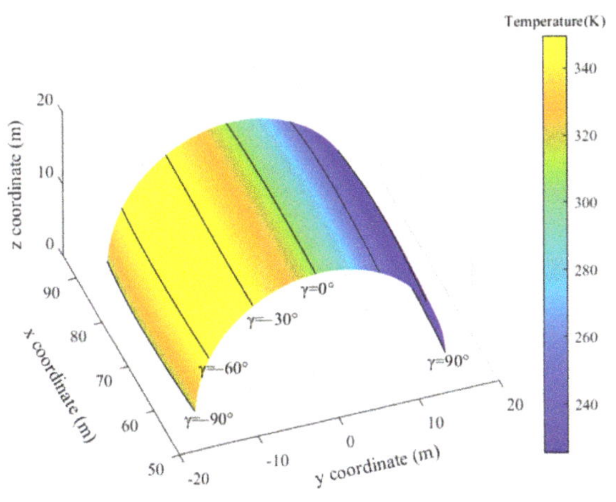

Figure 9. Distribution of solar cell temperature with traditional layout.

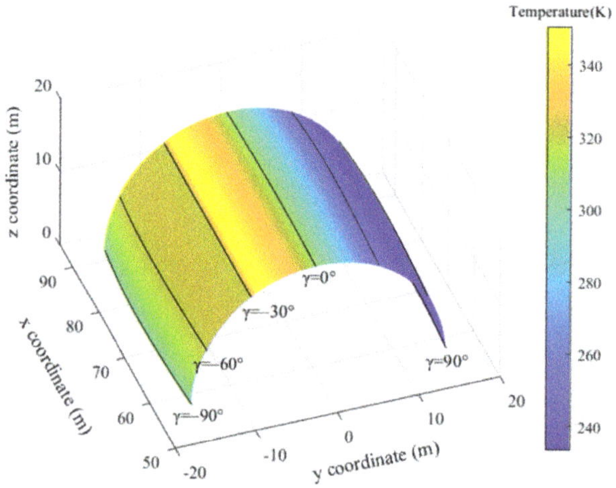

Figure 10. Distribution of solar cell temperature with optimized layout.

Figure 11 displays the distribution of the solar cell output power, with the traditional configuration of the insulation layer, at 12:00. The output power of cells with the central angle, γ, of $-90°$ to $-6°$ is basically the same. The output power of solar cells with γ of $-47°$ is not significantly higher than that of the other cells with γ of $-90°$ to $-6°$, despite the maximum illumination intensity. The output power of cells decreases as the γ changes from $-6°$ to $90°$.

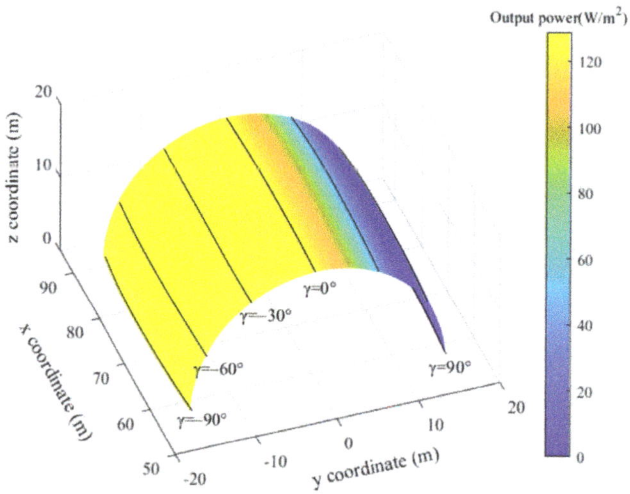

Figure 11. Distribution of output power with traditional layout.

Figure 12 demonstrates the distribution of the output power with the optimized configuration of insulations. The solar cells with a central angle γ of $-47°$ to $-30°$ have the maximum increasement of output power. The output power of solar cells with γ of $-90°$ to $-47°$ also increases. The output power cells with γ of $-30°$ to $0°$ decreases, but the magnitude of change is lower than that of the cells with γ of $-47°$ to $-30°$. The output power of the solar cells with γ of $0°$ to $90°$ has no obvious change.

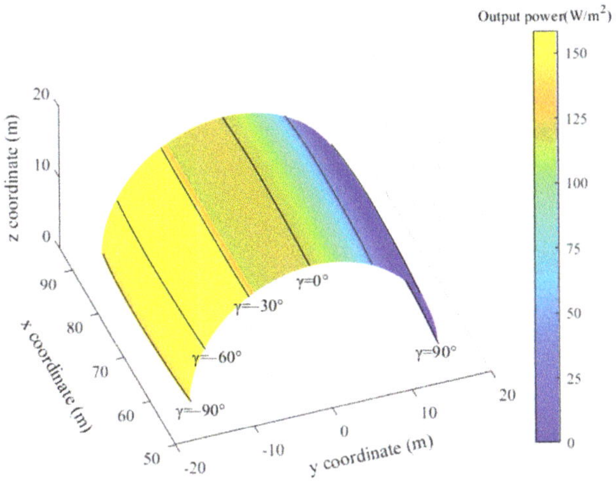

Figure 12. Distribution of output power with optimized layout.

Figure 13 depicts the output power and differential pressure, ΔP, during the day, with the traditional and optimized layouts of insulation layers. The output power of the solar array with optimized thermal insulations is higher than that of traditional insulations from 8:00 am to 16:00 pm. The maximum increase of the output power is 9.9% at 11:58 am. The values of the total energy of solar array with traditional and optimized configurations of insulations during the day are 4.32 and 4.60 GJ, respectively. The output energy is improved by 6.5%, with optimized layout of thermal insulations. The optimized configuration of the insulation layer results in the rise of differential pressure during the daytime, and the maximum differential pressure is 1190.2 Pa at 12:10 pm. The limit of differential pressure, ΔP_{limit}, of the investigated airship is 1254 Pa, so the structural integrity is guaranteed.

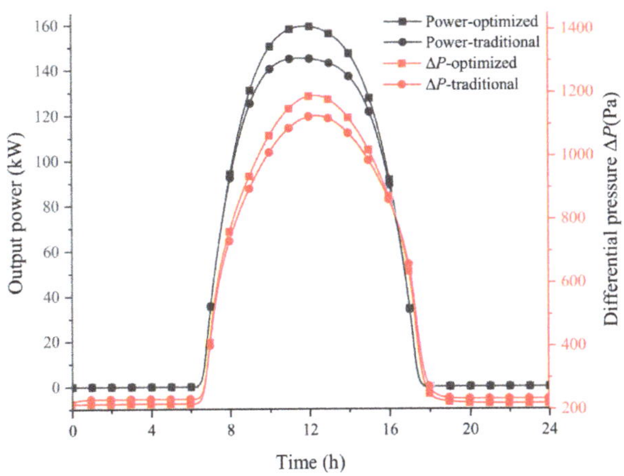

Figure 13. The output power and differential pressure, ΔP, with the traditional and optimized configurations.

Figure 14 compares the variation of output energy of solar array and maximum differential pressure, ΔP_{max}, on different dates of a year, with the traditional and optimized configurations of the insulation layers. The maximum values of the energy acquirement and ΔP_{max} occur on 21 June, when the solar irradiation is highest at the investigated location. It is also indicated that the optimized insulation layer functions to improve the energy output throughout the year. The maximum increasement of the total output energy during a day is 8.2%, on 21 December, when the energy acquirements of the solar powered airship with traditional and optimized insulations are 4.02 and 4.35 GJ, respectively. The ΔP_{max} of the airship with an optimized layout of the insulations is higher than that of the airship with traditional layout on most dates of the year, and the maximum derivation is 77.6 Pa.

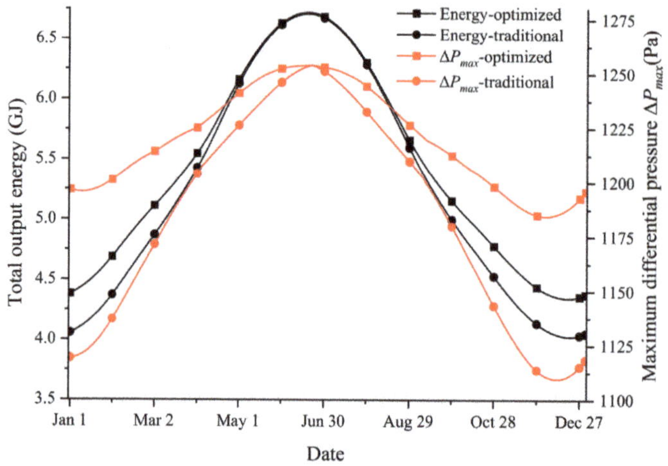

Figure 14. Variation of output power and total energy on different dates.

The results from the conducted analysis indicate that, compared with the traditional arrangement, with constant thickness of the insulation layer under each solar cell unit, the optimized configuration decides the thicknesses of the insulation layer under the solar cells according to their solar irradiation intensity. The thicknesses of the insulation layers under the solar cells with sufficient solar irradiation decrease, so as to improve the output power of the solar array. Meanwhile, the thicknesses under other cells increase to avoid an excessive rise of differential pressure of the airship. Based on the multidisciplinary optimization method, the everyday output energy of solar powered airships can be improved during flight in a year.

4. Conclusions

The arrangement of the insulation configuration plays an important role in the temperature and output power of solar array. This paper discusses an optimized layout of insulation to improve the total output energy of the solar array. The theoretical method, consisting of spatial geometry, thermal, and power models, is developed and validated. The optimized configuration of the insulation structure on the curved surface of the airship is obtained and compared with the traditional layout with constant thicknesses. The conclusions are drawn as follows:

1. The thickness of the insulation layer has a significant influence on the output power of the solar array and maximum differential pressure of the airship. Decreasing thickness is beneficial for reducing the average temperatures of the solar cells, hence increasing energy acquisition. However, a thin thermal insulation layer results in the temperature rise of the inner gas of the airship, causing an increase of differential pressure, which could lead to structural failure of the airship film.
2. The optimized configuration significantly reduces the total weight of the thermal insulation layers. Meanwhile, the limit of the maximum differential pressure is not exceeded. The thicknesses of the optimized insulation layers vary with the central angles. For the investigated airship, the thicknesses of the insulation layers with a central angle from 0° to ±30° are thickened, and the others are thinned. The decreased mass of insulations can be used to increase the capacity of carrying payload.
3. The optimized layout of the insulations improves the output energy of the solar array throughout the entire year. In the long time period, the optimized arrangement functions to increase the energy acquirement and maximum differential pressure, especially on dates when the solar radiation are insufficient (before and after the winter solstice). The maximum increase of the total output energy during a day is 8.2% on the winter solstice. The proposed method plays an essential role in the long endurance mission of stratospheric airships.

Author Contributions: Conceptualization, Y.L.; methodology, Y.L.; software, Y.L. and Z.X.; validation, Y.L., Z.X. and H.D.; investigation, M.L.; data curation, M.L.; writing—original draft preparation, Y.L.; writing—review and editing, H.D.; visualization, H.D.; supervision, M.L. All authors have read and agreed to the published version of the manuscript.

Funding: This work was funded by the National Natural Science Foundation of China, grant number 51775021; the China Postdoctoral Science Foundation, grant number 2021M700322; and the Postdoctoral Fellows of "Zhuoyue" Program of Beihang University.

Institutional Review Board Statement: Not applicable.

Informed Consent Statement: Not applicable.

Data Availability Statement: Not applicable.

Conflicts of Interest: The authors declare no conflict of interest.

References

1. Gonzalo, J.; López, D.; Domínguez, D.; García, A.; Escapa, A. On the capabilities and limitations of high altitude pseudo-satellites. *Prog. Aerosp. Sci.* **2018**, *98*, 37–56. [CrossRef]
2. Frezza, L.; Marzioli, P.; Santoni, F.; Piergentili, F. VHF Omnidirectional Range (VOR) Experimental Positioning for Stratospheric Vehicles. *Aerospace* **2021**, *8*, 263. [CrossRef]
3. Mano, S.; Ajay Sriram, R.; Vinayagamurthy, G.; Pillai, S.N.; Pasha, A.A.; Reddy, D.S.K.; Rahman, M.M. Effect of a Circular Slot on Hybrid Airship Aerodynamic Characteristics. *Aerospace* **2021**, *8*, 166. [CrossRef]
4. Pande, D.; Verstraete, D. Impact of solar cell characteristics and operating conditions on the sizing of a solar powered nonrigid airship. *Aerosp. Sci. Technol.* **2018**, *72*, 353–363. [CrossRef]
5. Xu, Y.; Zhu, W.; Li, J.; Zhang, L. Improvement of endurance performance for high-altitude solar-powered airships: A review. *Acta Astronaut.* **2020**, *167*, 245–259. [CrossRef]
6. Garg, A.; Burnwal, S.; Pallapothu, A.; Alawa, R.; Ghosh, A. Solar Panel Area Estimation and Optimization for Geostationary Stratospheric Airships. In Proceedings of the 11th AIAA Aviation Technology, Integration and Operations (ATIO) Conference, Virginia Beach, VA, USA, 20–22 September 2011.
7. Wang, H.; Song, B.; Zuo, L. Effect of High-Altitude Airship's Attitude on Performance of its Energy System. *J. Aircr.* **2007**, *44*, 2077–2080. [CrossRef]
8. Alam, M.I.; Pant, R.S. Multidisciplinary approach for solar area optimization of high altitude airships. *Energy Convers. Manag.* **2018**, *164*, 301–310. [CrossRef]
9. Li, J.; Lv, M.; Tan, D.; Zhu, W.; Sun, K.; Zhang, Y. Output performance analyses of solar array on stratospheric airship with thermal effect. *Appl. Therm. Eng.* **2016**, *104*, 743–750. [CrossRef]
10. Du, H.; Zhu, W.; Wu, Y.; Zhang, L.; Li, J.; Lv, M. Effect of angular losses on the output performance of solar array on long-endurance stratospheric airship. *Energy Convers. Manag.* **2017**, *147*, 135–144. [CrossRef]
11. Dai, Q.; Cao, L.; Zhang, G.; Fang, X. Thermal performance analysis of solar array for solar powered stratospheric airship. *Appl. Therm. Eng.* **2020**, *171*, 115077. [CrossRef]
12. Shi, H.; Chen, J.; Hu, L.; Geng, S.; Zhang, T.; Feng, Y. Multi-parameter sensitivity analysis on thermal characteristics of stratospheric airship. *Case Stud. Therm. Eng.* **2021**, *25*, 100902. [CrossRef]
13. Wang, Y.; Yang, C. Thermal Analysis of a Stratospheric Airship in Working Process. In Proceedings of the AIAA Balloon Systems Conference; American Institute for Aeronautics and Astronautics (AIAA): Reston, VA, USA, 2009.
14. Li, X.; Fang, X.; Dai, Q. Research on Thermal Characteristics of Photovoltaic Array of Stratospheric Airship. *J. Aircr.* **2011**, *48*, 1380–1386. [CrossRef]
15. Sun, K.; Yang, Q.; Yang, Y.; Wang, S.; Xie, Y.; Sun, M.; Chen, X.; Xu, J. Numerical Simulation and Thermal Analysis of Stratospheric Airship. *Procedia Eng.* **2015**, *99*, 763–772. [CrossRef]
16. Kim, H.K.; Han, C.Y. Analytical and numerical approaches of a solar array thermal analysis in a low-earth orbit satellite. *Adv. Space Res.* **2010**, *46*, 1427–1439. [CrossRef]
17. Sun, K.; Zhu, M.; Liu, Q. Membrane Material-Based Rigid Solar Array Design and Thermal Simulation for Stratospheric Airships. *Adv. Mater. Sci. Eng.* **2014**, *2014*, 192707. [CrossRef]
18. Li, J.; Lv, M.; Sun, K.; Zhu, W. Thermal insulation performance of lightweight substrate for solar array on stratospheric airships. *Appl. Therm. Eng.* **2016**, *107*, 1158–1165. [CrossRef]
19. Meng, J.; Liu, S.; Yao, Z.; Lv, M. Optimization design of a thermal protection structure for the solar array of stratospheric airships. *Renew. Energy* **2019**, *133*, 593–605. [CrossRef]
20. Tang, R.; Wu, T. Optimal tilt-angles for solar collectors used in China. *Appl. Energy* **2004**, *79*, 239–248. [CrossRef]
21. Atmeh, G.; Subbarao, K. Guidance, Navigation and Control of Unmanned Airships under Time-Varying Wind for Extended Surveillance. *Aerospace* **2016**, *3*, 8. [CrossRef]
22. Khoury, G.A. *Airship Technology*; Cambridge University Press: Cambridge, UK, 2012.

23. Gemignani, M.; Marcuccio, S. Dynamic Characterization of a High-Altitude Balloon during a Flight Campaign for the Detection of ISM Radio Background in the Stratosphere. *Aerospace* **2021**, *8*, 21. [CrossRef]
24. Yang, X.; Liu, D. Renewable power system simulation and endurance analysis for stratospheric airships. *Renew. Energy* **2017**, *113*, 1070–1076. [CrossRef]
25. Despotovic, M.; Nedic, V. Comparison of optimum tilt angles of solar collectors determined at yearly, seasonal and monthly levels. *Energy Convers. Manage.* **2015**, *97*, 121–131. [CrossRef]
26. Yang, X.; Zhang, W.; Hou, Z. Improved Thermal and Vertical Trajectory Model for Performance Prediction of Stratospheric Balloons. *J. Aerosp. Eng.* **2013**, *28*, 04014075. [CrossRef]
27. Gao, X.-Z.; Hou, Z.-X.; Guo, Z.; Liu, J.-X.; Chen, X.-Q. Energy management strategy for solar-powered high-altitude long endurance aircraft. *Energy Convers. Manag.* **2013**, *70*, 20–30. [CrossRef]
28. Yang, X. Prediction of thermal behavior and trajectory of stratospheric airships during ascent based on simulation. *Adv. Space Res.* **2016**, *57*, 2326–2336. [CrossRef]
29. Zhang, L.; Li, J.; Wu, Y.; Lv, M. Analysis of attitude planning and energy balance of stratospheric airship. *Energy* **2019**, *183*, 1089–1103. [CrossRef]
30. Yamawaki, T.; Mizukami, S.; Masui, T.; Takahashi, H. Experimental investigation on generated power of amorphous PV module for roof azimuth. *Sol. Energy Mater. Sol. Cells* **2001**, *67*, 369–377. [CrossRef]
31. Martin, N.; Ruiz, J.M. Calculation of the PV modules angular losses under field conditions by means of an analytical model. *Sol. Energy Mater. Sol. Cells* **2001**, *70*, 25–38. [CrossRef]
32. Khan, F.; Baek, S.-H.; Kim, J.H. Wide range temperature dependence of analytical photovoltaic cell parameters for silicon solar cells under high illumination conditions. *Appl. Energy* **2016**, *183*, 715–724. [CrossRef]
33. Dai, Q.; Fang, X.; Li, X.; Tian, L. Performance simulation of high altitude scientific balloons. *Adv. Space Res.* **2012**, *49*, 1045–1052. [CrossRef]
34. Anoop, S.; Velamati, R.K.; Oruganti, V. Aerodynamic characteristics of an aerostat under unsteady wind gust conditions. *Aerosp. Sci. Technol.* **2021**, *113*, 106684. [CrossRef]
35. Kennedy, J.; Eberhart, R. Particle Swarm Optimization. In Proceedings of the Icnn95-international Conference on Neural Networks, Perth, WA, Australia, 27 November–1 December 1995.
36. Gangadhar, A.; Manikandan, M.; Rajaram, D.; Mavris, D. Conceptual Design and Feasibility Study of Winged Hybrid Airship. *Aerospace* **2021**, *9*, 8. [CrossRef]
37. Harada, K.; Eguchi, K.; Sano, M.; Sasa, S. Experimental Study of Thermal Modeling for Stratospheric Platform Airship. In Proceedings of the AIAA's 3rd Annual Aviation Technology, Integration, and Operations (ATIO) Forum, Denver, CO, USA, 17–19 November 2003.
38. Hersbach, H.; Bell, B.; Berrisford, P.; Biavati, G.; Horány, A.; Muñoz Sabater, J.; Nicolas, J.; Peubey, C.; Radu, R.; Rozum, I.; et al. ERA5 Hourly Data on Single Levels from 1979 to Present. Copernicus Climate Change Service (C3S) Climate Data Store (CDS). 2018. Available online: https://cds.climate.copernicus.eu/cdsapp#!/dataset/reanalysis-era5-single-levels?tab=overview (accessed on 1 December 2021).

aerospace MDPI

rticle

On the Feasibility of a Launcher-Deployable High-Altitude Airship: Effects of Design Constraints in an Optimal Sizing Framework

Carlo E.D. Riboldi *,†, Alberto Rolando *,† and Gregory Regazzoni †

Dipartimento di Scienze e Tecnologie Aerospaziali, Politecnico di Milano, Via La Masa 34, 20156 Milan, Italy; gregory.regazzoni@mail.polimi.it
* Correspondence: carlo.riboldi@polimi.it (C.E.D.R.); alberto.rolando@polimi.it (A.R.)
† These authors contributed equally to this work.

Abstract: When ground observation or signal relaying in the vicinity of an unfriendly operative scenario are of interest, such as for military actions or disaster relief, high-altitude airships (HAA) offer some technical benefits. Featuring a milder cost and higher deployment flexibility with respect to lower-Earth orbit satellites, these platforms, often baptized as high-altitude pseudo-satellites (HAPS), operate sufficiently far from the ground to provide better imaging coverage and farther-reaching signal relaying than standard low-flying systems, such as aircraft or helicopters. Despite the atmospheric conditions in the higher atmosphere, they offer stable airstreams and highly-predictable solar energy density, thus ideally giving the chance of smooth operation for a prolonged period of time. The design of airships for the task is often conditioned by the need to go through the lower layers of the atmosphere, featuring less predictable and often unstable aerodynamics, during the climb to the target altitude. With the aim of simultaneously largely increasing the ease and quickness of platform deployment, removing most of the design constraints for the HAPS induced by the crossing of the lower atmosphere, and thus allowing for the design of a machine best suited to matching optimal performance at altitude, the deployment of the HAA by means of a missile is an interesting concept. However, since the HAA platform should take the role of a launcher payload, the feasibility of the mission is subject to a careful negotiation of specification, such that the ensuing overall weight of the airship is as low as possible. A preliminary design technique for high-altitude airships is therefore introduced initially, customized to some features typical to missile-assisted deployment, but with the potential for broader applications. The proposed procedure bends itself to the inclusion in an optimal framework, with the aim of seeking a design solution automatically. A validation of the adopted models and assumptions on existing HAPS is proposed first. The design of the airship is then carried out in a parameterized fashion, highlighting the impact of operative and technological constraints on the resulting sizing solutions. This allows for the marking of the boundaries of the space of design solutions for a launcher-deployable airship.

Keywords: high-altitude airship (HAA); high-altitude pseudo-satellite (HAPS); missile-assisted deployment; lighter-than-air (LTA); earth observation; signal relay; long endurance; airship design; optimal design; mission design; specifications negotiation

1. Introduction

Following the massive introduction of satellites to support the needs of imagery and signal relay missions, for commercial, humanitarian, and military purposes, platforms which feature a greater flexibility and meet the requirements for a lower-budget and quicker deployment have been in the scope of research [1,2].

The so-called high altitude pseudo-satellite (HAPS) concept, i.e., a flying craft stationed in the higher layers of the atmosphere where aerodynamic flight is still possible, typically between 18 and 25 km from the ground, can be managed by using very different crafts [3]. The high altitude allows one to take advantage of the good stability and predictability of airstreams [4], and the lower distance from the ground allows for operation with comparatively less sophisticated imagery and signal relay plants, while retaining or even surpassing the quality of the output of space satellites (for instance, a better absolute image resolution may be achieved with a lower-budget zoom lens). Furthermore, the more intense solar radiation at higher altitudes in the atmosphere allows solar power harvesting to be a very significant—even primary—contributor to the power balance of the flying craft, similarly to space platforms [5].

Since payload is at a premium when reaching an altitude that high, HAPS platforms are typically unmanned. Engineered realizations based on a winged configuration have been tested [6–8]. As it is known, winged aircraft require relative airspeed to remain airborne, and this creates the need for engines and the energy to feed them for the duration of the flight. This in turns limits the weight onboard which can be used for payload. Furthermore, the need for aircraft to keep moving constrains their ability to carry out fixed-point observation. Combined with limited—despite significant—endurance at the current technological level, this makes winged configurations largely sub-optimal for signal relay in support of disaster relief missions or continued fixed-point surveillance for military tasks.

Conversely, high-altitude airships (HAA) offer the advantage of a different method to counter weight, i.e., aerostatic force, which allows for a strong reduction in the power required for propulsion, thus enabling a comparatively larger payload weight for the same take-off weight [7,9]. Furthermore, provided they can face the airstream at altitude, HAA can be employed for fixed-point flight with good accuracy. A major shortcoming of every flying craft destined for station-keeping missions at higher altitudes (such as HAPS) is constituted by the need to go through more unstable and more dense layers of the atmosphere while flying towards the destination level [8,10]. To tackle this issue, missile-assisted deployment is an interesting alternative. However, the latter finds its more straightforward application when an HAA is considered being deployed at altitude, thanks to the ability to make the airship very compact when not inflated, thus making its size compatible with the typical size of candidate high-atmospheric missiles. Clearly, the adoption of a compact missile launcher for the deployment also adds to the quick response capability of the overall system, which is in good agreement with the typical requirements of disaster relief missions.

The chance to take the airship to the intended deployment altitude does impose a strict constraint, especially on weight, because the HAA has to be considered primarily as the payload of a missile. Consequently, the HAA needs to be conceived from preliminary sizing as a weight-minimal (i.e., weight-optimal) system. When the chosen missile is lighter, typically for the reason of increased transportability, that means a lower cost and increased ease of deployment. Furthermore, the constraint on the weight of the airship to be put onboard is more stringent.

Building on this concept, in this work the chance to cover a HAPS mission by means of a missile-deployed HAA is analyzed, with the aim of assessing its feasibility. With this goal, a focus has been put on a detailed preliminary design procedure of an airship for high-altitude operation. Besides the advantages of the low turbulence and good predictability of airstreams in the higher layers of the atmosphere, the high solar power harvesting opportunities, and the lack of any constraints due to the autonomous climb to the stationing altitude, comes the need to minimize weight and make it compatible with the payload of the lightest (i.e., cheapest and most easily transportable) low-atmospheric missiles.

The concept of a missile-assisted deployment of an HAA has been proposed in [11], showing its general validity, but making use of simplified models for energy balance and mass characterization without analyzing the effects of technology constraints on airship sizing.

Despite considerable efforts in the setup of design procedures for airships, leading to some consolidated approaches [12], in the current literature it is often pointed out that existing methodologies rarely allow a high degree of automation in the design process [13–15]. Furthermore, when analyzing the design of HAA, it is more generally treated as autonomously deployable (as opposed to launcher-deployable); the existing literature invariably makes use of strong assumptions and approximations on the environment and especially on the modeling of the wind and solar power radiation, both during ascent and at altitude [16–20]. When this is not crucial for the ascent phase, which in the concept of interest here is not faced by the airship itself (but by the missile instead), the effect of the choice of the stationing altitude, geographical position, and time of the year is extremely relevant for the resulting characteristics of the airship, including its weight, and this can potentially either allow or jeopardize the feasibility of the entire mission. Similarly, the geometry of the envelope and the technological features of the airship (including properties of materials) have been studied as sub-problems [21,22]. However, they were never thoroughly integrated into a complete design loop. Specifically, the weight was not looked at as a hard constraint and conclusions were not made on the feasibility of a launch-assisted deployment at altitude.

Therefore, despite previous research that is capable of providing useful direction, a comparative analysis of the feasibility of a launch-deployed HAA that gives a detailed description of the included realization technologies and of the environment at altitude needed for accurate predictions is still missing.

In this work, a complete preliminary sizing methodology is introduced first. This is based on two conceptual steps. In the first, a guess sizing is carried out based on the models described in the text, fed by a set of assigned variables, which need to be first guessed by the user. In a second step, the outcome of the sizing should be verified with respect to a set of major constraints. Thanks to the structure and information workflow envisaged in this procedure, the sizing loop can be fully automated and cast as the core of an optimization algorithm targeting the minimum weight of the airship. This has been considered as a natural choice, considering the entire HAA needs to fit in a missile and be lifted by its boosters. In an optimal approach, starting from a guess of a set of variable parameters, a sizing is carried out by running the loop, enabling the computation of the overall mass of the system (optimization target) and of the constraint violation, so that a new iteration can be started accordingly. The process is repeated until convergence to an optimal solution which satisfies all constraints.

The sizing loop has been validated in terms of the models and logic adopted, making use of the data of an existing HAA [14], accounting for the specifications of the chosen test bed with non-substantial amendments to the sizing procedure.

To the aim of understanding the effect of a negotiation of specifications on the feasibility of the missile-assisted deployment, the optimal sizing methodology has been employed to carry out the preliminary sizing of an HAA for a HAPS mission, in four selected test cases featuring very different geographical locations. This analysis highlights strong trends and dependencies, underlining the relevance of the choice of some of the design parameters, including the stationing altitude in combination with the geographical position. In particular, a strong effect on overall weight can be observed in some test cases. To better investigate mission feasibility, further analyses have been carried out, investigating the result of a choice of the technological parameters involved in the sizing. These show that a suitable combination of realistic technological choices enables advantages on overall weight, allowing for the fitting of the airship as a payload on a missile. However, other combinations of environmental conditions and technological choices do not bear feasible results, thus showing the paramount importance of the proposed systematic, critical approach to the analysis of design specifications.

2. Preliminary Sizing of a High-Altitude Airship: Introductory Concepts and Models for Design

A procedure for the preliminary sizing of a high-altitude airship (HAA) making use of solar power can be based on the simultaneous satisfaction of feasibility criteria, quantitatively defined according to three major drivers:

1. Buoyancy-weight balance. The buoyant lift force generated by the lifting gas enclosed in the envelope has to be equal to the weight of all elements making the airship, including payload.
2. Energy balance. The energy that is collected over the mission from the solar array has to be equal or higher than the total electrical energy required for the mission by the airship's own systems and for operating the on-board payload.
3. Limit stress compliance. The ultimate stress of the envelope material should not exceed the estimated limit load acting on it, amplified by a safety factor.

Similar to the case of fixed-wing aircraft, preliminary sizing is an iterative process, invariably requiring a starting guess.

The latter can be identified starting from a mission goal, identified in carrying an assumed payload (e.g., a sensor suite), which needs to work over an assigned time schedule. This produces a core component of the weight and energy/power requirements to be satisfied for the mission. Furthermore, the altitude (or altitude range) acceptable for the stationing of the airship will be similarly the result of the payload choice.

Further assumptions in the first guess involve the general arrangement of the airship envelope and a starting geometrical sizing. For an HAA based on solar cells, this step corresponds also to the definition of the surface of solar arrays. Merged with the time schedule, geographical location on Earth, and target altitude (or altitude range), this directly allows for the estimation of the ideal captured solar radiation. Combining geometrical information and mission schedule requirements—the latter especially with the aim of estimating the inside/outside pressure gradient, thus allowing for the assessment of the required thickness of the envelope material—the weight break-down of the entire airship can be computed. This allows for the estimation of the overall energy and power requirements. In turn, the first guess sizing allows for the triggering of the iterative procedure mentioned above, where criteria corresponding to the three drivers just introduced need to be simultaneously verified for an acceptable sizing.

Before going into detail concerning the sizing procedure just outlined, it should be pointed out that, in order to allow the quick definition of the preliminary sizing of an HAA, an automated, optimally oriented approach has been envisaged. Therefore, all considerations in the setup of the design loop have been formulated so as to avoid the need for human intervention in running the process. If on the one hand this may cut short some speculations, thus excluding some area in the space of design solutions a priori, on the other it also allows one to carry out sizing computations more quickly, so as to allow for the performance of sensitivity studies, understanding the effect of a change in the design specifications on the resulting characteristics of the airship.

2.1. Envelope Lofting

One of the first steps in the sizing of the airship is the definition of its geometric characteristics. The geometry of an airship bears a direct influence on all aspects governing its operation. Firstly, the power consumption is directly related to the experienced drag and thus to both the size and shape of the envelope, while solar power capture is affected by the shape of the envelope and the layout of the solar array laid on it, which determines the actual amount of incident radiation that is captured and converted into electric energy. Secondly, the buoyant lift force is a function of the envelope volume and is therefore affected by envelope geometry. Finally, the overall mass is the result of a number of contributions, with numerical values bound to the geometry and size of the airship.

2.1.1. Analytic Geometry of the Envelope

Considering HAA designed for a high-altitude pseudo-satellite (HAPS) mission, keeping the airship in a fixed position with respect to the ground, in the face of reasonably slow-varying (albeit intense) headwind, is performed by adopting an axial-symmetric geometry. As a matter of fact, for that type of mission the envelope geometry is typically based on a solid of revolution. In mathematical terms, these geometries can be assigned based on different modeling approaches [6,16,19,23]. In this work, in consideration of the applications which will be later presented (see Section 4), the low-drag bi-ellipsoidal shape designed by the NPL [19] is adopted, consisting of two semi-ellipsoids jointed at the maximum thickness longitudinal location. For this analytic geometry the envelope is completely defined by the overall length of the envelope L and its fineness ratio $FR = \frac{L}{2b}$, with b minor semi-axis of the ellipsoid. However, it should be remarked that whatever the mathematical modeling, once the geometry of a solid of revolution is assigned, the corresponding features required for aerostatic, power capture, and structural integrity computations can be uniquely evaluated—namely, the fineness ratio FR, envelope volume V_{env} and area A_{env} of external surface, area of lateral projection A_{side}, and position of the buoyancy center $\boldsymbol{BC}$.

2.1.2. Geometry of Solar Cells

In order to assess the solar power and energy capture, the relative positioning of solar cells with respect to the local radiation, as well as the absolute size of the panels, are needed inputs. Considering the seasonal change in the direction of sun radiation, as well as a specific positioning of the HAA at a lower or higher latitude, the problem of the positioning of solar cells arrays on the envelope surface—i.e., where to lay them—may not be solved with an always-valid solution. In other words, whether putting solar cells on top of the envelope or to the sides (i.e., the most commonly adopted solutions) would be an optimal choice that cannot be determined in general a priori. Therefore, a set of analytic quantities to define the coverage of the envelope with solar cells is introduced, keeping a further set of variables to be assigned in the sizing procedure. In a later section, it will be shown how to select the corresponding values, according to optimal criteria. It is assumed that the area covered with solar cells is always symmetric with respect to the vertical plane of symmetry of the airship. Furthermore, a set of two azimuthal ($\theta_{in}, \theta_{out}$) and two longitudinal ($x_{le}, x_{te}$) coordinates uniquely defines the portion of the envelope surface covered by cells. Clearly, as mentioned, an area obtained by symmetry with respect to the vertical plane will also be correspondingly covered by cells. This is illustrated in Figure 1 for an example geometry.

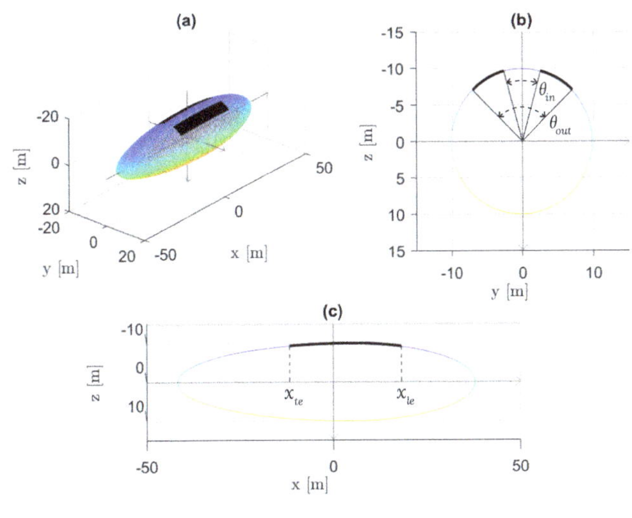

Figure 1. Modeling approach for the geometry of solar cells. (**a**) 3-D view. (**b**) y-z plane view. (**c**) x-z plane view.

Since the shape—in particular, the curvature—of the envelope may yield significant differences in the power-capturing ability of one portion or another of the overall surface of the envelope associated to solar cells, the latter is discretized according to the curved paving model [19,22], based on a suitable number of azimuthal and longitudinal elements. This should significantly increase the accuracy of the prediction of the capturing potential, with respect to more coarse methods [6,16,19,23].

2.2. Power Balance

Energy depletion on board the HAA can be imputed to three major power sinks, namely motion—or fixed-point keeping in the presence of wind—payload operation, and other plants (such as computers, sensors, actuators, etc.).

2.2.1. Power Required for Station Keeping

Concerning motion, for an HAA performing a HAPS mission it is typically desirable to stay in a near-fixed position with respect to a point on the ground, in spite of the airstream present at altitude in the stratosphere. In case this requirement is not strict, an optimal energetic strategy can be envisaged, where the airship moves upstream during the day, profiting from the higher solar power available, and drifts downstream during the night. This typically allows for the optimization of the battery mass required for energy storage and reduces the overall weight [2,24]. However, if precise positioning is at a premium, that can be ensured by a proper sizing of the batteries. This more stringent scenario is considered herein. For station keeping, the thrust F_T produced by the propulsion system should balance the aerodynamic drag. Therefore, it is associated with the propulsive power P_{prop} that can be computed according to Equation (1)

$$P_{prop} = F_T v = \frac{1}{2}\rho v^2 V_{env}^{2/3} C_D, \tag{1}$$

where v is the relative airspeed of the airstream with respect to the airship, ρ the density of air at altitude, V_{env} the volume of the envelope, and C_D the drag coefficient of the airship [12]. To estimate the latter, lumped parameter methods can be deployed, expressing its dependence on the actual geometry as a function of a few integral parameters, such as the fineness ratio [25,26], or it can be estimated with numerical codes, based on a discretization of the hull shape. The axial-symmetric envelope can be considered aligned with the airstream, thus working at a null angle of attack with respect to it. Considering a null dynamic lift component, the value of the lift coefficient is correspondingly null, and the drag coefficient value C_D needed for the computation is just one value (i.e., that for a null angle of attack), instead of a function.

Considering a null relative motion of the airship with respect to the ground, the relative airspeed v in Equation (1) can be estimated based on that of the airstream only. The latter is computed in this research according to the Horizontal Wind Model (HWM) [4,27], where the components of the airstream average velocity vector are assigned up to an altitude of 500 km as a function of geographic location (coordinates), altitude, day of the year, solar local time, and geomagnetic activity.

The power supply designed for the propulsion actually needed is obtained accounting for power losses, namely due to the electric motor (typically including an electronic controller), gearbox, and propeller. Electric motors come as a natural choice with respect to fuel-burning units, allowing for an increased mechanical and plant simplicity (e.g., no fuel system is required), higher energy efficiency [28], and the chance to reload the corresponding energy storage, thanks to solar power harvesting. The elements of the power train just introduced are represented, respectively, by the efficiencies η_m, η_g, and η_p, which in turn produce the power required for propulsion $P_{req,prop}$ in Equation (2)

$$P_{req,prop} = \frac{P_{prop}}{\eta_m \eta_g \eta_p}. \tag{2}$$

2.2.2. Power Required for Utilities

As pointed out, other power utilities onboard include the payload, associated to a power P_{pl}, and further electrical systems, associated to P_{other}. These can be modeled in a power balance by means of the nominal power required for their use, divided by corresponding efficiency factors η_{pl} and η_{other}, yielding the total power required, Equation (3)

$$P_{req} = P_{req,prop} + \frac{P_{pl}}{\eta_{pl}} + \frac{P_{other}}{\eta_{other}}. \tag{3}$$

2.2.3. Solar Power Harvesting

The power input to the system comes from the solar cells, geometrically described according to the assumptions in Section 2.1. In order to numerically compute the power input, it is first necessary to compute power capture. Considering the peculiarities of a HAPS station-keeping mission, where the attitude of the airship in 3D spaces can be considered constant with null pitch and roll, the only variable associated to a change is yaw ψ. The latter will be such to keep the airship aligned with the stratospheric airstream at altitude, and will be proportional to the local wind direction, an output of the HWM model [4,27] for a specific location.

With the attitude of the airship assigned with respect to the irradiation direction, for an assigned solar cell spacial geometry, the relative incidence and tilt angles of each discretized component on the assumed solar cells area can be computed. This can be cast as input to an irradiation model—the SMARTS model has been adopted in this research [5,29,30]—to obtain the spectral global irradiance E_g and the corresponding broadband value I_g on a component of a solar cell surface. Further inputs required by the SMARTS model include geographical inputs (altitude and zenith angle of the location) and time of the year (for eccentricity correction of extraterrestrial radiation).

The power coming from the i-th discretized component of a solar cell can be analytically expressed as in Equation (4),

$$P_{sc,i} = \eta_{MPPT}\eta_{pack}\eta_{sc,i}A_i I_{g,i}, \tag{4}$$

where η_{MPPT} is the efficiency of the maximum power point tracker (MPPT), $\eta_{sc,i}$ is the photoelectric conversion efficiency of the cell, A_i is the effective area exposed to solar radiation for the i-th discretized cell, and $I_{g,i}$ the broadband irradiance previously introduced, evaluated for the i-th discretized component of the solar array. The further efficiency η_{pack} appearing in Equation (4) represents the power decrease due to the wiring between adjoining cells [6].

The photoelectric conversion efficiency $\eta_{sc,i}$ depends on technological parameters, reflecting the technology of the solar cell, and environmental parameters. In this work, several technologies have been investigated, including CIGS, a-Si, and CdTe [31–33]. For each technology, nominal η_{sc} values can be obtained, which need to be corrected on account of a few environmental parameters. The latter include:

1. Temperature: η_{sc} is typically linearly influenced by the temperature of the panel, and the coefficients of the linear function are specific to the technology of the cell [34–36]. The computation of the temperature of the panel can be carried out based on models, such as NOCT [31,34] or HOMER [36]. The latter has been adopted in this research, bearing the actual temperature of the panel based on the assumption of a steady-state equilibrium between the solar energy absorbed by the module on one hand and the electrical output plus the heat transfer to the surroundings on the other.
2. Radiation spectrum: generally a weak dependence for top-tier solar cells technologies, it can be however simply taken into account according to the model presented in [37], where the pressure-corrected air mass is considered as an input.
3. Radiation intensity: the actual intensity of the radiation I_g affects photoelectric efficiency and may result in changes of the order of some percent units on its value. This

dependence, featuring an optimum for a specific radiation intensity, is technology-specific, and can be modeled according to the model presented in [37], suitably tuned based on the specific technology in use.

Once all components $P_{sc,i}$ have been computed, considering the overall discretized solar cells area, the total value of captured power can be obtained through a direct sum over i, yielding

$$P_{sc} = \sum_i P_{sc,i}. \tag{5}$$

2.3. Structural Integrity

Since HAA are of interest, the assumption is to design the airship as a super-pressure system, meaning that the pressure differential between the lifting gas and atmospheric air is left free to change, according to daily temperature cycles of the lifting gas, in turn resulting from the cyclic change of the exposure to solar radiation. Therefore, the airship is designed so that at operating altitude and at night, i.e., when the coldest internal temperature occurs and the lifting gas is most contracted, the gas fills the whole envelope, also providing a pressure differential sufficient for preserving shape and envelope rigidity. Conversely, the maximum design internal pressure can be estimated from the maximum expected super heat.

In order to compute the stress on the envelope, three geometric parameters, namely the airship length L, fineness ratio FR, and envelope volume V_{env}, and two environmental specifications, namely the operational altitude z and the maximum wind speed v, are required.

Stress computations on the envelope bear as a primary output a minimum and maximum pressure differential.

2.3.1. Minimum Pressure Differential

The minimum pressure differential is obtained in order to simultaneously satisfy three design objectives, namely:

1. Nose caving. A criterion for granting a sufficient rigidity of the envelope, to avoid nose caving while station keeping in an airstream, is based on the evaluation of the dynamic pressure of the airstream, assuming it turns into a static pressure in the vicinity of the nose cone (ideally on its tip). A maximum airstream speed $v_{\max}$ that the airship will encounter over its entire mission is introduced, and a safety factor of up to $s_{NC} = 50\%$ [12] is typically adopted, thus yielding for this criterion [12,17,38]

$$\Delta P_{nose\text{-}caving} = s_{NC}\frac{1}{2}\rho v_{\max}^2. \tag{6}$$

   It should be remarked that the choice of $v_{\max}$ cannot be known precisely a priori, so a conservative estimation based on the available experience is assumed.

2. Extreme aerodynamic bending. In order to ensure a sufficient rigidity all along the airship envelope, in the presence of aerodynamic loads pushing the ship from the sides, a minimum $\Delta P_{bending}$ can be obtained from an empirical relationship, accounting for the contributions of aerodynamics-induced bending and thrust. The former (aerodynamics-induced) can be obtained from the following relationship [12,17,39]

$$\Delta M_{lat} = 0.02155\Big(1 + (FR - 4)\Big(0.5492L^{0.02} - 0.5\Big)\rho u_g v_g V_{env} L^{0.25}\Big), \tag{7}$$

   where u_g is a gust speed, assigned as $u_g = 25$ ft/s when v_g is taken as the design maximum level flight airspeed (i.e., $v_g = v_{\max}$) or $u_g = 35$ ft/s when v_g is instead taken as the design airspeed for maximum gust intensity, defined as the minimum between 35 kt and $0.65v$. All calculations in this work have been performed in both cases, selecting the most demanding outcome.

The contribution due to thrust can be included directly in the model expressing the minimum pressure differential, yielding the overall side-load pressure differential threshold

$$\Delta P_{side\text{-}load} = \frac{2M_{lat} + F_T b}{\pi b^3}, \tag{8}$$

where F_T is thrust and b is the top radius in the envelope geometry. It should be remarked that the value obtained from Equation (8) is a conservative estimation, since the pressure difference may be reduced by a dynamic suction effect, as well as a hydrostatic effect due to the difference in the density of the lifting gas and surrounding air [17,39]. Such corrections are typically mild however, and Equation (8) captures most of the actual value according to this criterion.

3. Envelope shape consistency at the bottom. As a result of the vertical pressure gradient in the atmosphere, especially for larger airships, a vertical gradient of the pressure difference between the inner and outer sides of the envelope may be experienced as well, yielding in particular a lower pressure differential at the bottom of the envelope. A requirement on the pressure differential according to this criterion comes in the form

$$\Delta P_{hydrostatic} + \Delta P_{consistency} \geq 0 \rightarrow \left(-\rho + \rho_{lg}\right) gb + \Delta P_{consistency} \geq 0, \tag{9}$$

where g is gravitational acceleration, ρ is the density of air, and ρ_{lg} is the density of the lifting gas. The density of the lifting gas can be expressed as a function of the temperature, therefore yielding the following explicit expression for $\Delta P_{consistency}$,

$$\Delta P_{consistency} = s_C \frac{\left(\rho - \left(\frac{k_{lg}}{R_{lg}} + \frac{1-k_{lg}}{R_a}\right)\frac{P}{T_{\min}}\right) g_0\, b}{1 - \left(\frac{k_{lg}}{R_{lg}} + \frac{1-k_{lg}}{R_a}\right)\frac{1}{T_{\min}}\, g_0\, b}, \tag{10}$$

where R_{lg} and R_a are gas kinematic constants of the lifting gas and air, respectively, k_{lg} is the purity ratio of the lifting gas, and $T_{\min}$ the minimum temperature expected along the temperature cycle considered for the design. Finally, s_C is a safety factor.

The actual minimum pressure differential can be computed taking the most stringent outcome of the three criteria just introduced (Equations (6), (8) and (10)), thus yielding

$$\Delta P_{\min} = \max\left(\Delta P_{nose\text{-}caving}, \Delta P_{side\text{-}load}, \Delta P_{consistency}\right). \tag{11}$$

2.3.2. Maximum Pressure Differential and Breaking Stress

Concerning the maximum pressure differential, since the volume of the envelope is fixed, as the temperature of the lifting gas varies throughout the day, its density cannot change. Therefore, the static pressure of the lifting gas varies proportionally to its temperature. The maximum expected differential pressure across the envelope skin can be predicted based on a reasonable assumption of the maximum super heat experienced by the lifting gas, while the airship operates at altitude. Several experimental studies have investigated a prediction for the super-heat maximum temperature [1,40–45]. In consideration of the HAA mission, a conservative (i.e., high) change of temperature of +50 K is hypothesized. Accordingly, the maximum pressure differential can be computed based on Equation (12),

$$\Delta P_{\max} = \frac{T_{\max}}{T_{\min}}(P + \Delta P_{\min}) - P, \tag{12}$$

where P is the atmospheric pressure at altitude and $T_{\max}$ and $T_{\min}$ the maximum and minimum temperatures encountered over the time frame of the mission. From Equation (12), a corresponding tensile stress can be computed, translating the pressure differential $\Delta P_{\max}$

into a circumferential stress component (called hoop stress) and a longitudinal one, acting on the envelope skin. This is achieved through the following models [39]:

$$\begin{aligned}\sigma_{hoop,\max} &= \left(\Delta P_{\max} + C_p \frac{1}{2} \rho\, v_{\max}^2 + \left(\rho - \rho_{lg}\right) g_0\, b\right) b,\\ \sigma_{long,\max} &= \frac{\left(\Delta P_{\max} + C_p \frac{1}{2} \rho\, v_g^2 - \left(\rho - \rho_{lg}\right) g_0\, b\right) b}{2} + \frac{2\, M_{lat} + F_T\, b}{2\,\pi\, b^2},\end{aligned} \tag{13}$$

where C_p is the pressure coefficient, which can be estimated based on basic aerodynamic prediction models [17].

The top value among the two in Equation (13) should be taken into account to yield the constraining value in the sizing process, properly modulated with a safety factor s_σ. In analytic terms,

$$\sigma_{brk} = s_\sigma \max\left(\sigma_{hoop,\max}, \sigma_{long,\max}\right), \tag{14}$$

where σ_{brk} is the breaking stress computed based on the sizing. This quantity shall be verified in the sizing process with respect to the actual ultimate stress of the assumed material for the envelope, σ_{env}.

2.4. Estimation of Mass Components

The estimation of the mass of the components of the airship can be carried out considering the following breakdown,

$$\begin{aligned}M =& M_{pl} + \left(M_{env} + M_{fins} + M_{curt}\right) + \left(M_m + M_p\right)\\ &+ \left(M_{sc} + M_{MPPT} + M_{bat}\right) + \left(M_{lg} + M_{tank}\right) + M_{gon},\end{aligned} \tag{15}$$

where the components are, respectively, those pertaining to the payload, the envelope, fins and curtain suspensions, the motors and propellers, the solar cells, MPPT and batteries, those of the lifting gas and corresponding tank, and that of the gondola. Ballonets are not included since they are not considered in the current design, which calls for a fixed-altitude HAA, which does not need to go through atmospheric layers during deployment. Considering the concept of an HAA of interest here, to be deployed at altitude making use of a missile launcher, the estimation of the components in Equation (15) allows for the definition of the overall mass to be stored on board the launcher. Historical regressions and first-principle relationships allow one to carry out the computation of each component, and they have been selected according to the mission of interest in this research. They will be listed in the following paragraphs.

2.4.1. Envelope, Curtain Suspension, and Fins

The density of the material composing the envelope and fins has been hypothesized to be identical, i.e., $\rho_{env} = \rho_{fins}$ [6,17,21]. The value of the material specific density is a function of the technology of the fabric, and it is related to the tensile strength to be supported. Multiple layers of the same material produce an increase in the top stress supported, at the price of a higher density. The relationship between density and ultimate stress is linear [12]. The stress to be supported can be computed according to the points outlined in Section 2.3. The corresponding surface density is obtained from the relationship [6,17,21]

$$\rho_{env} = k_1 \frac{\sigma}{g} + k_2, \tag{16}$$

where k_1 and k_2 are typical to a material, and σ can be computed based on Equation (13). Clearly, the mass of the envelope can be computed as

$$M_{env} = s_{env} \rho_{env} A_{env}, \tag{17}$$

where s_{env} is a safety factor.

The size of the fin is typically a function of the volume of the airship. The historical regression method proposed in [6] is adopted in this research. It considers the relationship between the total fin area and airship volume of a number of existing airships, to establish a ratio of fin-area-to-airship-volume, which comes out as $R_{fa} = 0.0121\ \text{m}^{-1}$. This method has been compared with the outcome of the volume coefficient approach [12], finding a good accordance for envelope volumes ranging between 5000 m^3 and 40,000 m^3, when the distance between the airship center of gravity and the quarter chord of the vertical/horizontal tail mean aerodynamic chord is assumed to be between 30% and 40% of the length of the airship. Hence, the mass of the fins can be predicted as

$$M_{fin} = k_{fin}\rho_{fin}A_{fin} = k_{fin}\rho_{fin}R_{fa}V_{env}, \tag{18}$$

where k_{fin} is a factor which accounts for the internal reinforcements (cover tape, internal structure tape, spar fabric curtains, etc.) and the increase of mass due to manufacturing. According to the literature [6,17,19,21], a factor of $k_{fin} = 2$, a fairly conservative assumption, can be assumed for the inflatable fins of the HAA design of interest in this work.

Finally, the mass of the curtain load suspension system is given once again by the product between the fabric area density and the area of the curtain, which is estimated according to the method suggested for non-rigid airships in [12]. The curtain surface can be assumed equal to 20% of the envelope side projected area A_{side}, already introduced as a geometric specification for an assigned envelope geometry. Considering as usual a safety factor $k_{curt} = 1.06$ [12], to account for assembly joints, the mass of the suspension system is written as

$$M_{curt} = k_{curt}\rho_{curt}A_{curt} = 0.2k_{curt}\rho_{curt}A_{side}. \tag{19}$$

As previously observed, no ballonets are envisaged for deployment on the HAA of interest here. Most notable examples of HAA for HAPS missions actually do not make use of this system [46,47].

2.4.2. Motor and Propeller

A natural choice for HAA, which benefit from the high level of irradiance typical to the higher layers of the atmosphere, a purely electrical motor plant is assumed for the mission. The relationship between the power and weight of electric motors in the range of interest for aeronautical applications has been investigated in previous research [48,49]. The relationship between the two quantities can be assumed linear. In order to estimate the weight of the electric motor, the value of the maximum required power is therefore needed. This can be computed as

$$P_{motor,\max} = \frac{\frac{1}{2}\rho v_{\max}^3 V_{env}^{2/3} C_D}{\eta_g \eta_p}. \tag{20}$$

Correspondingly, the weight of the electric motor is

$$M_m = P_{motor,\max} d_m + c_m, \tag{21}$$

where coefficients d_m, c_m can be obtained from the literature or by proper regression of a pertinent motor database [49]. A similar approach has been adopted for the weight of the propeller, since a linear relationship exists between the logarithm of maximum propeller power and propeller mass. This is analytically modeled through coefficient w_p. Therefore,

$$M_p = P_{motor,\max} w_p. \tag{22}$$

Both the motor and propeller weights can be increased through multiplication by a factor $w_m > 1$ to account for engine mounting.

2.4.3. Solar Cells, MPPT, and Batteries

The estimation of the weight of the solar cells, and of the MPPT required for cells operation, can be simply carried out introducing a mass density for the former and a mass-to-power ratio for the latter. Consequently, for solar cells a mass of

$$M_{sc} = \rho_{sc} A_{sc} w_{sc}, \tag{23}$$

where ρ_{sc} is the surface mass density of the cells and w_{sc} a correction factor taking into account collectors, grid network, and other electronic and auxiliary components [17,21,50]. For the MPPT, the top power value associated to any of the cells $P_{sc,\max}$ is the driver for the corresponding mass. A mass-to-power ratio w_{MPPT} is introduced [50], such that

$$M_{MPPT} = w_{MPPT} P_{sc,\max}. \tag{24}$$

Concerning the batteries, these are required primarily to store a sufficient amount of energy for nighttime operations of the HAA, taking into account the need to feed the system in peak power conditions. An accurate computation in this sense is typically not carried out in airship design, leading to fairly inaccurate predictions of the energy storage mass. In the present work, an accurate closed-cycle model for the 24-h energy balance of the airship is proposed. According to this model, the required energy storage Q for continuous operation of the system is given by the integral in Equation (25)

$$Q = \int_0^{24\mathrm{h}} \omega_{dn} \frac{s_{pwr} P_{req} - P_{sc}}{\eta_{dsch} \eta_{tr} \eta_{sw}} dt, \tag{25}$$

where it is assumed that one will compute the maximum energy which should be stored in the batteries by computing an integral over a day of operations of the difference between the power captured by the solar cells and the power required by the system for work. The term ω_{dn} is a binary factor with the function of distinguishing between time instants when the solar array output is sufficient to power the airship from those when it is not; therefore, the contribution of the energy storage system is needed to feed the airship plants. Furthermore, s_{pwr} is a safety factor to slightly upscale the whole power system, making it able to cope with peaks in power demand or uncertainties, due for instance to lower-than-expected power generation, unpredictably excessive airstream speeds, or an airship orientation misaligned with respect to the stream direction. A value of $s_{pwr} = 1.1$ is assumed [12]. The terms which appear in the denominator in Equation (25) are the discharge efficiency η_{dsch} of the energy storage system and the transmission and power switch/step efficiencies, η_{tr} and η_{sw}, respectively, which account for the line losses downstream of the system.

Based on Equation (25), the mass of the energy storage system can be predicted by simultaneously taking into account two competing sizing drivers, namely the need to store the amount of electrical energy required for nighttime operation and the need to satisfy any power demand of the platform, meaning that the peak discharge power must be greater than or equal to the maximum power requirement. This way of reasoning, already applied to the field of electrically propelled fixed-wing aircraft [48,49,51], can be put in analytic terms as per Equation (26)

$$M_{bat} = \max\left(\frac{Q}{\eta_{dod}\, q_{bat}}, \frac{\omega_{dn}\left(s_{pwr} P_{req} - P_{sc}\right)}{p_{dsch}}\right) \tag{26}$$

where η_{dod} is the depth-of-discharge or the fraction of actual discharge capability relative to the overall capacity of the energy storage system, and q_{bat} and p_{dsch} are, respectively, the specific energy and the specific discharge power of the battery.

Having defined the required mass of the batteries, or in general of the energy storage system, in order to collect sufficient energy for continuous operation, the closed-cycle model for the 24-h energy balance of the airship can now be completed, by also evaluating the surplus energy available for storage during daytime. Not all the excess power produced by the solar array is taken into account for the energy balance, but consistently with the concepts just introduced, the actual maximum continuous charge rate of the system is considered. Stated another way, if the power production of the solar array presents a peak that exceeds the charge rate of the system, the latter will limit the energy actually added to the system. As a result, the surplus energy effectively available for storage is given by Equation (27),

$$Q_{exc} = \int_0^{24\text{h}} \min\left((1-\omega_{dn})\left(P_{sc} - s_{pwr}\,P_{req}\right)\eta_{dsch}\,\eta_{tr}\,\eta_{sw},\ M_{bat}\,p_{ch}\right)dt, \tag{27}$$

where p_{ch} is the battery specific charging power.

2.4.4. Gondola

The mass of the gondola, which is the part of the airship mainly devoted to the storage of the payload, batteries, and in general all subsystems required for control, navigation, and operation, can be put into a relationship with the corresponding masses, as per Equation (28),

$$M_{gon} = s_{gon} w_{gon}\left(M_{pl} + M_{bat}\right) \tag{28}$$

where the sensitivity $w_{gon} = 0.15$ is introduced [12]. In addition, a safety factor s_{gon} is included in order to oversize the gondola and take into account a possible increase of mass due to ancillary components, such as pipelines, wires, and cables, as well as attachments of the gondola to the hull. By default, a 20% increase is considered; thus, a $s_{gon} = 1.2$ is assumed [12].

It may be pointed out that, with respect to the existing literature, recent research on multi-functional materials [52], and in particular structural batteries, allows one to envisage an at least partial fusion between the gondola structure and the energy storage system. This would be especially interesting in this specific application, since the loads that are to be carried by the gondola should be lower than on the fuselage of aircraft of comparable weight, and they are expected to be generally quasi-steady and highly predictable in the specific application of an HAA. What advantages the inclusion of this technology would bear on the mass of the airship has to be assessed. This investigation has been left to a future development of the present work.

2.4.5. Lifting Gas and Tank

The lifting gas typically adopted on board airships is either helium or hydrogen. The evaluation of the actual mass of the lifting gas onboard is made straightforward by the knowledge of the volume required for inflation V_{env} and of the density of the lifting gas ρ_{lg}, which is determined as a function of temperature, similar to Equation (10). Consequently, the following definition can be applied

$$M_{lg} = \rho_{lg} V_{env}. \tag{29}$$

Since the airship is intended to be lifted to the stationing altitude by means of a missile, a set of pressurized tanks capable of storing the entire mass of the lifting gas M_{lg} up to the stationing altitude, where the HAA will be inflated and released, needs to be taken into account in the missile payload mass. A model for defining the corresponding mass M_{tank} has been taken from the consolidated design practice of hydrogen fuel cells, which need to similarly store relevant masses of hydrogen inside a tank, usually facing volume constraints (fuel cells are currently mostly adopted in transport vehicles) [53,54]. The gravimetric index K_{tank} can be defined as

$$K_{tank} = \frac{M_{lg}}{M_{tank} + M_{lg}}, \tag{30}$$

and is a technological parameter typical to the chosen gas. The value of K_{tank} for the lightest gases (hydrogen, helium) is unfavorable, ranging between 4.2% and 5.8%, thus implying a significant mass of the tank for storing a very limited mass of gas [53,54]. However, according to Equation (30), it is possible to find an explicit definition of M_{tank}. Therefore,

$$M_{tank} = s_{tank}\left(K_{tank}^{-1} - 1\right)M_{lg}. \tag{31}$$

A correction factor s_{tank} is added especially to take into account the presence of a residual lifting gas in the tank, due to the decrease in the pressure differential while the tank is emptied to inflate the airship envelope, leading to equalization and to the usually incomplete expulsion of the lifting gas from the tank.

3. Sizing Algorithm: An Optimal Approach

As outlined in the introduction (Section 1), based on the estimation models presented in the previous Section 2, it is now possible to envisage the complete structure of a sizing loop. As stated when introducing models correlating sizing parameters to one another, special attention has been devoted to the formulation of the sizing problem so as to allow an automatic search of an optimal solution, driven by a general requirement on the minimization of a cost function, while granting the satisfaction of a set of suitable constraints.

3.1. Iteration Workflow

The workflow of the sizing process is introduced in Figure 2. Correspondingly, the flow along a single sizing loop, to be ideally performed in the seek for the optimum of a certain merit function, can be retraced schematically. In the figure, cyan blocks refer to data which need to be guessed or assigned, blue blocks to quantities computed based on the models presented in Section 2, and purple blocks to relevant sizing outputs. The latter are the result of a sizing loop. Based on them, the double-circled pink block to the bottom of the figure ideally represents the comparison of the output of a sizing loop with respect to constraints. This and the chosen objective function (in the present case, total mass) produce an adjustment of the geometry and allow one to run a further iteration (see Section 3.2).

A set of quantities related to the mission (*Mission parameters* block in Figure 2) need to be assigned as constant, i.e., non-negotiable, parameters:

1. Mass and power adsorbed by the payload, M_{pl} and P_{pl}.
2. Geographical location, in terms of latitude λ, longitude ϕ, and stationing altitude z.
3. Solar dates for the mission.

The geographical coordinates and the solar dates (points 2 and 3) are the required input for the evaluation of the intensity and direction of the solar radiation at altitude, which is a major input to compute the irradiance I_g, and therefore power capture, through the model introduced in Equation (4). The computation of the solar radiation at altitude can be performed based on standard models and is not explicitly covered in this work [29]. In a similar fashion, given the geographical position of the airship and its altitude, an estimation of the airstream speed v and direction β_w can be obtained from standard models for the higher levels of the atmosphere [4,27]. These computations are represented by the block *Environmental conditions* in Figure 2.

Further environmental parameters which are required for some of the later computations are the atmospheric pressure and temperature, which depend on the global positioning and solar date. The average density of air ρ and the maximum and minimum temperature $T_{\min}$ and $T_{\max}$ expected during the mission are environmental characteristics and do not depend on the actual sizing of the airship.

In order to assign the geometry, a set of parameters is needed. These are specified in the block *Geometry parameters* in Figure 2. As stated in Section 2.1, in this work a bi-ellipsoidal solid of revolution has been adopted, bearing a description of the envelope geometry by means of its length L and fineness ratio FR. With the geometry completely assigned, it is possible to compute the corresponding envelope volume and areas V_{env}, A_{env}, and A_{side}, as well as the position of the buoyancy center ***BC*** (block *Envelope geometry* in Figure 2). As mentioned, the parameters chosen for assigning the envelope geometry may be different from those chosen in this work, according to an alternative modeling choice in terms of analytic geometry (for instance, more suitable to model a specific shape of the envelope, as is the case for teardrop designs). However, for an assigned geometry, the corresponding values just listed (as well as the fineness ratio, in case it is not a specification as in the present case) can be always computed.

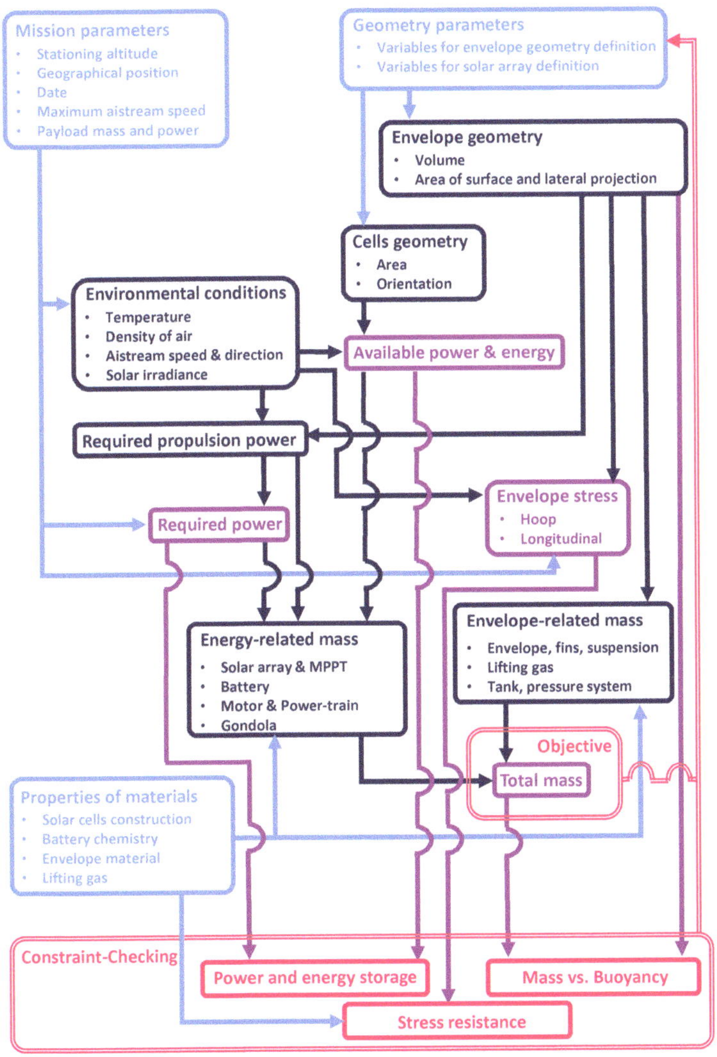

Figure 2. Workflow of the sizing loop and optimization logic.

In order to complete the assignment of the geometry, the parameters describing the extension of solar cells on the corresponding portion of the surface of the envelope are also required. As explained in Section 2.1, this is equal to assigning four further scalar quantities, namely θ_{in}, θ_{out}, x_{te}, and x_{le} (block *Cells geometry* in Figure 2).

Once the geometry of the envelope and cells has been assigned, and the atmospheric environment is quantitatively described, it is possible to compute the power captured and to complete the evaluation of the power needed. In mathematical terms, this translates into computing the power captured by the solar cells P_{sc} (block *Available power & energy*) and the power required P_{req} for keeping the airship in place in the airstream (block *Required propulsion power*), running the payload, and possibly any other subsystem onboard requiring a power expenditure (all components are included in the block *Required power* in Figure 2).

The quantities defined in the process of evaluating power balance allow for the triggering of the computation of the mass of several components in the design. Further required inputs for this step include technological features of the cells, battery, envelope, and lifting gas (block *Properties of materials* in Figure 2). With knowledge of the power captured and needed, it is possible to compute the instantaneous difference between the two, according to the schedule of the mission, i.e., how long it is going to last, and when it will take place in the solar calendar. This is an input to the computation of battery mass M_{bat} (Equation (26)). Similarly, when computing the power required for propulsion, the main input is obtained to compute the power of the motors, and therefore to size up the mass of the motors M_m and propeller M_{prop} (Equations (21) and (22)). The mass of the solar cells M_{sc} is obtained directly from the assignment of their extension, and similarly the mass of the MPPT, M_{MPPT}, can be computed from the power captured. These computations are included in block *Energy-related mass* in Figure 2.

The acquired knowledge of thrust F_T, as well as the definition of the environment, including the minimum and maximum expected temperatures and pressures, allows for the computation of the stress the envelope needs to sustain itself (Equation (13), block *Envelope stress*). This in turn allows for the completion of the computation of mass M_{env}, and similarly of those of the fin and of the suspension system, M_{fin} and M_{curt} (Equations (17)–(19); all computations are included in block *Envelope-related mass*).

Finally, the mass of the gondola M_{gon} and of the lifting gas and tank, M_{lg} and M_{tank}, can be computed based on the corresponding geometrical assignments, as explained in Section 2.4.5.

Table 1 summarizes the fixed technological parameters which are required for computing the mass components. The corresponding numerical values reported therein refer to the reference sizing example presented in Section 4.2. It should be remarked that the super-cold and super-hot temperature changes are such that $T_{min} = T + \Delta T_{cold}$ and $T_{max} = T + \Delta T_{hot}$, i.e., they should be added to the average atmospheric temperature at altitude.

Table 1. Fixed parameters for a sizing loop and assumed values for computations in Section 4.2.

Parameter	Value	Unit	Description
k_{lg}	0.98	-	Lifting gas purity.
s_{NC}	1.5	-	Pressure differential safety factor for nose-caving.
ΔT_{cold}	0	[K]	Super-cold temperature difference.
ΔT_{hot}	50	[K]	Super-hot temperature difference.
k_1	2.1×10^{-5}	[1/m]	Sensitivity of ρ_{env} to stress.
k_2	0.14	[kg/m^2]	ρ_{env} constant in stress dependency.
s_{env}	1.2	-	Factor for envelope mass increase.
σ_{env}	900	[N/cm]	Tensile strength (tenacity) of the envelope material.
s_σ	4	-	Breaking strength safety factor.
R_{fa}	0.0121	[m^{-1}]	Ratio of fin area to airship volume.
k_{fin}	2.0	-	Factor for fins mass increase.
k_{curt}	1.06	-	Factor for curtain suspension system mass increase.
η_p	0.80	-	Propeller efficiency.
η_m	0.90	-	Electric motor efficiency.
η_g	0.97	-	Gear transmission efficiency.
d_m	-	[kg/W]	Electric motor mass-to-power ratio (regression assumed with $c_m = 0$).
w_p	5×10^{-4}	[kg/W]	Propeller mass-to-power ratio.
w_m	1.2	-	Engine mounts to propulsion group mass ratio.
η_{sc}	0.13	-	Solar array conversion efficiency.

Table 1. *Cont.*

Parameter	Value	Unit	Description
η_{pack}	0.90	-	Solar array adjoining cells packing efficiency.
ρ_{sc}	0.150	[kg/m^2]	Solar array area density.
w_{sc}	1.3	-	Multiplicative factor for solar array mass increase.
η_{MPPT}	0.97	-	MPPT efficiency.
w_{MPPT}	1/2368	[kg/W]	MPPT mass to maximum power ratio.
η_{ch}	0.96	-	Energy storage system charge efficiency.
η_{dsch}	0.96	-	Energy storage system discharge efficiency.
η_{dod}	0.95	-	Energy storage system Depth-of-Discharge.
q_{bat}	250	[Wh/kg]	Energy storage system specific energy.
p_{ch}	NaN	[W/kg]	Energy stor. syst. specific charge power (computed from q_{bat} here).
p_{dsch}	NaN	[W/kg]	Energy stor. syst. specific discharge power (computed from q_{bat} here).
s_{pwr}	1.1	-	Power requirement safety factor.
η_{tr}	0.98	-	Transmission efficiency in/out the energy storage system.
η_{sw}	0.90	-	Power switch/step efficiency in/out the energy storage system.
w_{gon}	0.15	-	Gondola to payload mass ratio.
s_{gon}	1.2	-	Gondola mass safety factor.
K_{tank}^{-1}	18	-	Inverse of lifting gas storage system specific capacity.
s_{tank}	1.1	-	Tank mass safety factor.

3.2. Optimal Approach

The workflow described in the previous section can be adopted to run an optimal sizing problem, where the target of the optimization is that of minimizing the overall mass of the system. This is especially relevant for the intended mission profile of the HAA under study, since the entire airship should constitute the payload of a missile.

In analytic terms, the optimal problem can be formulated as

$$\begin{gathered}\min_{p} M\\ \text{s.t. } \boldsymbol{c}, \boldsymbol{b}\end{gathered} \tag{32}$$

where M has been defined in Equation (15), optimization parameters $\boldsymbol{p}$ are expressed by

$$\boldsymbol{p} = (L, FR, x_{le}, x_{te}, \theta_{in}, \theta_{out}) \tag{33}$$

and the sets of constraints and bounds, $\boldsymbol{c}$ and $\boldsymbol{b}$, are defined as follows,

$$\boldsymbol{c}: \begin{cases} M_{airship} - \rho_{env} V_{env} \leq 0 \\ Q - Q_{exc} \leq 0 \\ x_{te} - L + 2 \leq 0 \\ x_{le} - x_{te} \leq 0 \\ \theta_{in} - \theta_{out} \leq 0 \\ FR - FR_{\max} \leq 0 \\ -FR + FR_{\min} \leq 0 \end{cases}, \quad \boldsymbol{b}: \begin{cases} L_{\min} \leq L \leq L_{\max} \\ x_{le,\min} \leq x_{le} \leq L_{\max} \\ x_{te,\min} \leq x_{te} \leq L_{\max} \\ 0 \leq \theta_{in} \leq \pi \\ 0 \leq \theta_{out} \leq \pi \end{cases}. \tag{34}$$

The first two inequality constraints in $\boldsymbol{c}$ are non-linear with respect to the optimization parameters and require a minimum lifting action from envelope buoyancy and a minimum energy storage capacity from the batteries. As can be noticed from Equation (34), the mass that needs to be sustained by the airship at altitude is not that of the entire system M. The latter is of course targeted in the optimization since it is the payload of the missile launcher. However, it is hypothesized that the tanks of the lifting gas are left on the launcher following inflation, thus reducing the required lifting force by a due amount. Actually, as will be clear from the results, that reduction in the lifting requirement from M to $M_{airship}$—where the latter is the mass of the HAA, without the tanks—is rather dramatic, since the mass of the tanks M_{tank} is one of the primary components of the overall mass of the system in the presented study cases.

The other constraints in $\boldsymbol{c}$ (Equation (34)) are linear in the parameters and enforce consistency between different geometric variables defining the shape of the airship, consistency between longitudinal coordinates of the solar array segments and length of the airship, consistency between leading edge and trailing edge coordinates as well as inner and outer central angles of the same segment, and finally consistency between the longitudinal and angular coordinates of different segments, so as to avoid overlapping. The fineness ratio is constrained between values $FR_{\min}$ and $FR_{\max}$, chosen so as to avoid unrealistic geometries. Similarly, the bounds on the optimization variables in set $\boldsymbol{b}$ help to prevent unrealistic sizing solutions.

The continuity and regular shape of the merit function and constraining equations in Equation (32) make the optimal problem ideally trackable by means of a gradient-based algorithm. This was adopted with largely satisfactory results (both in terms of optimum resolution and machine-time) working on the test cases presented in the next sections.

4. Application Results

4.1. Validation of the Sizing Loop

Before launching the sizing algorithm within the body of an optimization, thus letting it search autonomously for a design solution, it is worth testing the models and assumptions which constitute one iteration of the sizing procedure on the data of an existing airship. The comparison between the actual sizing and the outcome of the design loop may support the validity of the latter for obtaining results in real field applications.

The test case chosen for this validation is the HiSentinel80. That was a single-use, non-rigid, fixed volume, single chamber airship, with inflatable tail fins and a recoverable equipment pod [13]. It was 199 ft (60.6 m) long and featured a 45.5 ft (13.9 m) diameter, for a total gas volume of 6846 m^3, and its envelope was made of a lightweight Vectran® based material, including Nylon as the gas barrier [13,46]. The non-gaseous mass of the airship without payload was 1068 lbs (484.4 kg), while the helium mass was 212 lbs (96.3 kg) [46]. It was designed to operate at 65,000 ft (19,812 m) at an average cruise speed of 18 kts (9.3 m/s) and for a mission duration up to 24-h [13,46]. Power was supplied by batteries supplemented by a non-pointed 1.2 kW thin-film, flexible photovoltaic array, mounted inside the hull near the nose (70% light transmission through the envelope). It was mounted as such to reach a horizontal configuration once at altitude, as depicted in Figure 3 [13,46]. Propulsion was provided by one tail-mounted electric-motor-driven propeller [13], and a minimum differential pressure exceeding 150 Pa was required across the envelope to eliminate buckling in the conical section at the propulsion strut ends [46].

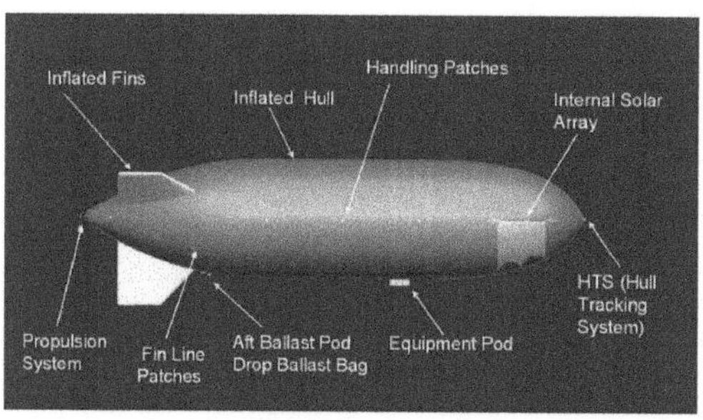

Figure 3. HiSentinel80 airframe configuration [46].

Worst case loading of the hull was expected to occur during the day at a temperature close to 0 °C [46], while minimum helium temperatures in the order of 200 K were expected at nighttime [55].

The HiSentinel80 mission payload was housed in an insulated, cuboid container with dimensions 23-23-30 in. Passive heating and active, electrical heating maintained internal temperature when the payload was not powered. Otherwise, power dissipated by the payload during operation maintained the internal temperature. The payload housed a repeater, an Iridium transceiver, a high-resolution camera system, data storage, a high data rate transceiver, a GPS, an environmental monitoring package, a computer stack, and a power management and control subsystem, resulting in a payload mass of 86.2 lbs (39.1 kg) and a required power of 50 W [13,46].

Mission specifications for the design of the HiSentinel80 considered for the validation of the methodology are listed in Table 2. The location is assumed to be Page, Arizona, where the launch of the flight test performed on 10 November 2010 took place, and winter solstice is considered, as it is the most critical day of the year concerning solar energy harvesting. The airship had no ballonets.

Table 2. Parameters for the sizing tool validation on the HiSentinel80.

Parameter	Value	Unit
z	19.812	[km]
Δz	0	[km]
ϕ	36.9142	[°]
λ	−111.4600	[°]
Date	21 December	-
M_{pl}	39.1	[kg]
P_{pl}	50	[W]
$v_{\max}$	13.38	[m/s]

Due to the lack of information on the HiSentinel80 concerning the technological part, as well as some specific features of this airship, assumptions were made in some cases, slightly modifying the procedure shown in the previous sections, justified *a posteriori* by the good results of the validation. In order to obtain a good agreement with the shape of the HiSentinel80 shown in Figure 3, a shape with a cylindrical mid-section is considered, instead of a bi-ellipsoid one. A specific value of prescribed minimum pressure differential has been taken into account, required to prevent buckling, an issue inherent to the HiSentinel80, thus substituting the computation of the corresponding threshold value. Concerning the envelope material, since it is a light-weight Vectran® based fabric, its sizing is performed considering the model derived by Carichner and Nicolai for Vectran® laminate [12]. Finally, further amendments pertain to the technological parameters of batteries and solar array. On account of the relatively older age of the airship, a low specific energy value of $q_{bat} = 200$ Wh/kg is considered for the batteries. Concerning the solar array, it is assumed that a-Si cells were employed, similar to other airships with a similar mission [15]. Consistently, proper values of area density and efficiency are assumed based on product data-sheets. Furthermore, in order to take into account the increase in mass due to the internal solar array assembly, which allows its horizontal deployment during ascent, also a higher value of the multiplicative factor w_{sc} is assumed. The parameters modified with respect to the values in Table 1 for the purpose of validation are listed in Table 3.

Table 3. Parameters assigned specifically for the validation on the HiSentinel80.

Parameter	Value	Unit	Comment
ΔP	150	[Pa]	for stress computation, assigned, not computed
ΔT_{cold}	−15	[K]	wrt. temperature at altitude
ΔT_{hot}	70	[K]	wrt. temperature at altitude
q_{bat}	200	[Wh/kg]	
ρ_{sc}	0.120	[kg/m^2]	
η_{sc}	0.08	-	

Concerning geometrical sizing, the length and fineness ratio of the HiSentinel80 are known, but the remaining information is not. Assuming, as stated, a geometry more resembling the actual one, proper values for the semi-axes of the rear and front semi-ellipsoids are then determined with a trial and error approach, with the goal of obtaining an envelope volume sufficiently close to the real one. Concerning the solar array, since it is not located on the hull surface as typically done in stratospheric airships, slight modifications are required with respect to the normal sizing methodology. The solar array arrangement can be modeled more simply as a single horizontal, rectangular array inside the hull. Consequently, θ_{in} and θ_{out} are not assigned, and the solar array layout can be defined by the two longitudinal coordinates x_{le} and x_{te}, plus a half-width Δy. Since the array is horizontal, its actual location inside the hull is not relevant to the incident solar radiation and power production calculations, so proper values for the three geometrical parameters are determined based on the required solar array area, which is known, and again considering the goal to obtain a good agreement with Figure 3. Indeed, knowing that the solar array has a nominal power of 1.2 kW, once efficiencies are assumed, the area of the solar cells array can be determined. A value of $\eta_{sc} = 8\%$ is assumed, and an area $A_{sc} = 15.002\ \text{m}^2$ is obtained here, which corresponds with the actual value ($A_{sc} = 15\ \text{m}^2$). Values considered for the parameters defining the geometry of the airship and the arrangement of the solar array are listed in Table 4 (where a_1 and a_2 represent the characteristic dimensions of the envelope section), and the resulting layout considered for the validation is shown in Figure 4.

Table 4. HiSentinel80 assumed design parameters.

Parameter	Value	Unit
L	60.65	[m]
FR	4.374	-
a_1	19.05	[m]
a_2	26.94	[m]
x_{le}	5.50	[m]
x_{te}	10.00	[m]
Δy	1.67	[m]

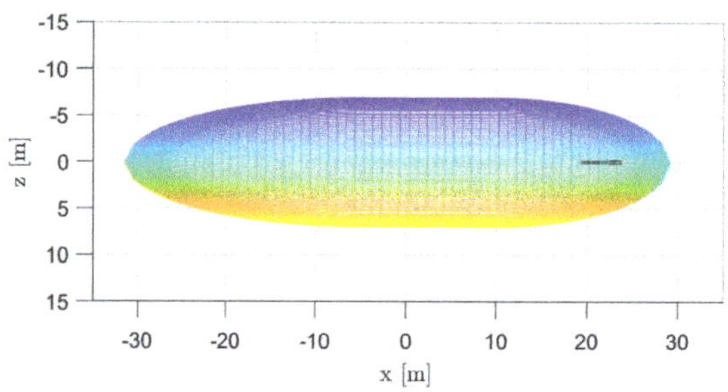

Figure 4. HiSentinel80 modeled geometry, considered for validation.

It is worth making a few more comments about some differences in how validation is here performed, with respect to the sizing loop introduced in Section 3.1.

Firstly, it is assumed that the system is sized in order to operate on a specific day and location, without satisfying the 24-h energy balance, that is, without respecting the constraint that the surplus energy generated by the solar array during daytime is sufficient to recharge batteries for nighttime operation, as required for more-than-one-day missions. In fact, since the HiSentinel80 is designed for a mission duration of up to one day, the need for battery recharge is not considered, and it is assumed that the system is sized such that the batteries, starting from a fully charged state, together with the solar array, can power the airship for 24 h.

Secondly, the wind speed profile is not computed from models, but an average cruise speed of 18 kts, reported by references for this specific airship, is considered to compute the required propulsive power. It was however verified that this airstream speed value is close to the prediction of the adopted HWM model. Next, in order to take into account the fact that the solar array is located inside the hull and only 70% of the radiation is transmitted through it, after the broadband global radiation is computed as explained in this work, its value is multiplied by a 0.7 safety factor.

A comparison between the known details about the HiSentinel80 design and the predictions produced by the sizing methodology is shown in Table 5.

Table 5. Comparison between HiSentinel80 and methodology predictions.

Parameter	HiSentinel80	Prediction	Unit	Error
V_{env}	6846.02	6846.68	[m^3]	0.01%
M	-	627.43	[kg]	-
M_{solid}	484.44	480.88	[kg]	−0.73%
M_{lg}	-	107.45	[kg]	-
M_{He}	96.16	93.62	[kg]	−2.64%

In Table 5, M_{solid} is the total mass of the airship without gases, and M_{He} is the mass of helium onboard.

Despite the many uncertainties in the technological parameters, coped with by making reasonable assumptions and in some cases taking values from airships with a comparable mission, the results obtained confirm the reliability of the sizing approach and of the specific models which have been adopted.

4.2. Optimal Sizing and Parameter Analysis

The optimal approach introduced in Section 3.2 has been applied to a sizing scenario where an assigned HAPS mission payload of $M_{pl} = 10$ kg, with a consumed power of $P_{pl} = 100$ W, needs to be positioned and operated for one year at a fixed altitude of $z = 20$ km.

To better show how the automatic procedure can be exploited, four sizing problems have been solved, where the geographical position is chosen corresponding to four different locations, namely Pontianak (Indonesia), Port-au-Prince (Haiti), Houston (Texas), and L'Aquila (Italy), stricken by natural disasters over the last two decades. The corresponding coordinates vary significantly, as shown in Table 6.

Table 6. Mission design specifications for operations in the considered case studies.

Parameter	Pontianak	Port-au-Prince	Houston	L'Aquila	Unit
z	20	20	20	20	[km]
ϕ	−0.0206	18.5754	29.7499	42.3540	[°]
λ	109.3414	−72.2947	−95.3584	13.3920	[°]
Date	26 January	5 August	3 August	19 January	-
M_{pl}	10	10	10	10	[kg]
P_{pl}	100	100	100	100	[W]
$v_{\max}$	30	30	30	30	[m/s]

Two remarks concern the settings of the sizing problem solved here.

The first is about the day of the year specified in the table. The sizing of a HAPS should be such to sustain operations over an unlimited time frame. This would require in principle the analysis of power balance over the intended time frame, taking into account the worst conditions of solar energy harvesting, due to seasonal change in irradiance especially at higher latitudes, combined with the worst wind intensity, which produces a higher power needed to keep the airship in position. In order to save on machine time, instead of analyzing the performance over the entire yearly calendar, the worst day is selected based on the worst (most intense) wind condition, which is typically the effect producing the more intense effect on sizing. A posteriori, the so-obtained sizing is checked on the day with the lowest overall solar power harvesting, in order to make sure that that sizing allows for the satisfaction of the power balance also in that case. If the latter turns out to require more than the one initially hypothesized, the representative day for the sizing is changed correspondingly. However, it should be noted that this way of proceeding is not substantial—it is of course possible to simply extend the time frame of the analysis from one day to one full year, without the need to do any crosschecks, at the price of some increase in the required machine time.

The second remark in Table 6 concerns the choice of the maximum airstream speed for station keeping $v_{\max}$ which is chosen equal for all the sizing problems at hand, similarly to the stationing altitude. Specifically, the value chosen for speed is very conservative considering all four locations. This choice was made to allow a fairer comparison among the scenarios, which will be analyzed next.

4.2.1. Optimal Sizing at 20 km Stationing Altitude in a Single Location—Pontianak, Indonesia

In this paragraph, the complete sizing for the case of Pontianak will be shown, whereas in the next a summary and comparison of the results for all four considered geographical locations will be proposed.

The first location, Pontianak, Indonesia, is located very close to the equator. The most critical day on which the design of the system is performed is identified according to the procedure illustrated in the previous paragraph. Figure 5 shows how wind speed varies throughout the year at 20 km altitude above Pontianak. Based on this annual wind profile, 3 February is initially selected as the day of the year that most requires the design.

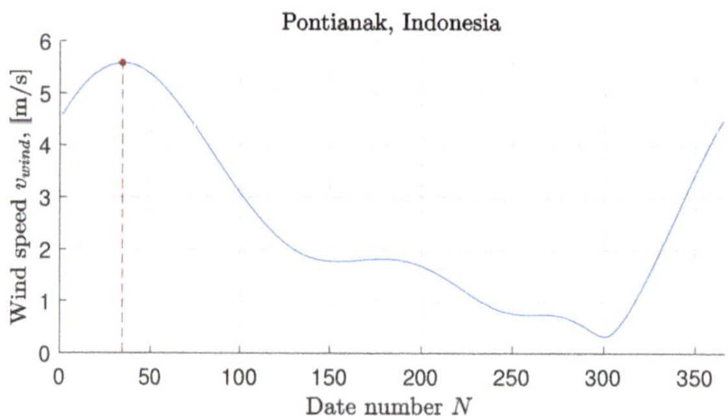

Figure 5. Airstream speed over the year at a 20 km altitude above Pontianak, Indonesia (0°1′14″ S, 109°20′29″ E). Values measured at midnight.

The resulting sizing solution, however, proves to be incapable of meeting the 24-h energy demand on 26 January, when the energy available for storage is 2.37% less than what is required. Thus, 26 January is selected as a new design date. The definitive mission specifications for the design of the HAPS platform to operate continuously year-round at 20 km altitude above Pontianak have been reported in Table 6.

The values of the design variables output by the optimization algorithm are reported in Table 7, whereas Figure 6 shows the layout of the corresponding optimal design for the airship.

Table 7. Design variables defining the HAPS optimal layout for operation at 20 km altitude above Pontianak, Indonesia.

Parameter	Value	Unit
L	48.84	[m]
FR	3.667	-
x_{le}	34.09	[m]
x_{te}	40.04	[m]
θ_{in}	1.331	[rad]
θ_{out}	1.697	[rad]

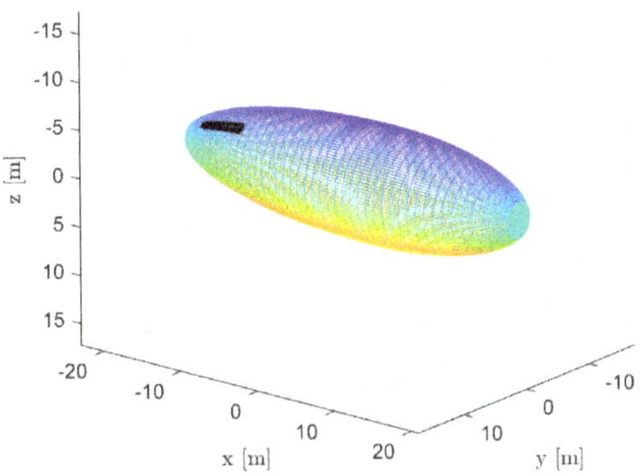

Figure 6. HAPS optimal layout for operation at 20 km altitude above Pontianak, Indonesia.

The details of the optimal solution and its components, as output by an implementation of the optimal design algorithm, are presented in Figures 7 and 8, which provide, respectively, the mass breakdown and the constraints evaluation obtained from the methodology.

```
MASS BREAKDOWN

- TOTAL LAUNCHER PAYLOAD: 682.501 kg
    - SOLIDS MASS W/O PAYLOAD: 235.809 kg
        - HULL: 177.834 kg
            - Envelope: 158.957 kg
            - Fins: 7.889 kg
            - Septum/curtain suspension system: 10.988 kg
        - GONDOLA: 4.844 kg
        - PROPULSION GROUP: 34.090 kg
            - Electric motors: 7.640 kg
            - Propellers: 7.856 kg
            - Mounts: 18.594 kg
        - POWER SYSTEM: 19.042 kg
            - Solar array: 1.758 kg
            - Batteries: 16.909 kg
            - Mppt: 0.375 kg
    - PURE LIFTING GAS FOR ENVELOPE FILLING: 39.923 kg
    - PAYLOAD: 10.000 kg
    - LIFTING GAS STORAGE SYSTEM: 396.769 kg
        - Excess of pure lifting gas to be stored: 1.857 kg
        - Lifting gas storage system (tanks, valves, etc.): 328.298 kg
        - Lifting gas storage system pod: 66.614 kg
- LIFTING GAS IMPURITIES WITHIN THE ENVELOPE: 5.896 kg
```

Figure 7. Mass breakdown of the HAPS optimal design for operation at 20 km altitude above Pontianak, Indonesia.

Finally, Figure 9 shows the feasibility of the obtained solution for year-round operation, reporting both the evaluation of the 24-h energy balance and the required battery capacity throughout the year. In particular, having selected January 26 as the date for the design, the 24-h energy balance is strictly satisfied on every day of the year even considering a 10%

increase in power consumption assumed in the design for safety ($s_{pwr} = 1.1$). Concerning the energy storage requirement, instead, a maximum additional 0.19% mass of batteries is still required on 29 January. However, considering the actual power consumption predicted by the methodology ($s_{pwr} = 1.0$) instead of that increased with a 10% margin adopted as mentioned, the capacity of batteries is also sufficient for every day of the year, and a good margin is maintained even on 29 January.

```
WEIGHT-BUOYANCY BALANCE CONSTRAINT:

- Total weight: 2859.893 N
- Total buoyancy: 2859.894 N

ENERGY BALANCE CONSTRAINT:

- Surplus energy available for storage: 4.016 kWh
- Energy storage requirement: 4.016 kWh

ENVELOPE MATERIAL STRESS CONSTRAINT:

- Envelope material breaking strength: 322.675 N/cm
- Envelope material limit load (accounting for safety factor): 322.675 N/cm
```

Figure 8. Constraints check for HAPS optimal design for operation at 20 km altitude above Pontianak, Indonesia.

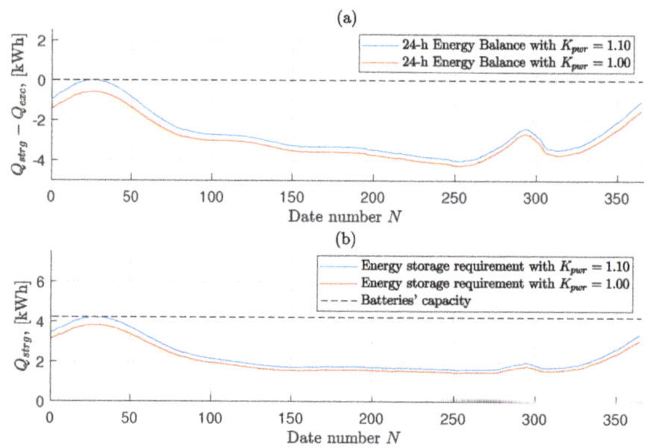

Figure 9. Energy balance evaluation (**a**) and energy storage requirement (**b**) for operation throughout the year at 20 km altitude above Pontianak, Indonesia.

4.2.2. Comparison of Optimal Sizing Results for Different Geographical Locations at 20 km Altitude

The main parameters output by the optimization process for the four case studies just introduced above (Section 4.2) are reported in Table 8.

The influence of the specific stratospheric environment over the different geographical locations on airship sizing is evident from Table 8. Required propulsion power increases with the third power of the wind speed, increasing the need to store energy to ensure continuous operation. This results in an increase of the energy storage system mass, as well as that of the solar array. Furthermore, power consumption is also dependent on the geometrical size of the airship. Increasing airship volume provides increased lift, required to balance the increased weight of the overall system. However, the increased size induces an increase in power required, due to the larger drag of the airship. This in turn increases the mass of the airship again. Because of this connection between airship size, lifting capacity, and power production and consumption, converging to a feasible solution with a sizing such to generate the required lift and capable of producing and storing enough energy suggests an iterative process, in which alterations in mass or power consumption can translate into large changes in airship size. This stands in support of the choice of an automatic solver in the search for a sizing solution.

Table 8. Comparison between HAPS optimal designs for operation at 20 km altitude above the four considered locations.

Param.	Unit	Pontianak	Port-au-Prince	Houston	L'Aquila
L	[m]	43.84	128.53	68.92	117.42
FR	-	3.667	5.185	4.422	5.158
x_{le}	[m]	34.09	2.00	2.00	38.37
x_{te}	[m]	40.04	88.24	55.55	104.59
θ_{in}	[rad]	1.331	0.000	2.121	2.070
θ_{out}	[rad]	1.697	0.462	2.443	2.876
V_{env}	[m^3]	3280.043	41,365.970	8765.707	31,858.069
A_{env}	[m^2]	1332.873	7991.302	2708.574	6703.987
A_{sc}	[m^2]	9.014	435.005	118.360	544.908
σ_{brk}	[N/cm]	322.675	668.870	420.618	614.189
ρ_{env}	[kg/m^2]	0.099	0.156	0.116	0.147
Q	[kWh]	4.016	213.159	26.544	148.232
M	[kg]	682.501	8607.296	1823.940	6628.923
$M_{airship}$	[kg]	291.628	3677.840	779.357	2832.494
M_{tank}	[kg]	396.769	5003.812	1060.339	3853.694
M_{solid}	[kg]	235.809	3090.002	646.910	2377.471
M_{env}	[kg]	158.957	1499.369	375.410	1185.444
M_{sc}	[kg]	1.758	84.826	23.080	106.257
M_{bat}	[kg]	16.909	897.510	111.764	624.135
M_{prop}	[kg]	34.090	155.061	60.377	132.036
M_{gon}	[kg]	4.844	163.352	21.917	114.144
M_{lg}	[kg]	45.819	577.838	122.448	445.023

The coupling of the required and available power with the environment makes the sizing dependent on both where and when it is to be flown. Hence, the extremely low wind speed in Pontianak, combined with the fact that, lying on the equator, the number of daylight hours is practically unchanged throughout the year, result in a system with a size and weight which are smaller compared to those resulting for other locations—even by an order of magnitude comparing the cases of Port-au-Prince and L'Aquila.

Airstream speed, in particular, is a main driver of the sizing. Actually, the system designed for operation above Port-au-Prince, where wind speeds are the highest at 20 km altitude, is characterized by the largest size and energy storage requirement, and consequently also by the highest total mass, as well as the highest mass of many components. Similarly, the maximum wind speed at 20 km altitude above L'Aquila is higher than above Houston, and consequently the system designed for the former location features a higher power consumption and larger size and mass.

However, where the difference in maximum wind speed are comparable, the increase in overall mass between L'Aquila and Houston is much larger compared to that between Port-au-Prince and L'Aquila. The reason, which can be understood by looking at the energy storage requirements, lies in the critical conditions that drove the design in the three cases. On the one hand, at 20 km altitude Port-au-Prince and Houston have their critical day of the year in summer, when there are more daylight hours and thus the energy storage requirement is made less stringent by the shorter time for which the solar array is not capturing power. The opposite scenario is encountered in L'Aquila, where the most critical day at 20 km altitude occurs in the middle of the winter, when less than 10 daylight hours are available, and thus the energy storage requirement is more stringent. This difference is also reflected by the required area and mass of the solar array. Despite the lower energy storage requirement, the optimal solution for operation above L'Aquila is characterized by a larger solar array than the one designed to operate above Port-au-Prince. In fact, due to the fewer daylight hours on the critical design date, in the former case the energy for night operation must be collected in a shorter time, meaning that a higher output power from the solar array, and hence a larger solar cell surface, are required.

Broadly speaking, it should be observed that for locations such as Port-au-Prince and L'Aquila, and to a lesser extent Houston, a system with a large mass is required to carry a payload which is relatively limited in both mass and power consumption. This would practically result in the unfeasibility of a missile-deployed HAPS concept in such locations, at the considered altitude and with the adopted baseline technologies. This is due to the inability of such a heavy system to fit into any small enough missile that would allow the ease of transport and handling and the fast deployment required to meet the time responsiveness goal underlying the proposed concept. By carefully analyzing the weight distribution among the various components listed in Table 8, where envelope and batteries are the major contributors to the mass of the airship (except in the Pontianak case, in which the energy storage requirement is minimal), what actually contributes most to the excessive weight of the system is not the airship itself but rather the lifting gas storage system, which due to the low gravimetric capacity of state-of-the-art tanks takes a portion around 58% of the total weight of the airship, and therefore of the missile payload.

A further consideration applies, which supports the use of an automatic approach to the sizing, considering the trade-off between length and fineness ratio to obtain the volume required for the airship in order to balance its own weight. In fact, such volume can be obtained either with a combination of increased length and fineness ratio, or, on the contrary, with a combination of reduced length and fineness ratio. A number of aspects relevant to airship design and operation are influenced by its geometry, meaning that manually selecting such combination would require taking into account several different considerations. For instance, volume and area of the airship do not vary in the same way for changing length and fineness ratio, but the more the shape deviates from the spherical one, the larger the surface area of the hull becomes. Then, the volumetric drag coefficient depends on the shape of the airship, and specifically it is a function of the fineness ratio, with an optimal value depending on the operational Reynolds number range of the airship. In addition, the stress acting on the envelope skin is influenced by the shape of the hull, both the hoop and the longitudinal stresses being influenced by the diameter and the fineness ratio. Finally, a higher length or a larger diameter may be needed to place the solar array such that it is better radiated by sunlight.

The optimization algorithm can conveniently manage all these conflicting factors and determine the optimal layout depending on the required size of the airship, the specific conditions in which it should operate, and the characteristics of the components employed (strength-to-weight ratio of the envelope material, efficiency of the solar array, etc.), resulting in different combinations of length and fineness ratio defining the geometry of the platform in different case studies.

4.3. Studying the Effect of Stationing Altitude

The purpose of this section is to show how the selection of an appropriate operational altitude can considerably influence the sizing and design of the HAA and specifically grant the potential feasibility of the missile-deployed HAPS concept, for the reasons bound to the missile payload outlined in the final remarks of Section 4.2.2. The goal of the proposed analysis is not to determine the optimal altitude at which the airship should be flown over a certain location, i.e., the stationing altitude is not considered as a further optimization parameter, but it is treated as a parameter to investigate how sensitive the design is to it.

The approach presented here is based on a model trying to predict the relative drag acting on the airship, as a function of airstream speed and air density only. Actually, Equation (35) is used here, inspired by [6],

$$P_{rel} = \frac{v_{wind}^3}{\rho^{2/3}} \quad (35)$$

where the relative drag has been multiplied by the airstream speed v_{wind}, since the required propulsive power is assumed to be the quantity to be minimized in order to effectively reduce the size and weight of the airship. For a given location, Equation (35) is used to build a three-dimensional surface representing the relative propulsive power P_{rel} required for station keeping as a function of both altitude and day of the year. This way, for each altitude the most critical day of the year can be identified based on this relative propulsive power only. The altitude at which the relative propulsive power on the worst day is the smallest can be assumed as the most suitable for airship operation in that geographical location.

Clearly, the stationing altitude obtained with this method will be sub-optimal due to a number of reasons. Firstly, Equation (35) provides just an estimate, which obviously cannot capture all aspects that influence the power requirement of the airship. Furthermore, as in the previous section, airstream speed and thus relative propulsive power are evaluated at a given time (midnight), which might result in an underestimation of the top value adopted for sizing. Finally, airstream speed and related propulsive power are not the only aspects relevant to the operation of the airship.

The plots in Figure 10 show the three-dimensional surfaces obtained for the four considered locations, representing the relative propulsive power required for station keeping evaluated at midnight of each day of the year, for altitudes between 17.5 km and 21 km. Pontianak and L'Aquila are characterized by a specific time of the year, namely winter, in which wind speed and thus relative propulsive power are higher compared to the rest of the year. The most advisable altitude for the operation of the airship is on the lowest point of the only ridge on the surface (marked by a red dot on the plots). Conversely, Port-au-Prince and Houston are characterized by a higher variability of the airstream speed with altitude throughout the year, resulting in two distinct ridges, one in winter, which decreases with altitude, and the other in summer, which instead increases with altitude. The most convenient altitude for year-round operation is where the two ridges are equally high (marked again by a red dot on the winter ridge only).

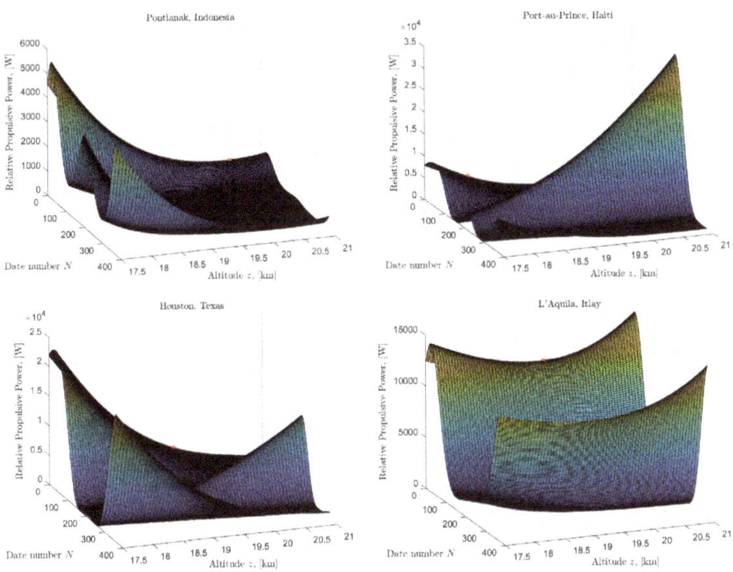

Figure 10. Three-dimensional surface representing the relative propulsive power required for station keeping on each day of the year (at midnight), at altitudes between 17.5 km and 21 km. (**Top row**): (**left**), Pontianak, (**right**), Port-au-Prince. (**Bottom row**): (**left**), Houston, (**right**), L'Aquila.

The stationing altitude suggested for Pontianak and L'Aquila is, respectively, 20.38 km and 19.41 km, but in both cases the slope of the ridge for altitudes around the minimum is small, meaning there is no significant difference in considering these values or a 20 km altitude as previously assumed. On the contrary, for Port-au-Prince and Houston, for which the most convenient altitudes resulting from the considered approach are 18.06 km and 19.46 km, respectively, due to the considerable slope of the two ridges, i.e., the sensitivity of required propulsive power with altitude, a significant difference between the new altitude instead of the 20 km previously assumed shall be encountered.

Re-Design for a Different Altitude, Port-au-Prince Case

Considering the case of Port-au-Prince as an example, in order to assess the convenience of the estimated 18.06 km stationing altitude compared to the 20 km previously considered, the design of the system has been re-performed for such location, changing only the operating altitude and design day. Concerning the most critical day for the design, the date corresponding to the red dot in Figure 10, namely February 17, is selected as a first guess. The annual wind profile at the new fixed altitude of 18.06 km above Port-au-Prince is shown in Figure 11. This can be extracted from Figure 10 in correspondence to the considered altitude.

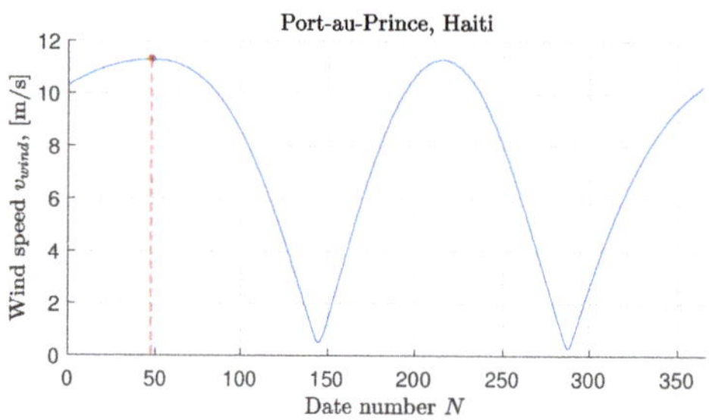

Figure 11. Yearly variability in airstream speed (evaluated at midnight of each day) at 18.059 km altitude above Port-au-Prince, Haiti (18°34′31″ N, 72°17′41″ W).

The selection of the design date is refined with the same approach previously explained (Section 4.2). With the optimal solution resulting from the design performed on 17 February, both a maximum 12.67% violation of the 24-h energy balance and a maximum 18.53% additional energy storage requirement occur on 5 August. Therefore, the design of the HAPS platform for continuous operation throughout the year at 18.06 km altitude above Port-au-Prince is repeated considering this new date, resulting in a set of mission specifications reported in Table 9.

Table 9. HAPS mission design specifications for operation at 18.06 km altitude above Port-au-Prince, Haiti.

Parameter	Value	Unit
z	18.06	[km]
ϕ	18.5754	[°]
λ	−72.2947	[°]
Date	5 August	-
M_{pl}	10	[kg]
P_{pl}	100	[W]
$v_{\max}$	30	[m/s]

The optimal layout resulting from the design methodology is given in Table 10 and Figure 12.

Table 10. Design variables defining the optimal sizing for operation at 18.06 km altitude above Port-au-Prince, Haiti.

Parameter	Value	Unit
L	67.71	[m]
FR	4.791	-
x_{le}	2.00	[m]
x_{te}	40.98	[m]
θ_{in}	1.703	[rad]
θ_{out}	2.184	[rad]

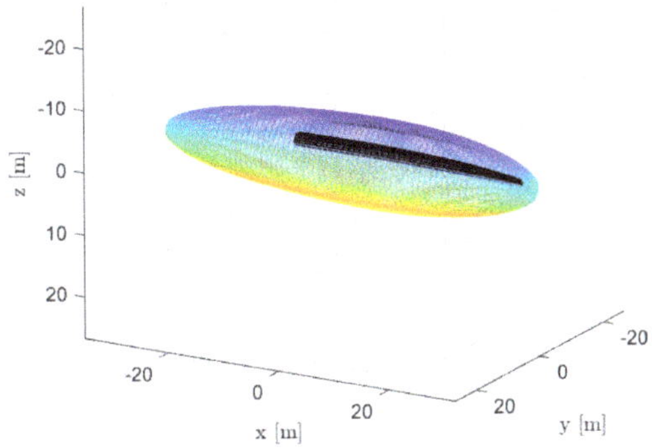

Figure 12. Optimal layout for operation at 18.06 km altitude above Port-au-Prince, Haiti.

The feasibility of the designed optimal solution is assessed by means of Figure 13, where it is possible to visually check the fulfillment of the 24-h energy balance and the required energy storage throughout the year. The capacity of the batteries resulting from the design is sufficient, even considering a 10% margin over power consumption. Concerning the 24-h energy balance instead, the only violation occurs on the same day for which the design has been conducted, falling within the constraint tolerance of the optimization algorithm.

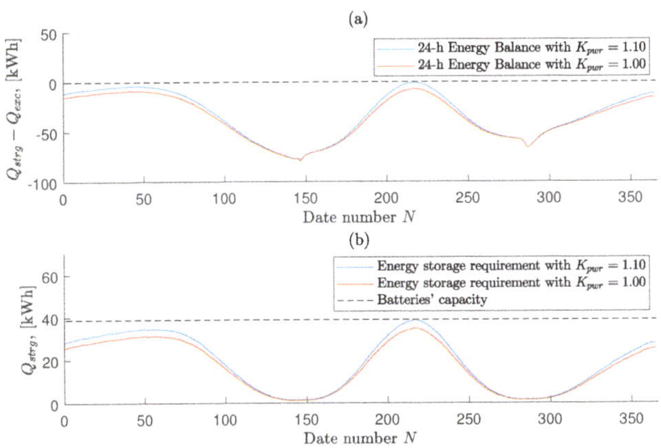

Figure 13. Energy balance evaluation (**a**) and energy storage requirement (**b**) for operation throughout the year at 18.06 km altitude above Port-au-Prince, Haiti.

Finally, a comparison between the optimal solutions obtained for year-round operation at 20 and 18.06 km altitudes above Port-au-Prince is reported in Table 11.

Table 11. Comparison between HAPS optimal designs for operation at 20 and 18.06 km altitudes above Port-au-Prince, Haiti.

Parameter	Unit	20 km	18.059 km	Reduction
L	[m]	128.53	67.71	-
FR	-	5.185	4.791	-
x_{le}	[m]	2.00	2.00	-
x_{te}	[m]	88.24	40.98	-
θ_{in}	[rad]	0.000	1.703	-
θ_{out}	[rad]	0.462	2.184	-
V_{env}	[m^3]	41,365.970	7083.912	82.88%
A_{env}	[m^2]	7991.302	2406.371	69.89%
A_{sc}	[m^2]	435.005	117.567	72.98%
σ_{brk}	[N/cm]	668.870	517.170	22.68%
ρ_{env}	[kg/m^2]	0.156	0.131	16.03%
Q	[kWh]	213.159	36.938	82.67%
M	[kg]	8607.296	1998.161	76.79%
$M_{airship}$	[kg]	3677.840	853.801	76.79%
M_{tank}	[kg]	5003.812	1161.621	76.79%
M_{solid}	[kg]	3090.002	709.657	77.03%
M_{env}	[kg]	1499.369	379.407	74.70%
M_{sc}	[kg]	84.826	22.926	72.97%
M_{bat}	[kg]	897.510	155.529	82.67%
M_{prop}	[kg]	155.061	67.701	56.34%
M_{gon}	[kg]	163.352	29.795	81.76%
M_{lg}	[kg]	577.838	134.144	76.79%

The advantage of reducing the operating altitude of the HAPS platform is evident from Table 11. The significant reduction of the maximum wind speed that the airship is expected to face when it is sized for the new altitude, together with the increase in air density (that appears at the denominator in Equation (35)), and that allows a given amount of lift to be generated with a smaller volume), leads to a substantial decrease in power consumption, and therefore energy harvesting and storage requirement, as well as in the size of the airship. As a result, the mass of the batteries and all power system components are greatly reduced and similarly is the mass of all other components onboard. The final result is a dramatic 82.88% reduction in HAPS size and a 76.79% reduction in its total mass. Considered as the payload for a missile responsible for positioning the HAPS at altitude, the new values constitute a feasible solution, according to the technology and size of existing launchers.

4.4. Effects of Technology on Optimal Sizing

In this section, the potential of some technological advancements on the outcome of the optimal design phase is analyzed. In particular, understanding whether such technological advances may ensure the feasibility of the missile-launched HAPS, by reducing its overall weight, is of interest here. The case study of L'Aquila is taken into account in this section, since according to the result of the sizing previously presented, a missile deployment is hardly feasible in that case. Each change investigated in the HAPS baseline design is introduced individually, in order to ease the evaluation of its specific impact. In addition, the effect of combining first only some, and finally all changes, is investigated.

In the following, a list of the combinations of alterations to the technological settings for HAPS optimal design is presented. Firstly, potential advancements in the three main key technologies for an HAA—namely envelope material, batteries, and photovoltaic film—are taken into account. Furthermore, some alternatives concerning the safety factors used for structural sizing, as well as lifting gas properties, are explored.

1. **Case #1:** envelope's laminated material with higher strength-to-weight ratio. In particular, the relationship provided by Carichner and Nicolai for Dyneema® laminate is employed instead of that for Vectran® [12]. More in general, the resulting lower value of surface density for a given minimum required breaking strength could also refer to a material with a different load-bearing component, such as Zylon®, which is characterized by a strength-to-weight ratio very similar to that of Dyneema®.
2. **Case #2:** batteries with higher gravimetric energy density. A specific energy of the batteries of $q_{bat} = 400$ Wh/kg can be considered a target within reach for a number of commercial and non-commercial available technologies, such as Li-ion with silicon nanowire anode (Amprius), Li-metal (SionPower, SolidEnergy), and Li-S (Oxis), once issues related to safety, instability, and limited cycle life are completely solved.
3. **Case #3:** thin photovoltaic film with higher conversion efficiency. Based on the constantly increasing performance of best-research laboratory solar cells, $\eta_{sc} = 17\%$ is considered a reasonable target for efficiency, already achieved by MiaSolé thin-film CIGS solar cells on a stainless steel substrate.
4. **Case #4:** simultaneous application of Cases #1, #2, and #3.
5. **Case #5:** lower safety factor over the maximum expected loads for sizing of the envelope skin material. A lower safety factor than that prescribed in FAA Airship Design Criteria for non-rigid airships may probably be used for an unmanned system, where mass and size minimization are a priority. A safety factor $s_\sigma = 3$, higher than that used in the design of the HiSentinel80, is considered.
6. **Case #6:** hydrogen instead of helium as lifting gas. Safety concerns related to the flammability of hydrogen when mixed with air could take second place in an unmanned system. Hence, using hydrogen, lighter and cheaper than helium, is an interesting choice to explore.
7. **Case #7:** simultaneous application of Cases #5 and #6.
8. **Case #8:** simultaneous application of Cases #4 and #7.

Mission specifications are the same in all cases and are reported in Table 6. The actual feasibility for year-round operation of the solution resulting from the optimization process is then assessed as usual, and a confirmation is obtained in each case, meaning that 9 January actually remains the most critical day for the sizing of the airship and its power system, even when the considered modifications are introduced.

Tables 12–15 summarize the values taken by the optimal design variables and the output of the optimization process when the eight alternatives listed above are considered, comparing the obtained values with those of the baseline, labeled as #0.

Concerning technological advancements, as expected, improvements in the envelope material and energy storage system characteristics have the greatest impact on the size and weight of the system. In fact, they allow a considerable lightening of the two corresponding components that contribute the most to the airship mass. On the contrary, improving the conversion efficiency (or, more in general, the output power-to-weight ratio) of the solar array is not so effective in enhancing the sizing of a stratospheric airship, due to the small fraction of total weight it takes. Altogether, the three considered technological advancements, all based on developments that are absolutely within reach, would make it possible to reduce the size and weight of the system by 76.11%, resulting in a payload for the missile weighing 1584 kg against the 6629 kg of the baseline design solution.

Table 12. Effect of technological advancements and design improvements on optimal design for operation at 20 km altitude above L'Aquila, Italy (Part 1). For each case, the first column gives the absolute value of the parameter, while the second column gives the percent reduction compared to the baseline #0.

Param.	Unit	#0	#1		#2	
L	[m]	117.42	94.37	-	92.30	-
FR	-	5.158	5.334	-	4.870	-
x_{le}	[m]	38.37	30.31	-	29.52	-
x_{te}	[m]	104.59	84.71	-	81.81	-
θ_{in}	[rad]	2.070	2.063	-	2.046	-
θ_{out}	[rad]	2.876	2.882	-	2.899	-
V_{env}	[m^3]	31,858.07	15,466.27	51.45%	17,355.40	45.52%
A_{env}	[m^2]	6703.99	4183.52	37.60%	4394.96	34.44%
A_{sc}	[m^2]	544.91	352.23	35.36%	380.22	30.22%
σ_{brk}	[N/cm]	614.19	477.44	22.27%	511.38	16.74%
ρ_{env}	[kg/m^2]	0.147	0.088	40.14%	0.130	11.56%
Q	[kWh]	148.23	95.81	35.37%	103.37	30.26%
M	[kg]	6628.92	3218.17	51.45%	3611.26	45.52%
$M_{airship}$	[kg]	2832.49	1375.10	51.45%	1543.06	45.52%
M_{tank}	[kg]	3853.69	1870.87	51.45%	2099.39	45.52%
M_{solid}	[kg]	2377.47	1149.06	51.67%	1290.63	45.71%
M_{env}	[kg]	1185.44	443.67	62.57%	687.92	41.97%
M_{sc}	[kg]	106.26	68.68	35.36%	74.14	30.22%
M_{bat}	[kg]	624.13	403.40	35.37%	272.03	56.42%
M_{prop}	[kg]	132.04	84.24	36.20%	91.14	30.97%
M_{gon}	[kg]	114.14	74.41	34.80%	50.77	55.53%
M_{lg}	[kg]	445.02	216.05	51.45%	242.44	45.52%

Table 13. Effect of technological advancements and design improvements on optimal design for operation at 20 km altitude above L'Aquila, Italy (Part 2). For each case, the first column gives the absolute value of the parameter, while the second column gives the percent reduction compared to the baseline #0.

Param.	Unit	#0	#3		#4	
L	[m]	117.42	120.71	-	68.29	-
FR	-	5.158	5.510	-	4.680	-
x_{le}	[m]	38.37	33.88	-	24.37	-
x_{te}	[m]	104.59	75.86	-	52.73	-
θ_{in}	[rad]	2.070	2.068	-	1.959	-
θ_{out}	[rad]	2.876	2.947	-	2.855	-
V_{env}	[m^3]	31,858.07	30,330.85	4.79%	7612.13	76.11%
A_{env}	[m^2]	6703.99	6619.82	1.26%	2507.02	62.60%
A_{sc}	[m^2]	544.91	396.34	27.27%	175.07	67.87%
σ_{brk}	[N/cm]	614.19	591.09	3.76%	393.81	35.88%
ρ_{env}	[kg/m^2]	0.147	0.144	2.04%	0.078	46.94%
Q	[kWh]	148.23	143.41	3.25%	63.52	57.15%
M	[kg]	6628.92	6311.14	4.79%	1583.91	76.11%
$M_{airship}$	[kg]	2832.49	2696.71	4.79%	676.79	76.11%
M_{tank}	[kg]	3853.69	3668.95	4.79%	920.80	76.11%
M_{solid}	[kg]	2377.47	2263.02	4.81%	560.46	76.43%
M_{env}	[kg]	1185.44	1140.36	3.80%	235.19	80.16%
M_{sc}	[kg]	106.26	77.29	27.27%	34.14	67.87%
M_{bat}	[kg]	624.13	603.83	3.25%	167.15	73.22%
M_{prop}	[kg]	132.04	127.64	3.33%	54.81	58.49%
M_{gon}	[kg]	114.14	110.49	3.20%	31.89	72.10%
M_{lg}	[kg]	445.02	423.69	4.79%	106.33	76.11%

Table 14. Effect of technological advancements and design improvements on optimal design for operation at 20 km altitude above L'Aquila, Italy (Part 3). For each case, the first column gives the absolute value of the parameter, while the second column gives the percent reduction compared to the baseline #0.

Param.	Unit	#0	#5		#6	
L	[m]	117.42	100.05	-	109.22	-
FR	-	5.158	5.020	-	5.478	-
x_{le}	[m]	38.37	32.23	-	27.10	-
x_{te}	[m]	104.59	88.99	-	81.72	-
θ_{in}	[rad]	2.070	2.056	-	1.309	-
θ_{out}	[rad]	2.876	2.889	-	2.197	-
V_{env}	[m^3]	31,858.07	20,814.52	34.66%	22,734.69	28.64%
A_{env}	[m^2]	6703.99	5006.33	25.32%	5452.55	18.67%
A_{sc}	[m^2]	544.91	422.75	22.42%	461.39	15.33%
σ_{brk}	[N/cm]	614.19	403.40	34.32%	538.26	12.36%
ρ_{env}	[kg/m^2]	0.147	0.113	23.13%	0.135	8.16%
Q	[kWh]	148.23	114.96	22.44%	120.58	18.65%
M	[kg]	6628.92	4331.02	34.66%	4958.51	25.20%
$M_{airship}$	[kg]	2832.49	1850.61	34.66%	2021.33	28.64%
M_{tank}	[kg]	3853.69	2517.82	34.66%	2978.04	22.72%
M_{solid}	[kg]	2377.47	1549.86	34.81%	1830.82	22.99%
M_{env}	[kg]	1185.44	676.86	42.90%	882.40	25.56%
M_{sc}	[kg]	106.26	82.44	22.42%	89.97	15.33%
M_{bat}	[kg]	624.13	484.06	22.44%	507.72	18.65%
M_{prop}	[kg]	132.04	101.71	22.97%	106.83	19.09%
M_{gon}	[kg]	114.14	88.93	22.09%	93.19	18.36%
M_{lg}	[kg]	445.02	290.76	34.66%	180.52	59.44%

Table 15. Effect of technological advancements and design improvements on optimal design for operation at 20 km altitude above L'Aquila, Italy (Part 4). For each case, the first column gives the absolute value of the parameter, while the second column gives the percent reduction compared to the baseline #0.

Param.	Unit	#0	#7		#8	
L	[m]	117.42	90.54	-	60.72	-
FR	-	5.158	5.001	-	5.020	-
x_{le}	[m]	38.37	28.92	-	20.96	-
x_{te}	[m]	104.59	80.60	-	53.31	-
θ_{in}	[rad]	2.070	2.050	-	2.107	-
θ_{out}	[rad]	2.876	2.895	-	2.844	-
V_{env}	[m^3]	31,858.07	15,539.87	51.22%	4649.91	85.40%
A_{env}	[m^2]	6703.99	4115.38	38.61%	1843.26	72.50%
A_{sc}	[m^2]	544.91	354.93	34.86%	129.79	76.18%
σ_{brk}	[N/cm]	614.19	366.59	40.31%	244.90	60.13%
ρ_{env}	[kg/m^2]	0.147	0.107	27.21%	0.060	59.18%
Q	[kWh]	148.23	96.51	34.89%	47.16	68.19%
M	[kg]	6628.92	3389.30	48.87%	1014.16	84.70%
$M_{airship}$	[kg]	2832.49	1381.65	51.22%	413.42	85.40%
M_{tank}	[kg]	3853.69	2035.58	47.18%	609.10	84.19%
M_{solid}	[kg]	2377.47	1248.26	47.50%	366.50	84.58%
M_{env}	[kg]	1185.44	526.49	55.59%	132.74	88.80%
M_{sc}	[kg]	106.26	69.21	34.86%	25.31	76.18%
M_{bat}	[kg]	624.13	406.34	34.89%	124.10	80.12%
M_{prop}	[kg]	132.04	84.88	35.71%	39.89	69.79%
M_{gon}	[kg]	114.14	74.94	34.34%	24.14	78.85%
M_{lg}	[kg]	445.02	123.39	72.27%	36.92	91.70%

In addition, the proposed design improvements also have the potential to allow a significant enhancement of the system and contribute to its feasibility. In particular, lowering the envelope material sizing requirement by reducing the safety factor, while still retaining a sizable 66% margin over the maximum expected in-flight loads, is another effective method of lightening the airship envelope, leading to an overall gain in size and weight of the system. On the other hand, filling the envelope with hydrogen instead of helium would allow to more than halve the required mass of lifting gas—thanks to the lower density of hydrogen and the reduction in size following the reduced lift requirement—and consequently to reduce the weight of the system.

Jointly, the three assumed technological advancements and these two changes in the design of the platform would lead to a 85.40% reduction in airship size and a 84.70% lightening of the missile total payload, from 6629 kg to only 1014 kg. Multiple combinations of advancements could be used to achieve the same or even better results, and improvements could also be assumed for other components (gondola material, efficiencies, and power-to-weight ratios of other power-train components) or other design aspects (shape and resulting drag coefficient model, more optimistic predictions of temperature, and hence pressure variations inside the envelope). In particular, considering the dramatic influence of the lifting gas storage system on the overall weight, assuming an even only slightly better gravimetric capacity of high-pressure tanks would greatly contribute to its reduction. However, the obtained results, based on reliable and attainable assumptions only, represent the proof of the potential to obtain, by means of a fair level of improvement, a system with a size and weight small enough to enable the missile-deployed HAPS concept, even for year-round operation within a high wind area such as that at 20 km altitude in the test case of L'Aquila.

5. Conclusions

The aim of the present work was that of investigating the feasibility of the concept of a launcher-deployable high altitude airship (HAA) for a high-altitude pseudo-satellite (HAPS) mission. This mission puts the weight of the entire HAA at a premium, since it should be intended as the payload of a missile launcher. In order to perform such analysis, the limited fidelity of environmental models for the wind and solar radiation at altitude typically adopted for preliminary sizing is not sufficient. Therefore, high-fidelity modeling of the environment, and of the interaction with a specific shape of the envelope and of solar arrays, has been taken into account.

The first-principle or semi-empirical models required to obtain a sizing, starting from mission requirements, have been introduced and implemented in an optimal framework, where the loop is iterated in the search for a design solution with the minimum achievable take-off mass while simultaneously capable of verifying a set of constraints.

The sizing procedure has been validated in the case of an existing airship, showing a very good agreement with a real field test case.

The sizing methodology has been used intensively to carry out a parameter analysis in four testing scenarios, corresponding to largely different geographical locations, and hence to different atmospheric and radiation conditions. In this phase, stationing altitude has been assigned equal to all analysis conditions. The complete energy balance of the mission over a yearly time frame has been considered, accounting for changing values of radiation and wind over a one year period, as well as on a daily basis. The results have shown a significant difference in the optimal sizing in the respective geographical cases, highlighting a generally strong dependence on the choice of the coordinates of deployment on the resulting HAA weight-optimal sizing. Where generally this allows one to carry out the deployment of the airship at altitude by means of a missile, the outcome of the comparative analysis shows that there exist critical conditions (i.e., critical sets of requirements) that jeopardize the adoption of the corresponding design solutions as payloads of a missile launcher.

Concurrently, further parameter analysis has been carried out, again computing sizing solutions through the methodology introduced in this work, to ascertain whether mission specifications, such as the stationing altitude, or technological parameters, including qualities of the materials, may be tuned to bear a satisfying design solution, i.e., an optimal solution satisfying mission requirements, and to be possibly loaded on a missile to be employed for deployment. The results of this parameter analysis have shown that such a solution indeed exists, with a proper choice of the mission and technological parameters. However, the relationship between technological assumptions and environmental conditions for a specific mission strongly influences the feasibility of a missile-deployable HAA for HAPS missions.

Future development of the present work will include the analysis of further technologies in the design process, including technologies for energy storage, in particular structural batteries, as well as altogether different power-trains, such as hydrogen-based fuel cells. These may further increase the efficiency in the usage of the masses on board, thus potentially allowing for more payload to be included in the airship design.

Author Contributions: C.E.D.R. and A.R. synthesized and structured the outcomes of the research, composing the present article. G.R. contributed to the refinement of the formulation and carried out the quantitative analyses. All authors participated equally in the development of the body of the work, contributing discussions and critical comments to the methodology and results. All authors have read and agreed to the published version of the manuscript.

Funding: This research received no external funding.

Informed Consent Statement: Not applicable.

Conflicts of Interest: The authors declare no conflict of interest.

References

1. Chu, A.; Blackmore, M.; Oholendt, R.G.; Welch, J.V.; Baird, G.; Cadogan, D.P.; Scarborough, S.E. A Novel Concept for Stratospheric Communications and Surveillance: Star Light. In Proceedings of the AIAA Baloon Systems Conference, Williamsburg, VA, USA, 21–24 May 2007. [CrossRef]
2. Miller, S.H.; Fesen, R.; Hillenbrand, L.; Rhodes, J.; Baird, G.; Blake, G.; Booth, J.; Carlile, D.E.; Duren, R.; Edworthy, F.G.; et al. *Airships: A New Horizon for Science*; Technical Report; Keck Institute for Space Studies: Pasadena, CA, USA, 2014.
3. Romeo, G.; Frulla, G.; Cestino, E. Design of a high-altitude long-endurance solar-powered unmanned air vehicle for multi-payload and operations. *J. Aerosp. Eng.* **2007**, *221*, 199–216. [CrossRef]
4. Drob, D.P.; Emmert, J.T.; Crowley, G.; Picone, J.; Shepherd, G.G.; Skinner, W.; Hays, P.; Niciejewski, R.J.; Larsen, M.; She, C.Y.; et al. An empirical model of the Earth's horizontal wind fields: HWM07. *J. Geophys. Res. Space Phys.* **2008**, *113*, A12304. [CrossRef]
5. Gueymard, C. *SMARTS2, a Simple Model of the Atmospheric Radiative Transfer of Sunshine: Algorithms and Performance Assessment*; Technical Report FSEC-PF-270-95; Florida Solar Energy Center: Cocoa, FL, USA, 1995.
6. Colozza, A. *Initial Feasibility Assessment of a High Altitude Long Endurance Airship*; Technical Report NASA/CR–2003-212724; Analex Corp.: Brooks Park, OH, USA, 2003.
7. Gonzalo, J.; López, D.; Domínguez, D.; García, A.; Escapa, A. On the capabilities and limitations of high altitude pseudo-satellites. *Prog. Aerosp. Sci.* **2018**, *98*, 37–56. [CrossRef]
8. d'Oliveira, F.A.; de Melo, F.C.L.; Devezas, T.C. High-Altitude Platforms—Present Situation and Technology Trends. *J. Aerosp. Technol. Manag.* **2016**, *8*, 249–262. [CrossRef]
9. Everaerts, J.; Lewyckyj, N.; Fransaer, D. Pegasus: Design of a Stratospheric Long Endurance UAV System for Remote Sensing. In Proceedings of the 20th Congress of the International Society for Photogrammetry and Remote Sensing (ISPRS), Istanbul, Turkey, 12–23 July 2004; Vito, Flemish Institute for Technological Research: Mol, Belgium, 2004.
10. Baraniello, V.R.; Persechino, G.; Borsa, R. *Tools for the Conceptual Design of a Stratospheric Hybrid Platform*; SAE Technical Paper; SAE: Warrendale, PA, USA, 2020. [CrossRef]
11. Jenkins, J.E.; Samsundar, J.; Neradka, V.F. A Design Methodology for Optimal Power Generation in High Altitude Airships Using Genetic Algorithms. In Proceedings of the 3rd International Energy Conversion Engineering Conference, San Francisco, CA, USA, 15–18 August 2005. [CrossRef]
12. Carichner, G.E.; Nicolai, L.M. *Fundamentals of Aircraft and Airship Design*; AIAA Education Series; American Institute of Aeronautics and Astronautics, Inc.: Reston, VA, USA, 2013.
13. Lobner, P. *HiSentinel Stratospheric Airships*; Technical Report; Lyncean Group of San Diego: Point Loma, CA, USA, 2020.
14. Lobner, P. *DARPA Integrated Sensor is Structure (ISIS) Airship*; Technical Report; Lyncean Group of San Diego: Point Loma, CA, USA, 2020.

15. Lobner, P. *Lockheed Martin HALE-D*; Technical Report; Lyncean Group of San Diego: Point Loma, CA, USA, 2020.
16. Wang, Q.; Chen, J.; Fu, G.; Duan, D.; Zhao, H. A methodology for optimisation design and analysis of stratosphere airship *Aeronaut. J.* **2009**, *113*, 533–540. [CrossRef]
17. Liang, H.; Zhu, M.; Guo, X.; Zheng, Z. Conceptual Design Optimization of High Altitude Airship in Concurrent Subspace Optimization. In Proceedings of the 50th AIAA Aerospace Sciences Meeting including the New Horizons Forum and Aerospace Exposition, Nashville, TN, USA, 9–12 January 2012. [CrossRef]
18. Chen, Q.; Zhu, M.; Sun, K. Analysis to Effects on Conceptual Parameters of Stratospheric Airship with Specified Factors. *J. Comput.* **2011**, *6*, 1055–1062. [CrossRef]
19. Yang, X.; Liu, D. Conceptual Design of Stratospheric Airships Focusing on Energy Balance. *J. Aerosp. Eng.* **2018**, *31*, 04017094. [CrossRef]
20. Zhang, L.; Zhu, W.; Du, H.; Lv, M. Multidisciplinary design of high altitude airship based on solar energy optimization. *Aerosp. Sci. Technol.* **2021**, *110*, 106440. [CrossRef]
21. Pant, R.; Shaik, S. Conceptual Design, Initial Sizing and Sensitivity Analyses of a High Altitude Airship using revised Volumetric Drag. In Proceedings of the 57th Israel Annual Conference on Aerospace Sciences, Tel Aviv-Haifa, Israel, 15–16 March 2017.
22. Li, J.; Lv, M.; Sun, K. Optimum area of solar array for stratospheric solar-powered airship. *J. Aerosp. Eng.* **2016**, *231*, 2654–2665. [CrossRef]
23. Wang, Q.; Chen, J.; G.; Duan, D. An approach for shape optimization of stratosphere airships based on multidisciplinary design optimization. *J. Zhejiang Univ.-A* **2009**, *10*, 1609–1616. [CrossRef]
24. Yang, X.; Liu, D. Renewable power system simulation and endurance analysis for stratospheric airships. *Renew. Energy* **2017**, *113*, 1070–1076. [CrossRef]
25. Hoerner, S.F. *Fluid-Dynamic Drag*; Hoerner Fluid Dynamics: Bakersfield, CA, USA, 1965.
26. Jones, S.P.; DeLaurier, J.D. Aerodynamic estimation techniques for aerostats and airships. *J. Aircr.* **1982**, *20*, 120–126. [CrossRef]
27. Drob, D.P.; Emmert, J.T.; Meriwether, J.W.; Makela, J.J.; Doornbos, E.; Conde, M.; Hernandez, G.; Noto, J.; Zawdie, K.A.; McDonald, S.E.; et al. An update to the Horizontal Wind Model (HWM): The quiet time thermosphere. *Earth Space Sci.* **2015**, 2, 301–319. [CrossRef]
28. Riboldi, C.E.D.; Trainelli, L.; Rolando, A.; Mariani, L. Predicting the effect of electric and hybrid electric aviation on acoustic pollution. *Noise Mapp.* **2020**, *7*, 35–36. [CrossRef]
29. Gueymard, C.A. *SMARTS Code, Version 2.9.5 for Windows, User's Manual*; Technical Report; Solar Consulting Services: Colebrook, NH, USA, 2005.
30. SMARTS: Simple Model of the Atmospheric Radiative Transfer of Sunshine. Available online: www.nrel.gov/grid/solar-resource/smarts.html (accessed on 1 December 2021).
31. Schopferer, S.; Liersch, C.; Brizon, M.; Froese, S. Evaluating the Energy Balance of High Altitude Platforms at Early Design Stages. In Proceedings of the 2016 International Conference on Unmanned Aircraft Systems (ICUAS), Arlington, VA, USA, 7–10 June 2016; pp. 170–177. [CrossRef]
32. Bücher, K. Site dependence of the energy collection of PV modules. *Sol. Energy Mater. Sol. Cells* **1997**, *47*, 85–94. [CrossRef]
33. Micheli, L.; Caballero, J.A.; Fernandez, E.F.; Smestad, G.P.; Nofuentes, G.; Mallick, T.K.; Almonacid, F. Correlating Photovoltaic Soiling Losses to Waveband and Single-Value Transmittance Measurements. *Energy* **2019**, *180*, 376–386. [CrossRef]
34. Brizon, M. *Solar Energy Generation Model for High Altitude Long Endurance Platforms*; Technical Report; KTH—Royal Institute of Technology: Stockholm, Sweden, 2015.
35. Fesharaki, V.J.; Dehghani, M.; Fesharaki, J.J.; Tavasoli, H. The Effect of Temperature on Photovoltaic Cell Efficiency. In Proceedings of the 1st International Conference on Emerging Trends in Energy Conservation (ETEC), Tehran, Iran, 20–21 November 2011.
36. How HOMER Calculates the PV Cell Temperature. Available online: www.homerenergy.com/products/pro/docs/latest/how_homer_calculates_the_pv_cell_temperature.html (accessed on 1 December 2021).
37. Keidel, B. Auslegung und Simulation von Hochfliegenden, Dauerhaft Stationierbaren Solardrohnen. Ph.D. Thesis, Lehrstuhl für Flugmechanik und Flugregelung Technische Universität München, Munich, Germany, 2000.
38. Gawale, A.; Raina, A.; Pant, R.; Jahagirdar, Y. Design, Fabrication and Operation of Remotely Controlled Airships in India. In Proceedings of the 26th Congress of International Council of the Aeronautical Sciences (ICAS), Anchorage, AK, USA, 14–19 September 2008. [CrossRef]
39. Khoury, G.A. *Airship Technology*; Cambridge Aerospace Series; Cambridge University Press: Cambridge, UK, 2012.
40. Wu, J.; Fang, X.; Wang, Z.; Hou, Z.; Ma, Z.; Zhang, H.; Dai, Q.; Xu, Y. Thermal modeling of stratospheric airships. *Prog. Aerosp. Sci.* **2015**, *75*, 26–37. [CrossRef]
41. Zheng, W.; Zhou, Q.; Li, Z.; Zhang, X.; Liu, Y.; Li, M. Performance analysis of a novel stratospheric airship concept based on gas-liquid phase change. *Appl. Therm. Eng.* **2019**, *159*, 113748. [CrossRef]
42. Liu, Q.; Yang, Y.; Cui, Y.; Cai, J. Thermal performance of stratospheric airship with photovoltaic array. *Adv. Space Res.* **2017**, *59*, 1486–1501. [CrossRef]
43. Shi, H.; Geng, S.; Qian, X. Thermodynamics analysis of a stratospheric airship with hovering capability. *Appl. Therm. Eng.* **2019**, *146*, 600–607. [CrossRef]
44. Dai, Q.; Fang, X.; Li, X.; Tian, L. Performance simulation of high altitude scientific balloons. *Adv. Space Res.* **2012**, *49*, 1045–1052. [CrossRef]

5. Zhang, T.; Geng, S.; Mu, X.; Chen, J.; Wang, J.; Wu, Z. Thermal Characteristics of a Stratospheric Airship with Natural Convection and External Forced Convection. *Int. J. Aerosp. Eng.* **2019**, *2019*, 4368046. [CrossRef]
6. Smith, S.; Fortenberry, M.; Lee, M.; Judy, R. HiSentinel80: Flight of a High Altitude Airship. In Proceedings of the 11th AIAA Aviation Technology, Integration, and Operations (ATIO) Conference, Virginia Beach, VA, USA, 20–22 September 2011. [CrossRef]
7. ISISPACE S-Band Transceiver. Available online: www.satcatalog.com/component/s-band-transceiver/ (accessed on 1 December 2021).
8. Riboldi, C.E.D.; Gualdoni, F. An integrated approach to the preliminary weight sizing of small electric aircraft. *Aerosp. Sci. Technol.* **2016**, *58*, 134–149. [CrossRef]
9. Riboldi, C.E.D. An optimal approach to the preliminary design of small hybrid-electric aircraft. *Aerosp. Sci. Technol.* **2018**, *81*, 14–31. [CrossRef]
0. Noth, A. Design of Solar Powered Airplanes for Continuous Flight. Ph.D. Thesis, ETH Zürich, Zürich, Switzerland, 2008. [CrossRef]
1. Riboldi, C.E.D. Energy-optimal off-design power management of hybrid-electric aircraft. *Aerosp. Sci. Technol.* **2019**, *95*, 105507. [CrossRef]
2. Riboldi, C.E.D.; Trainelli, L.; Biondani, F. Structural batteries in aviation: A preliminary sizing methodology. *J. Aerosp. Eng.* **2020**, *3*, 040200311–040200315. [CrossRef]
3. Barral, K.; Barthélémy, H. *Hydrogen High Pressure Tanks Storages: Overview and New Trends Due to H2 Energy Specifications and Constraints*; Technical Report INIS-FR–6812; IAEA: Vienna, Austria, 2006.
4. Rivard, E.; Trudeau, M.; Zaghib, K. Hydrogen Storage for Mobility: A Review. *Materials* **2019**, *12*, 1973. [CrossRef]
5. Smith, I.S. HiSentinel & Stratospheric Airship Design Sensitivity. In Proceedings of the Keck Institute for Space Studies (KISS) Workshop, Pasadena, CA, USA, 30 April–3 May 2013.

aerospace

MDPI

Article

Modelling and Stabilisation of an Unconventional Airship: A Polytopic Approach

Said Chaabani [1] and Naoufel Azouz [2,*]

1 Department of Maths-Info, Unité de Recherche en Matériaux Avancés et Nanotechnologies UR16ES03, Institute of Applied Sciences and Technology of Kasserine, Kairouan University, BP 471, Kasserine 1200, Tunisia; chaabani.said@yahoo.fr

2 Laboratoire de Mécanique et Energétique d'Evry, UnivEvry, University of Paris-Saclay, 40 rue du Pelvoux, 91025 Evry, France

* Correspondence: naoufel.azouz@univ-evry.fr

Abstract: The paper presents the modelling and stabilisation of an unconventional airship. The complexity of such a new design requires both proper dynamic modelling and control. A complete dynamic model is built here. Based on the developed dynamic model, a nonlinear control law is proposed for this airship to evaluate its sensitivity during manoeuvres above a loading area. The proposed stabilisation controller derives its source from a polytopic quasi-Linear Parameter varying (qLPV) model of the nonlinear system. A controller, which takes into account certain modelling uncertainties and the stability of the system, is analysed using Lyapunov's theory. Finally, to facilitate the design of the controller, we express the stability conditions using Linear Matrix Inequalities (LMIs). Numerical simulations are presented to highlight the power of the proposed controller.

Keywords: dynamic modelling; nonlinear systems; quasi-LPV systems; LMI formulation; stabilisation

1. Introduction

In recent decades, the use of large airships has attracted increasing attention over the world because of their potential in many application areas, such as surveillance, advertising, exploration, and in the medium term, heavy load carrying; see for example Liao et al. [1] and Li et al. [2] for more details. Ellipsoidal shapes are usually used for airships as noted by Jex et al. [3] and Hygounec et al. [4]. In order to explore new ways towards achieving the holy grail of aerodynamic performance, a number of original shapes were subjected to experiments in recent decades. This has become possible through the increased reliability of numerical aerodynamics, advanced textile technology, and control theory.

Here, we investigate a craft that differs from conventional airships. The airship presented here (Figure 1) is more like a flying wing. Hybrid airships have an exciting future, due to their increased lift compared with a conventional airship. In this sense, we can cite the pioneering airship LMH-1 from Lockheed Martin or the Airlander 10, the first Large Capacity Airship to have flown in this millennium. The flying wing shape presented here could minimise aerodynamic drag and has the advantage of presenting a larger flat (or quasi-flat), surface allowing the installation of photovoltaic cells necessary to provide a significant part of the thruster's energy. We first set about building a dynamic model for this airship: a Newton–Euler approach was taken. Note that the Lagrangian method has also been considered in other dynamic models of airships, for example El Omari et al. [5] and Bennaceur et al. [6].

In order to compensate for the airship's great sensitivity to gusts of wind, it is over-actuated. Moreover, in order to make these aircraft accessible and interesting for different countries and parties, even those which lack infrastructure, it is necessary to consider handling at altitude. This requires the development of a precise autopilot to stabilise the airship around a point of loading or unloading. Various techniques have been proposed for the control and stabilisation of aerial vehicles, see for example Yang et al. [7] where a backstepping method was used, or Moutinho [8] where a path-tracking gain-scheduling controller is designed and its performance and robustness evaluated in a simulation environment. Note that the dynamic models of aerial vehicles are often assumed to be linear, obtained from the linearization of the nonlinear system around a specific operating point. In the same way, linear models are extensively studied in the literature. Even though they solve many problems, the applicability of such developed approaches may fail to stabilise the system or lead to degraded performances when the system operates far from the operating point where the linear model is valid. It is, therefore, interesting in certain cases to take into account the nonlinear behaviours in the system. This allows the enhancing of the control performances and enlarges the domain of applicability of the controllers.

Figure 1. The studied Airship.

Nonlinear behaviours are often present in practical systems, especially in flight systems, such as airships. Everyone agrees that the nonlinear systems are complex and difficult to study; they nonetheless allow the expression of the system behaviour with more accuracy and in a large domain of operation. Several classes of nonlinear systems are studied in the literature, such as Lipschitz systems, Linearizable systems by output injection, etc. Among the most interesting structures of nonlinear systems, we find the Quasi-Linear Parameter Varying (qLPV) models, as seen in Onat et al. [9]. An important advantage of these models is the possibility of scaling up tools designed for linear to non-linear systems. Many control and observation themes are, in fact, applied to this class of systems, in particular stability or stabilisation. Note that this class of systems can be represented in a polytopic form which converges to the form of the well-known Takagi–Sugeno Systems, as seen in Takagi et al. [10]. Stabilisation and control of nonlinear systems with qLPV representation is largely studied. We can cite the works of Tanaka et al. [11,12] and Guerra et al. [13], where a quadratic Lyapunov function was used to establish Linear Matrix Inequalities (LMIs) for controller design. For the sake of conservatism of the proposed solutions, relaxed stability conditions were obtained using, for example, the Tuan's Lemma [14] or Polya's theorem in Sala et al. [15]. This last study provides sufficient and asymptotic necessary conditions for stability and stabilisation by the use of quadratic Lyapunov functions seen in Hayat et al. [16]. Another way dealing with the conservatism reduction is the use of non-quadratic Lyapunov functions, as seen in Tanaka et al. [17], which produce more relaxed LMI conditions.

Concerning uncertain systems, the stabilisation problem still remains open in the case of uncertain Quasi-LPV model. For this class of systems, we find some works dealing only with the stabilisation problem and stability analysis, for example Cao et al. [18] and Ibrir et al. [19]. Hence, this paper presents a new approach based on the Quasi-LPV model representation with the aim of proposing a control for nonlinear dynamic systems with uncertainties. To the best of our knowledge, this method has never been tested for aerial vehicles. We have exploited the particularities of the airship in order to establish a stabilisation around an operating point within a polytope, which guarantees the stability of the flying object in a domain, and which takes into account possible uncertainties concerning the model or concerning the stresses which could impact the airship.

In this approach, designs based on the LMI method are used to find the feedback gains of a controller, as well as common P matrices (defined positive) satisfying a stability criterion derived in terms of the Lyapunov direct method.

We have organised this paper as follows: Section 2 presents the dynamic modelling of the airship; in Section 3, we present the elaborated controller based on the q-LPV representation; and numerical results are presented and discussed in Section 4.

2. Dynamics Model of the Airship

2.1. Kinematics

One can see in Figure 2 a reduced model of the airship. This last is assumed to be a rigid flying object. As usual in aeronautics, we use the parameterisation in yaw, pitch and roll (ψ, θ, φ) to define the orientation of the moving frame $R_m = (O, X_m, Y_m, Z_m)$ fixed to the centre of inertia G of the airship; with regard to the inertial frame $R_0 = (O, X_0, Y_0, Z_0)$ and is a NED Cartesian frame, its origin being on the ground (Figure 2).

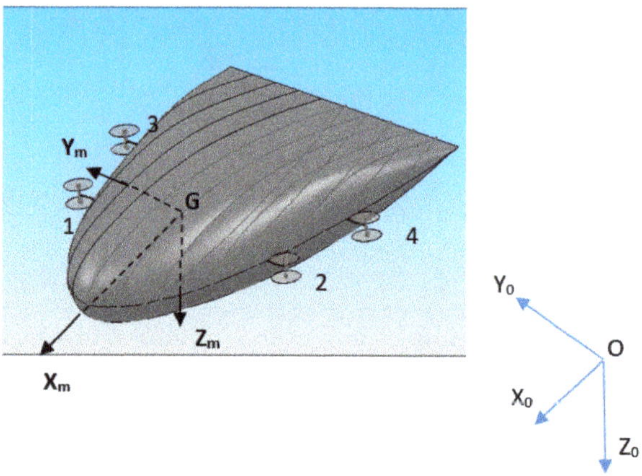

Figure 2. The different reference frames.

The airship subject of our study is propelled by four rotors. Each rotor consists of two counter-rotating propellers. A given rotor i can rotate around two axes: the Y_m-axis (β_i), $(-180° \le \beta_i \le 180°)$ and the Z_{Ri}-axis (γ_i) normal to Y_m and which matches the Z_m-axis (Figure 3) when $\beta_i = 0$ (cruise flight); γ_i has these limitations: $-30° \le \gamma_i \le 30°$. A fictive axis X_{iR} completes the rotor frame.

Let $\|F_i\|$ be the thrust of rotor i; this force is positioned in the point T_i in R_m. The location of the different thrusters is established as follows:

$$T_1 = (a, b_1, c_0)^T;\ T_2 = (a, -b_1, c_0)^T;\ T_3 = (-a, b_3, c_0)^T;\ T_4 = (-a, -b_3, c_0)^T$$

The transformation matrix from R_m to R_0 is defined by Goldstein [20]:

$$J_1^T = \begin{pmatrix} c\psi \cdot c\theta & s\psi \cdot c\theta & -s\theta \\ s\phi \cdot c\psi \cdot s\theta - c\phi \cdot s\psi & s\phi \cdot s\psi \cdot s\theta + c\psi \cdot c\phi & s\phi \cdot c\theta \\ c\phi \cdot c\psi \cdot s\theta + s\phi \cdot s\psi & c\phi \cdot s\theta \cdot s\psi - c\psi \cdot s\phi & c\theta \cdot c\phi \end{pmatrix} \quad (1)$$

For simplicity, we will note by: $c(\cdot) = \cos(\cdot)$ and $s(\cdot) = \sin(\cdot)$.

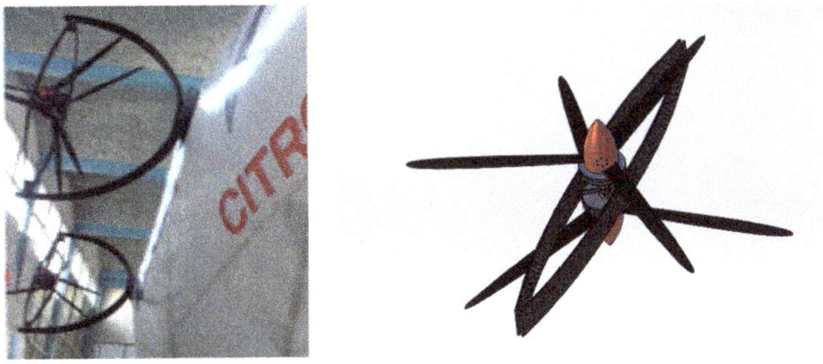

(a) (b)

Figure 3. (**a**) Thrusters of the airship; (**b**) CAD details of the thrusters.

The speed of the airship in R_0 is denoted by $\dot{\eta}_1 = [\dot{x}_0, \dot{y}_0, \dot{z}_0]^T$. It can be expressed with respect to the speed $\nu_1 = [u, v, w]^T$, which represents the speed expressed with respect to the local reference frame R_m, as is often the case in aeronautics.

$$\dot{\eta}_1 = J_1 \cdot \nu_1 \tag{2}$$

Note that $\eta_1 = [x_0, y_0, z_0]^T$ represents the position of the centre of inertia G. We also suppose that the distribution of masses is balanced between the top side and the bottom side of the airship, and thus the centre of volume coincides with the centre of gravity G.

Similarly, the speed of rotation $\dot{\eta}_2 = \left[\dot{\phi}, \dot{\theta}, \dot{\psi}\right]^T$ with respect to the fixed reference frame could be expressed with respect to the angular speed of the airship in R_m $\nu_2^T = [p, q, r]$ as:

$$\nu_2 = \begin{pmatrix} 1 & 0 & -s\theta \\ 0 & c\phi & c\theta \cdot s\phi \\ 0 & -s\phi & c\theta \cdot c\phi \end{pmatrix} \cdot \begin{pmatrix} \dot{\phi} \\ \dot{\theta} \\ \dot{\psi} \end{pmatrix} \tag{3}$$

reciprocally:

$$\dot{\eta}_2 = J_2 \cdot \nu_2 \tag{4}$$

The matrix J_2 is defined as follows:

$$J_2 = \begin{pmatrix} 1 & \tan\theta \cdot s\phi & \tan\theta \cdot c\phi \\ 0 & c\phi & -s\phi \\ 0 & \frac{s\phi}{c\theta} & \frac{c\phi}{c\theta} \end{pmatrix} \tag{5}$$

The overall kinematic relation of the airship can, therefore, be written in the following compact form:

$$\begin{pmatrix} \dot{\eta}_1 \\ \dot{\eta}_2 \end{pmatrix} = \begin{pmatrix} J_1 & 0 \\ 0 & J_2 \end{pmatrix} \begin{pmatrix} \nu_1 \\ \nu_2 \end{pmatrix} \tag{6}$$

2.2. *Dynamical Model*

The dynamic model of the flying body can be written in the following compact form, as seen in Shabana et al. [21]:

$$M \cdot \dot{\nu} = \mathbf{Q}_G + \tau \tag{7}$$

Here, $\tau = \begin{pmatrix} \tau_1 \\ \tau_2 \end{pmatrix}$ is the vector of forces and torques applied on the airship. This includes the thrust of the rotors, the weight of the flying machine, the buoyancy B_u, as well as the lift and the aerodynamic drag.

Using the definitions of the thrusters, this vector will be written as:

$$\tau_1 = \begin{pmatrix} \sum\limits_{i=1}^{4} c\gamma_i \cdot c\beta_i \cdot \|F_i\| - (-B_u + mg) \cdot s\theta \\ \sum\limits_{i=1}^{4} s\gamma_i \cdot \|F_i\| + (-B_u + mg) \cdot s\phi \cdot c\theta \\ -\sum\limits_{i=1}^{4} c\gamma_i \cdot s\beta_i \cdot \|F_i\| + (-B_u + mg) c\phi \cdot c\theta \end{pmatrix} \tag{8}$$

and:

$$\tau_2 = -\begin{pmatrix} c_0 \sum\limits_{i=1}^{4} s\gamma_i \|F_i\| + b_1 (c\gamma_1 \cdot s\beta_1 \|F_1\| - c\gamma_2 \cdot s\beta_2 \|F_2\| \\ +b_3 (c\gamma_3 \cdot s\beta_3 \|F_3\| - c\gamma_4 \cdot s\beta_4 \|F_4\|) + B_u z_B c\theta \cdot s\phi. \\ -c_0 \sum\limits_{i=1}^{4} c\gamma_i \cdot c\beta_i \|F_i\| + a (c\gamma_4 \cdot s\beta_4 \|F_4\| + c\gamma_3 \cdot s\beta_3 \|F_3\| \\ -c\gamma_1 \cdot s\beta_1 \|F_1\| - c\gamma_2 \cdot s\beta_2 \|F_2\|) + B_u z_B \cdot s\theta \\ b_1 (c\gamma_1 \cdot c\beta_1 \|F_1\| - c\gamma_2 \cdot c\beta_2 \|F_2\|) \\ +b_3 (c\gamma_3 \cdot c\beta_3 \|F_3\| - c\gamma_4 \cdot c\beta_4 \|F_4\|) \\ +a (s\gamma_4 \|F_4\| + s\gamma_3 \|F_3\| - s\gamma_1 \|F_1\| - s\gamma_2 \|F_2\|) \end{pmatrix} \tag{9}$$

Remark 1. *In this study, we are only interested in the stabilisation of the airship around a fixed point of loading and unloading; the drag and lift forces will, therefore, not be taken into consideration. This is also the case for the effect of the elevator and rudders.*

On the other hand, $M = \begin{pmatrix} M_{TT} & 0 \\ 0 & M_{RR} \end{pmatrix}$ is the inertia matrix, according to the choice of the mobile reference frame, and is block-diagonal. The indices TT and RR correspond, respectively, to the inertias in translation and in rotation. This 6×6 matrix includes both the terms of inertia and the terms of added masses.

The latter reflect the increase in inertia of the solid under the effect of the resistance force of the fluid which surrounds the airship. This force depends on the acceleration of the moving solid body. Although these added masses can be neglected for airplanes or helicopters, this cannot be tolerated when the density of the solid body and that of the fluid which surrounds it are comparable, as in the case of airships or submarines, reported in Bennaceur et al. [6]. The calculation of these added masses can be performed experimentally or analytically as was done, for example, for the airship in Chaabani et al. [22].

Note that the over-actuation of the airship can generate certain complications in the calculation of the response of the actuators. The specific problem of control allocation was solved in a recent study by Azouz et al. [23]. In the aforementioned study, a control vector of dimension six was used as input, corresponding to the six degrees of freedom of the airship; the twelve degrees of freedom of the actuators ($\|F_i\|$, β_i and γ_i) were used as output. The system to be solved is rectangular. If this is done by means of numerical inversions, this can be penalizing for real-time control. The developed technique is based on energy concepts; this allows a quick and precise solution to the previous problem. This technique gave similar results to those obtained from other known techniques.

The components of the inertia matrix are denoted by M_{ij}.

The gyroscopic forces and moments are grouped together within the vector: $Q_G = \begin{pmatrix} (M_{TT}\nu_1) \wedge \nu_2 \\ (M_{RR}\nu_2) \wedge \nu_2 + (M_{TT}\nu_1) \wedge \nu_1 \end{pmatrix}$. $\wedge$ is the cross product.

Taking into account Equations (8) and (9), the development of Equation (7) gives:

$$\begin{cases} \dot{u} = \frac{1}{M_{11}} \sum_{i=1}^{4} c\gamma_i \cdot c\beta_i \|F_i\| - \frac{1}{M_{11}} s\theta(-B_u + mg) + rv - qw \\ \dot{v} = \frac{1}{M_{22}} \sum_{i=1}^{4} s\gamma_i \cdot \|F_i\| + \frac{1}{M_{22}} s\phi \cdot c\theta(-B_u + mg) - ru + qw \\ \dot{w} = -\frac{1}{M_{33}} \sum_{i=1}^{4} c\gamma_i \cdot s\beta_i \|F_i\| + \frac{1}{M_{33}} c\phi \cdot c\theta(-B_u + mg) - vp + uq \end{cases} \tag{10}$$

and:

$$\begin{cases} \dot{p} = -\frac{M_{66}}{(M_{44}M_{66} - M^2{}_{46})} c_0 \sum_{i=1}^{4} s\gamma_i \|F_i\| + \frac{(M_{46} - M_{66})}{(M_{44}M_{66} - M^2{}_{46})} b_1 (c\gamma_1 \cdot s\beta_1 \|F_1\| - c\gamma_2 \cdot s\beta_2 \|F_2\|) \\ + \frac{(M_{46} - M_{66})}{(M_{44}M_{66} - M^2{}_{46})} b_3 (c\gamma_3 \cdot s\beta_3 \|F_3\| - c\gamma_4 \cdot s\beta_4 \|F_4\|) + \frac{M_{46}}{(M_{44}M_{66} - M^2{}_{46})} a(s\gamma_4 \|F_4\| + s\gamma_3 \|F_3\| - s\gamma_1 \|F_1\| - s\gamma_2 \|F_2\|) \\ - \frac{M_{66}}{(M_{44}M_{66} - M^2{}_{46})} B_u z_G s\phi \cdot c\theta - \frac{M_{46}(M_{44} - M_{55} + M_{66})}{(M_{44}M_{66} - M^2{}_{46})} pq + \frac{(M_{55}M_{66} - M^2{}_{46} - M^2{}_{66})}{(M_{44}M_{66} - M^2{}_{46})} qr \\ \dot{q} = -\frac{c_0}{M_{55}} \sum_{i=1}^{4} c\gamma_i \cdot c\beta_i \|F_i\| + \frac{a}{M_{55}} (c\gamma_4 \cdot s\beta_4 \|F_4\| + c\gamma_3 \cdot s\beta_3 \|F_3\| - c\gamma_1 \cdot s\beta_1 \|F_1\| - c\gamma_2 \cdot s\beta_2 \|F_2\|) \\ - \frac{B_u z_G}{M_{55}} \cdot s\theta + \frac{M_{46}}{M_{55}} p^2 + \frac{(M_{66} - M_{44})}{M_{55}} pr + \frac{M_{46}}{M_{55}} r^2 \\ \dot{r} = \frac{M_{46}}{(M_{44}M_{66} - M^2{}_{46})} c_0 \sum_{i=1}^{4} s\gamma_i \|F_i\| + \frac{(M_{46} - M_{44})}{(M_{44}M_{66} - M^2{}_{46})} b_1 (c\gamma_1 \cdot s\beta_1 \|F_1\| - c\gamma_2 \cdot s\beta_2 \|F_2\|) \\ + \frac{(M_{46} - M_{44})}{(M_{44}M_{66} - M^2{}_{46})} b_3 (c\gamma_3 \cdot s\beta_3 \|F_3\| - c\gamma_4 \cdot s\beta_4 \|F_4\|) - \frac{M_{44}}{(M_{44}M_{66} - M^2{}_{46})} a(s\gamma_4 \|F_4\| + s\gamma_3 \|F_3\| - s\gamma_1 \|F_1\| - s\gamma_2 \|F_2\|) \\ + \frac{M_{46}}{(M_{44}M_{66} - M^2{}_{46})} B_u z_G s\phi \cdot c\theta + \frac{(M^2{}_{44} + M^2{}_{46} - M_{44}M_{55})}{(M_{44}M_{66} - M^2{}_{46})} pq + \frac{M_{46}(M_{44} - M_{55} + M_{66})}{(M_{44}M_{66} - M^2{}_{46})} qr \end{cases} \tag{11}$$

This model will serve as a basis for the application of our control laws.

3. Quasi-LPV Representation for the Dynamic Model

3.1. Vector Control

To control the airship in hovering flight, the voltages at the terminals of the electric motors of the thrusters are varied, thus modifying the thrust forces F_i of the latter; the angles of inclination of the axes of these rotors are also varied by modifying the β_i and γ_i. The airship, in this configuration, is fully actuated.

We must point out that we are interested in a case of stabilisation of the airship around a loading point. We assume that in this configuration the reference frames R_0 and R_m are parallel and that the rotations of the airship are small.

The following control vector is proposed for the stabilisation of the airship:

$$\begin{cases} u_1 = \frac{1}{M_{11}} \sum_{i=1}^{4} c\gamma_i \cdot c\beta_i \|F_i\| \\ u_2 = \frac{1}{M_{22}} \sum_{i=1}^{4} s\gamma_i \|F_i\| \\ u_3 = -\frac{1}{M_{33}} \sum_{i=1}^{4} c\gamma_i \cdot s\beta_i \|F_i\| \end{cases} \tag{12}$$

and:

$$\begin{cases} u_4 = -\frac{M_{66}}{(M_{44}M_{66} - M^2{}_{46})} c_0 \sum_{i=1}^{4} \|F_i\| s\gamma_i + \frac{(I_{xz} - M_{66})}{(M_{44}M_{66} - M^2{}_{46})} b_1 (\|F_1\| c\gamma_1 \cdot s\beta_1 - \|F_2\| c\gamma_2 \cdot s\beta_2) \\ + \frac{(M_{46} - M_{66})}{(M_{44}M_{66} - M^2{}_{46})} b_3 (\|F_3\| c\gamma_3 \cdot s\beta_3 - \|F_4\| c\gamma_4 \cdot s\beta_4) + \frac{M_{46}}{(M_{44}M_{66} - M^2{}_{46})} a(\|F_4\| s\gamma_4 + \|F_3\| s\gamma_3 - \|F_1\| s\gamma_1 - \|F_2\| s\gamma_2) \\ u_5 = -\frac{c_0}{M_{55}} \sum_{i=1}^{4} \|F_i\| c\gamma_i \cdot c\beta_i + \frac{a}{M_{55}} (\|F_4\| c\gamma_4 \cdot s\beta_4 + \|F_3\| c\gamma_3 \cdot s\beta_3 - \|F_1\| c\gamma_1 \cdot s\beta_1 - \|F_2\| c\gamma_2 \cdot s\beta_2) \\ u_6 = \frac{M_{46}}{(M_{44}M_{66} - M^2{}_{46})} c_0 \sum_{i=1}^{4} \|F_i\| s\gamma_i + \frac{(M_{46} - M_{44})}{(M_{44}M_{66} - M^2{}_{46})} b_1 (\|F_1\| c\gamma_1 \cdot s\beta_1 - \|F_2\| c\gamma_2 \cdot s\beta_2) \\ + \frac{(M_{46} - M_{44})}{(M_{44}M_{66} - M^2{}_{46})} b_3 (\|F_3\| c\gamma_3 \cdot s\beta_3 - \|F_4\| c\gamma_4 \cdot s\beta_4) - \frac{M_{44}}{(M_{44}M_{66} - M^2{}_{46})} a(\|F_4\| s\gamma_4 + \|F_3\| s\gamma_3 - \|F_1\| s\gamma_1 - \|F_2\| s\gamma_2) \end{cases} \tag{13}$$

The dynamic model (10) and (11) becomes:

$$\left\{\begin{array}{lcl} \dot{u} & = & u_1 + c_1 \cdot s\theta + rv - qw \\ \dot{v} & = & u_2 + c_2 \cdot s\phi \cdot c\theta - ru + qw \\ \dot{w} & = & u_3 + \frac{1}{M_{33}} c\phi \cdot c\theta(-B_u + mg) - vp + uq \\ \dot{p} & = & u_4 + a_1 \cdot c\theta \cdot s\phi + a_2 \cdot pq + a_3 \cdot qr \\ \dot{q} & = & u_5 + a_5 \cdot s\theta + a_6 \cdot p^2 + a_4 \cdot pr + a_6 \cdot r^2 \\ \dot{r} & = & u_6 + a_7 \cdot s\phi \cdot c\theta + a_8 \cdot pq + a_9 \cdot qr \end{array}\right. \tag{14}$$

With:

$$\left\{\begin{array}{lcl} a_1 & = & -\frac{B_u z_G M_{66}}{(-M_{46}^2 + M_{66} M_{44})} \\ a_2 & = & -\frac{(M_{44} - M_{55} + M_{66}) M_{46}}{(-M_{46}^2 + M_{66} M_{44})} \\ a_3 & = & -\frac{(M_{55} M_{66} - M_{46}^2 - M_{66}^2)}{(-M_{46}^2 + M_{66} M_{44})} \\ a_4 & = & \frac{-M_{44} + M_{66}}{M_{55}} \\ a_5 & = & -\frac{B_u z_G}{M_{55}} \\ a_6 & = & -\frac{M_{46}}{M_{55}} \end{array}\right. \text{ and } \left\{\begin{array}{lcl} a_7 & = & \frac{B_u z_G M_{46}}{(-M_{46}^2 + M_{66} M_{44})} \\ a_8 & = & \frac{(M_{44}^2 + M_{46}^2 - M_{44} M_{55})}{(-M_{46}^2 + M_{66} M_{44})} \\ a_9 & = & \frac{(M_{44} + M_{66} - M_{55}) M_{46}}{(-M_{46}^2 + M_{66} M_{44})} \\ c_1 & = & \frac{-B_u + mg}{M_{11}} \\ c_2 & = & \frac{-B_u + mg}{M_{22}} \end{array}\right. \tag{15}$$

In matrix form, the system (14) becomes:

$$\dot{X} = AX + BU + E \cdot d \tag{16}$$

with: $X = \left[u, v, w, \phi, \dot{\phi}, \theta, \dot{\theta}, r\right]^T$ being the vector of states, $d = \cos\theta \cdot \cos\varphi$ is considered as a disturbance, and $U = [u_1, \ldots, u_6]^T$ being the vector of control. Since we are stabilising the airship around a given point, and given the assumption made concerning the reference frames, we use the following kinematics relations: $\dot{\phi} \approx p; \dot{\theta} \approx q; \dot{\psi} \approx r$.

We should mention here that, for the specific problem at hand, the quadratic terms of type $q \cdot w$ are small and will be neglected later in the approach. The matrices A, B and E could be written as:

$$A = \begin{pmatrix} 0 & r & 0 & 0 & 0 & c_1\left(1 - \frac{\theta^2}{6}\right) & 0 & 0 \\ -r & 0 & 0 & c_2 c\theta\left(1 - \frac{\theta^2}{6}\right) & 0 & 0 & 0 & 0 \\ 0 & -p & 0 & 0 & 0 & 0 & 0 & 0 \\ 0 & 0 & 0 & 0 & 1 & 0 & 0 & 0 \\ 0 & 0 & 0 & a_1\left(1 - \frac{\theta^2}{6}\right)c\theta & 0 & 0 & a_3 r + a_2 p & 0 \\ 0 & 0 & 0 & 0 & 0 & 0 & 1 & 0 \\ 0 & 0 & 0 & 0 & a_4 r + a_6 p & a_5\left(1 - \frac{\theta^2}{6}\right) & 0 & a_6 r \\ 0 & 0 & 0 & a_7\left(1 - \frac{\theta^2}{6}\right)c\theta & 0 & 0 & a_8 r + a_9 p & 0 \end{pmatrix} \tag{17}$$

and:

$$B = \begin{pmatrix} 1 & 0 & 0 & 0 & 0 & 0 \\ 0 & 1 & 0 & 0 & 0 & 0 \\ 0 & 0 & 1 & 0 & 0 & 0 \\ 0 & 0 & 0 & 0 & 0 & 0 \\ 0 & 0 & 0 & 1 & 0 & 0 \\ 0 & 0 & 0 & 0 & 0 & 0 \\ 0 & 0 & 0 & 0 & 1 & 0 \\ 0 & 0 & 0 & 0 & 0 & 1 \end{pmatrix}; E = \begin{pmatrix} 0 \\ 0 \\ \frac{mg - B_u}{M_{33}} \\ 0 \\ 0 \\ 0 \\ 0 \\ 0 \end{pmatrix} \tag{18}$$

One can choose the points of non-linearity as follows:

$$\left\{\begin{array}{l} z_1 = c\theta \cdot \left(1 - \frac{\varphi^2}{6}\right) \\ z_2 = p \end{array}\right. \text{ and } \left\{\begin{array}{l} z_3 = r \\ z_4 = \left(1 - \frac{\theta^2}{6}\right) \end{array}\right. \tag{19}$$

It is easy to check that all these points are bounded.

$$\left\{\begin{array}{l} {z_1}^{min} \leq z_1 \leq {z_1}^{max} \\ {z_2}^{min} \leq z_2 \leq {z_2}^{max} \end{array}\right. \text{ and } \left\{\begin{array}{l} {z_3}^{min} \leq z_3 \leq {z_3}^{max} \\ {z_4}^{min} \leq z_4 \leq {z_4}^{max} \end{array}\right. \tag{20}$$

with:

$$\left\{\begin{array}{l} z_1^{min} = -1; z_2^{min} = -1; z_3^{min} = -1; z_4^{min} = -\frac{2}{\text{ß}} \\ z_1^{max} = 1; z_2^{max} = 1; z_3^{max} = 1; z_4^{max} = \frac{2}{\text{ß}} \end{array}\right. \tag{21}$$

Then, this matrix A becomes:

$$A_{z_i}(z) = \begin{pmatrix} 0 & z_3 & 0 & 0 & 0 & c_1z_4 & 0 & 0 \\ -z_3 & 0 & 0 & c_2z_1 & 0 & 0 & 0 & 0 \\ 0 & -z_2 & 0 & 0 & 0 & 0 & 0 & 0 \\ 0 & 0 & 0 & 0 & 1 & 0 & 0 & 0 \\ 0 & 0 & 0 & a_1z_1 & 0 & 0 & a_3z_3 + a_2z_2 & 0 \\ 0 & 0 & 0 & 0 & 0 & 0 & 1 & 0 \\ 0 & 0 & 0 & 0 & a_4z_3 + a_6z_2 & a_5z_4 & 0 & a_6z_3 \\ 0 & 0 & 0 & a_7z_1 & 0 & 0 & a_8z_2 + a_9z_3 & 0 \end{pmatrix} \tag{22}$$

With z being the vector collection of the different z_i.

The quasi-LPV method allows the expression of the behaviour of a system as a set of linear models. Each sub-model contributes to this total representation according to a weight function to value in the interval. The multi-model structure is written in the following form:

$$\dot{X} = \sum_{i=1}^{16} h_i(X)A_{zi}X + BU + Ed \tag{23}$$

where $A_i \in R^{8\times 8}$. An example of a matrix A_i is detailed in Appendix A.1 in Appendix A.

The weighting functions h_i are nonlinear, are given in Appendix A.1, and satisfy the following properties:

$$\left\{\begin{array}{l} \sum_{i=1}^{16} h_i(X) = 1 \\ \text{with } 0 \leq h_i(X) \leq 1 \; \forall i \in \{1, 2, \ldots, 16\} \end{array}\right. \tag{24}$$

In summary, the proposed control strategy can be represented as follows (Figure 4):

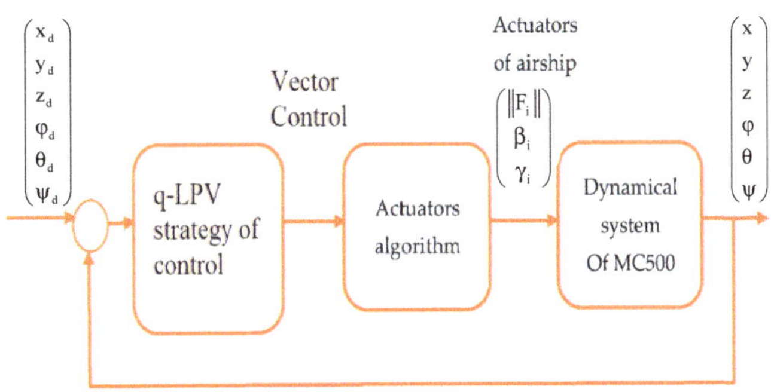

Figure 4. Closed-loop architecture of control.

3.2. The Uncertain Matrix of the Airship

To test the reliability and the robustness of our method, we introduce uncertainty into many parameters of the airship. Mathematical models of Large Capacity Airships always contain uncertain elements, which model the designer's lack of knowledge about certain parameter values, disturbances, and unmodelled dynamics. These uncertain parameters have values that are known only inside a given compact bounding set.

Let ΔA_z be the uncertain matrices having the same structure as $A(z)$.

We consider the following uncertain system:

$$\dot{X} = \sum_{i=1}^{16} h_i(X)\{A_i + \Delta A_z\}X + BU + Ed \quad (25)$$

where uncertainty is structured in the following way:

$$\Delta A_z = \widetilde{A}_1 \cdot F_1(t) \cdot \widetilde{E}_1 + \widetilde{A}_2 \cdot F_2(t) \cdot \widetilde{E}_2 + \widetilde{A}_3 \cdot F_3(t) \cdot \widetilde{E}_3 \quad (26)$$

Such as $\widetilde{E}_1, \widetilde{E}_2$, $\widetilde{E}_3, \widetilde{A}_1, \widetilde{A}_2$ and $\widetilde{A}_3$ are constant matrices defined later, and $F_1(t), F_2(t), F_3(t)$ satisfy the following condition:

$$\begin{cases} F_1^T(t) \cdot F_1(t) \leq I \\ F_2^T(t) \cdot F_2(t) \leq I, \ \forall t \\ F_3^T(t) \cdot F_3(t) \leq I \end{cases} \quad (27)$$

with I representing the identity matrix.

The uncertain matrix ΔA_z is given by:

$$\Delta A_z = \begin{pmatrix} 0 & 0 & 0 & 0 & 0 & \Delta c_1 z_4 & 0 & 0 \\ 0 & 0 & 0 & \Delta c_2 z_1 & 0 & 0 & 0 & 0 \\ 0 & 0 & 0 & 0 & 0 & 0 & 0 & 0 \\ 0 & 0 & 0 & 0 & 0 & 0 & 0 & 0 \\ 0 & 0 & 0 & \Delta a_1 z_1 & 0 & 0 & \Delta s_1 & 0 \\ 0 & 0 & 0 & 0 & 0 & 0 & 0 & 0 \\ 0 & 0 & 0 & 0 & \Delta s_2 & \Delta a_5 z_4 & 0 & \Delta a_6 z_3 \\ 0 & 0 & 0 & \Delta a_7 z_1 & 0 & 0 & \Delta s_3 & 0 \end{pmatrix} = \sum_{i=1}^{3} \widetilde{A}_i F_i(t) \widetilde{E}_i \quad (28)$$

One can see Appendix A.2 for more details regarding uncertain terms Δa_i.

With:

$$\widetilde{A}_1 F_1(t)\widetilde{E}_1 = \begin{pmatrix} 0&0&0&0&0&0&0&0\\ 0&0&0&0&0&0&0&0\\ 0&0&0&0&0&0&0&0\\ 0&0&0&0&0&0&0&0\\ 0&0&0&\Delta a_1 z_1&0&0&\Delta s_1&0\\ 0&0&0&0&0&0&0&0\\ 0&0&0&0&\Delta s_2&\Delta a_5 z_4&0&\Delta a_6 z_3\\ 0&0&0&0&0&0&0&0 \end{pmatrix}$$

$$\widetilde{A}_2 F_2(t)\widetilde{E}_2 = \begin{pmatrix} 0&0&0&0&0&\Delta c_1 z_4&0&0\\ 0&0&0&0&0&0&0&0\\ 0&0&0&0&0&0&0&0\\ 0&0&0&0&0&0&0&0\\ 0&0&0&0&0&0&0&0\\ 0&0&0&0&0&0&0&0\\ 0&0&0&0&0&0&0&0\\ 0&0&0&\Delta a_7 z_1&0&0&\Delta s_3&0 \end{pmatrix} \quad (29)$$

$$\widetilde{A}_3 F_3(t)\widetilde{E}_3 = \begin{pmatrix} 0&0&0&0&0&0&0&0\\ 0&0&0&\Delta c_2 z_1&0&0&0&0\\ 0&0&0&0&0&0&0&0\\ 0&0&0&0&0&0&0&0\\ 0&0&0&0&0&0&0&0\\ 0&0&0&0&0&0&0&0\\ 0&0&0&0&0&0&0&0\\ 0&0&0&0&0&0&0&0 \end{pmatrix}$$

Using the expression (28), we can decompose the matrixes $\widetilde{A}_i$, $F_i(t)$ and $\widetilde{E}_i$ $(i = 1,2,3)$, in the following way:

$$\widetilde{A}_1 = \begin{pmatrix} 0&0&0&0&0&0&0&0\\ 0&0&0&0&0&0&0&0\\ 0&0&0&0&0&0&0&0\\ 0&0&0&0&0&0&0&0\\ 0&0&0&1&0&0&1&0\\ 0&0&0&0&0&0&0&0\\ 0&0&0&0&1&1&0&1\\ 0&0&0&0&0&0&0&0 \end{pmatrix}; \ \widetilde{E}_1 = \begin{pmatrix} 0&0&0&0&0&0&0&0\\ 0&0&0&0&0&0&0&0\\ 0&0&0&0&0&0&0&0\\ 0&0&0&\sigma_1&0&0&0&0\\ 0&0&0&0&\sigma_2&0&0&0\\ 0&0&0&0&0&\sigma_3&0&0\\ 0&0&0&0&0&0&\sigma_4&0\\ 0&0&0&0&0&0&0&\sigma_5 \end{pmatrix} \quad (30)$$

$$F_1(t) = \begin{pmatrix} 0&0&0&0&0&0&0&0\\ 0&0&0&0&0&0&0&0\\ 0&0&0&0&0&0&0&0\\ 0&0&0&\frac{\Delta a_1 z_1}{\sigma_1}&0&0&0&0\\ 0&0&0&0&\frac{\Delta s_2}{\sigma_2}&0&0&0\\ 0&0&0&0&0&\frac{\Delta a_5 z_4}{\sigma_3}&0&0\\ 0&0&0&0&0&0&\frac{\Delta s_1}{\sigma_4}&0\\ 0&0&0&0&0&0&0&\frac{\Delta a_6 z_3}{\sigma_5} \end{pmatrix} \quad (31)$$

Two further decompositions are as follows:

$$\widetilde{A}_2(8,7) = \widetilde{A}_2(1,6) = \widetilde{A}_2(8,4) = 1 \quad (32)$$

$$\widetilde{E}_2(6,6) = \delta_2;\ \widetilde{E}_2(7,7) = \delta_3;\ \widetilde{E}_2(4,4) = \delta_1 \quad (33)$$

$$F_2(t)(6,6) = \frac{\Delta c_1 z_4}{\delta_2};\ F_2(t)(4,4) = \frac{\Delta a_7 z_1}{\delta_1};\ F_2(t)(7,7) = \frac{\Delta s_3}{\delta_3} \quad (34)$$

$$\widetilde{A}_3(2,4) = 1 \quad (35)$$

$$\widetilde{E}_3(4,4) = h_1 \quad (36)$$

$$F_3(t)(4,4) = \frac{\Delta c_2 z_1}{h_1} \quad (37)$$

Other terms are all zero: $F_2(t)(i,j) = \widetilde{A}_2(i,j) = \widetilde{E}_2(i,j) = 0$ and $F_3(t)(i,j) = \widetilde{A}_3(i,j) = \widetilde{E}_3(i,j) = 0$ for $i,j \in \{1,2,\ldots,8\}$, with:

$$\begin{cases} \Delta s_1 = \Delta a_2 z_2 + \Delta a_3 z_3; \Delta s_2 = \Delta a_6 z_2; \Delta s_3 = \Delta a_8 z_2 + \Delta a_9 z_3 \\ \sigma_1 = \max(\Delta a_1 z_1^{max});\ \sigma_2 = \max(\Delta a_6 z_2^{max});\ \sigma_3 = \max(\Delta a_5 z_4^{max}) \\ \sigma_4 = \max(\Delta a_2 z_2^{max} + \Delta a_3 z_3^{max}); \sigma_5 = \max(\Delta a_6 z_3^{max}); \\ \delta_1 = \max(\Delta a_7 z_1^{max}); \delta_2 = \max(\Delta c_1 z_4^{max}); \\ \delta_3 = \max(\Delta a_8 z_2^{max}); h_1 = \max(\Delta c_2 z_1^{max}) \end{cases} .$$

To eliminate the term E·d of disturbance, we can suppose that:

$$U = -\sum_{i=1}^{16} h_i(X) K_i X - S \cdot d \quad (38)$$

Using the control expression (38), the Quasi-LPV structure of the airship is given by:

$$\dot{X} = \sum_{i=1}^{16} h_i(X)(A_i + \Delta A - BK_i)X + (E - BS) \cdot d \quad (39)$$

With:

$$\begin{aligned} A &= \sum_{i=1}^{16} h_i(X) A_i;\ \Delta A = \widetilde{A}_2 F_2(t) \widetilde{E}_2 + \widetilde{A}_1 F_1(t) \widetilde{E}_1 \\ \text{and } U &= -\sum_{i=1}^{16} h_i(X) K_i X - S \cdot d \end{aligned} \quad (40)$$

where $A_i \in R^{8\times 8}$; $K_i \in R^{6\times 8}$

We can choose the matrix S as: $S = B^+ E$, where B^+ is the pseudo inverse matrix of the matrix B.

3.3. Robust Stabilisation of the Airship

Theorem 1. *The uncertain dynamic system (39) would be stable if there are definite, positive and symmetric matrixes* P, *and three positive diagonal matrixes* Ω_1, Ω_2 *and* Ω_3, *such that the following LMIs and LME hold:*

$$\begin{cases} \begin{bmatrix} \widetilde{Q}A_i^T + A_i\widetilde{Q} - M_i^T B^T - BM_i + \widetilde{E}_1^T \Omega_1 \widetilde{E}_1 + \widetilde{E}_2^T \Omega_2 \widetilde{E}_2 + \widetilde{E}_3^T \Omega_3 \widetilde{E}_3 & \widetilde{A}_1^T & \widetilde{A}_2^T & \widetilde{A}_3^T \\ \widetilde{A}_1 & -\Omega_1 & 0 & 0 \\ \widetilde{A}_2 & 0 & -\Omega_2 & 0 \\ \widetilde{A}_3 & 0 & 0 & -\Omega_3 \end{bmatrix} \\ E - BS = 0 \end{cases} \quad (41)$$

with: $\widetilde{Q} = P^{-1}$, *and the matrix gains* K_i *are defined by:* $K_i = M_i \widetilde{Q}^{-1}$, $i = 1, \ldots, 16$ *and*

$$\Omega_1 = \begin{bmatrix} \varepsilon_1^1 & 0 & 0 & 0 & 0 \\ 0 & \varepsilon_2^1 & 0 & 0 & 0 \\ 0 & 0 & \varepsilon_3^1 & 0 & \vdots \\ \vdots & \vdots & \vdots & \ddots & 0 \\ 0 & 0 & 0 & 0 & \varepsilon_{16}^1 \end{bmatrix}; \Omega_2 = \begin{bmatrix} \varepsilon_1^2 & 0 & 0 & 0 & 0 \\ 0 & \varepsilon_2^2 & 0 & 0 & 0 \\ 0 & 0 & \varepsilon_3^2 & 0 & \vdots \\ \vdots & \vdots & \vdots & \ddots & 0 \\ 0 & 0 & 0 & 0 & \varepsilon_{16}^2 \end{bmatrix}; \Omega_3 = \begin{bmatrix} \varepsilon_1^3 & 0 & 0 & 0 & 0 \\ 0 & \varepsilon_2^3 & 0 & 0 & 0 \\ 0 & 0 & \varepsilon_3^3 & 0 & \vdots \\ \vdots & \vdots & \vdots & \ddots & 0 \\ 0 & 0 & 0 & 0 & \varepsilon_{16}^3 \end{bmatrix} \quad (42)$$

such as: $\varepsilon_i^1, \varepsilon_i^2, \varepsilon_i^3 \succ 0$ *for* $i = 1, \ldots, 16$.

To prove the above theorem, we must first recall the following lemmas which are useful in the proof:

Lemma 1. *Peterson [24]: Suppose F(t) satisfying* $F^T(t)F(t) \leq I, \forall t, Q = Q^T, E, H$ *are matrixes of appropriate dimension, the inequality below:*

$$Q + HF(t)E + E^T F^T(t) H^T \prec 0 \tag{43}$$

is regarded as met only when the following inequality is true for:

$$Q + \varepsilon^{-1} HH^T + \varepsilon E^T E \prec 0 \tag{44}$$

Lemma 2. *Let* Ω *be a positive definite matrix; for* Y *and* X *matrixes with compatible dimensions, the property below is true:*

$$XY^T + X^T Y \prec X^T \Omega^{-1} X + Y \Omega Y^T, \; \Omega \succ 0 \tag{45}$$

Lemma 3. *(Congruence) Consider two matrixes* P *and* Q*; if* P *is positive definite and if* Q *is a full column rank matrix, then the matrix is positive definite.*

If P is a positive definite matrix, and if Q is a full column rank matrix, then the matrix product $Q \cdot P \cdot Q^T$ is definite positive.

This lemma is called the congruence lemma

Lemma 4. *Boyd* [25] *(Schur Complement). Given constant matrixes*M_1, M_2 *and* M_3 *with appropriate dimensions, where:* $M_1 = M_1^T$ *and* $M_2 = M_2^T$*, then:* $M_1 + M_3^T M_2^{-1} M_3 \prec 0$ *if and only if:*

$$\begin{bmatrix} M_1 & M_3^T \\ * & -M_2 \end{bmatrix} \prec 0 \text{ or } \begin{bmatrix} -M_2 & M_3 \\ * & M_1 \end{bmatrix} \prec 0 \tag{46}$$

Using these lemmas, the proof of Theorem 1 is presented as follows:

Proof. Consider the Lyapunov function expressed as follows:

$$V = X^T P X \tag{47}$$

with $P = P^T \geq 0$ being a definite positive matrix. □

The derivative with respect to time of V will then be:

$$\begin{aligned} \dot{V} &= X^T P \dot{X} + \dot{X}^T P X \\ &= X^T \sum_{i=1}^{16} h_i(X) \left[(A_i + \Delta A - BK_i)^T P + P(A_i + \Delta A - BK_i) \right] X + 2X^T P(E - BS)d \end{aligned} \tag{48}$$

And since we choose $E - BS = 0$, the Equation (48) becomes:

$$\dot{V} = X^T \sum_{i=1}^{16} h_i(X) \left[(A_i - BK_i + \Delta A)^T P + P(A_i - BK_i + \Delta A) \right] X \tag{49}$$

As the weighting functions h_i satisfy the property of convex sum, such that $\dot{V}(X)$ is negative, it is sufficient to check that:

$$(A_i + \Delta A - BK_i)^T P + P(A_i + \Delta A - BK_i) \prec 0, \; i = 1, \ldots, 16 \tag{50}$$

We replace ΔA by its value (27), and we obtain:

$$\begin{aligned}&\left(A_i+\widetilde{A}_2F_2(t)\widetilde{E}_2+\widetilde{A}_1F_1(t)\widetilde{E}_1+\widetilde{A}_3F_3(t)\widetilde{E}_3-BK_i\right)^TP\\&+P\left(A_i+\widetilde{A}_3F_3(t)\widetilde{E}_3+\widetilde{A}_1F_1(t)\widetilde{E}_1+\widetilde{A}_2F_2(t)\widetilde{E}_2-BK_i\right)\prec 0;\ i=1,\ldots,16\end{aligned}\tag{51}$$

Thereafter, and by relying on Lemma 1:

$$\begin{aligned}&A_i^TP+PA_i-PBK_i-(BK_i)^TP+P\widetilde{A}_1\Omega_1\widetilde{A}_1^TP+\widetilde{E}_1^TF_1^T(t)\Omega_1^{-1}F_1(t)\widetilde{E}_1\\&+P\widetilde{A}_2\Omega_2\widetilde{A}_2^TP+\widetilde{E}_2^TF_2^T(t)\Omega_2^{-1}F_2(t)\widetilde{E}_2+P\widetilde{A}_3\Omega_3\widetilde{A}_3^TP+\widetilde{E}_3^TF_3^T(t)\Omega_3^{-1}F_3(t)\widetilde{E}_3\prec 0;\\&i=1,\ldots,16\end{aligned}\tag{52}$$

and as: $F_i^T(t)F_i(t)\le I, \forall t$ and i = 1,2,3, then the inequality (51) becomes:

$$\begin{aligned}&A_i^TP+PA_i-PBK_i-(BK_i)^TP+P\widetilde{A}_1\Omega_1^{-1}\widetilde{A}_1^TP+\widetilde{E}_1^T\Omega_1\widetilde{E}_1\\&+P\widetilde{A}_2\Omega_2^{-1}\widetilde{A}_2^TP+\widetilde{E}_2^T\Omega_2\widetilde{E}_2+P\widetilde{A}_3\Omega_3^{-1}\widetilde{A}_3^TP+\widetilde{E}_3^T\Omega_3\widetilde{E}_3\prec 0;\\&i=1,\ldots,16\end{aligned}\tag{53}$$

Based on Lemma 2, and if we suppose $P^{-1}=\widetilde{Q}$ and $M_i=K_i\widetilde{Q}$, we obtain:

$$\begin{aligned}&P^{-1}[A_i^TP+PA_i-PBK_i-(BK_i)^TP+P\widetilde{A}_1\Omega_1^{-1}\widetilde{A}_1^TP+\widetilde{E}_1^T\Omega_1\widetilde{E}_1\\&+P\widetilde{A}_2\Omega_2^{-1}\widetilde{A}_2^TP+\widetilde{E}_2^T\Omega_2\widetilde{E}_2+P\widetilde{A}_3\Omega_3^{-1}\widetilde{A}_3^TP+\widetilde{E}_3^T\Omega_3\widetilde{E}_3]P^{-1}\prec 0;\\&i=1,\ldots,16\end{aligned}\tag{54}$$

We can rewrite the system (52) as:

$$\begin{aligned}&\widetilde{Q}A_i^T+A_i\widetilde{Q}-M_i^TB^T-BM_i+\widetilde{A}_1\Omega_1^{-1}\widetilde{A}_1^T+\widetilde{Q}\widetilde{E}_1^T\Omega_1\widetilde{E}_1\widetilde{Q}\\&+\widetilde{A}_2\Omega_2^{-1}\widetilde{A}_2^T+\widetilde{Q}\widetilde{E}_2^T\Omega_2\widetilde{E}_2\widetilde{Q}+\widetilde{A}_3\Omega_3^{-1}\widetilde{A}_3^T+\widetilde{Q}\widetilde{E}_3^T\Omega_3\widetilde{E}_3\widetilde{Q}\prec 0,\\&\forall i\in\{1,\ldots,16\}\end{aligned}\tag{55}$$

Finally, by using the Schur complement (Lemma 4), we will produce the following LMIs:

$$\begin{bmatrix}\widetilde{Q}A_i^T+A_i\widetilde{Q}-M_i^TB^T-BM_i+\widetilde{E}_1^T\Omega_1\widetilde{E}_1+\widetilde{E}_2^T\Omega_2\widetilde{E}_2+\widetilde{E}_3^T\Omega_3\widetilde{E}_3 & \widetilde{A}_1^T & \widetilde{A}_2^T & \widetilde{A}_3^T\\ \widetilde{A}_1 & -\Omega_1 & 0 & 0\\ \widetilde{A}_2 & 0 & -\Omega_2 & 0\\ \widetilde{A}_3 & 0 & 0 & -\Omega_3\end{bmatrix}\prec 0\tag{56}$$

4. Simulation Results

We present here some numerical examples demonstrating the power of our formulation. As support for our development, we will use the following characteristics of the studied airship:

Mass: m = 500 Kg; gravity: $g = 9.81\ m{\cdot}s^{-2}$;
Position of the centre of inertia $z_G = 0.5$ m;
Buoyancy $B_u = 5000$ N;
Components of the total mass matrix M:
$M_{11} = 630$ kg; $M_{22} = 712$ kg; $M_{33} = 1723$ kg
$M_{44} = 9415$ kg; $M_{55} = 10{,}455$ kg;
$M_{66} = 18{,}700$ kg; $M_{46} = 161$ kg.
Uncertainty values:

$$\Delta a_1 = \Delta\left(\frac{-M_{66}B_u z_G}{(-M_{46}^2+M_{44}M_{66})}\right) = 0.146$$
$$\Delta a_2 = \Delta\left(\frac{-(M_{46}M_{44}-M_{55}+M_{66})}{(-M_{46}^2+M_{44}M_{66})}\right) = 0.0083$$
$$\Delta a_3 = \Delta\left(\frac{(M_{55}M_{66}-M_{46}^2+M_{66}^2)}{(-M_{46}^2+M_{44}M_{66})}\right) = 1.2393$$
$$\Delta a_4 = \Delta\left(\frac{M_{66}-M_{44})}{M_{55}}\right) = 0.3577$$
$$\Delta a_5 = \Delta\left(-\frac{B_u z_G}{M_{55}}\right) = 0.0789,\ \Delta a_6 = \Delta\left(\frac{M_{46}}{M_{55}}\right) = 0.037$$
$$\Delta a_7 = \Delta\left(\frac{M_{46}B_u z_G}{(-M_{46}^2+M_{44}M_{66})}\right) = 0.013$$
$$\Delta a_8 = \Delta\left(\frac{M_{44}^2+M_{46}^2-M_{44}M_{55}}{(-M_{46}^2+M_{44}M_{66})}\right) = 0.2237$$
$$\Delta a_9 = \Delta\left(\frac{M_{46}(M_{44}-M_{55}M_{66})}{(-M_{46}^2+M_{44}M_{66})}\right) = 0.083$$

The envelope:
Axis: a = 2.5 m; c = 2 m; Volume: V= 500 m^3;
Position of the actuators: b_1 = 5.4 m; b_3 = 6.5 m;
The numerical developments were carried out using the MATLAB code.

As an illustration, we present a typical manoeuvre of a blimp over a loading area, which includes an attitude change due to a squall of wind. This change is estimated to $\frac{\pi}{6}$ rad in both pitch and roll motion.

Figures 5–12 show the stabilisation of the angles of roll and pitch, as well as their derivatives, and the convergence of the linear velocities u, v and w, in a relatively short period of time.

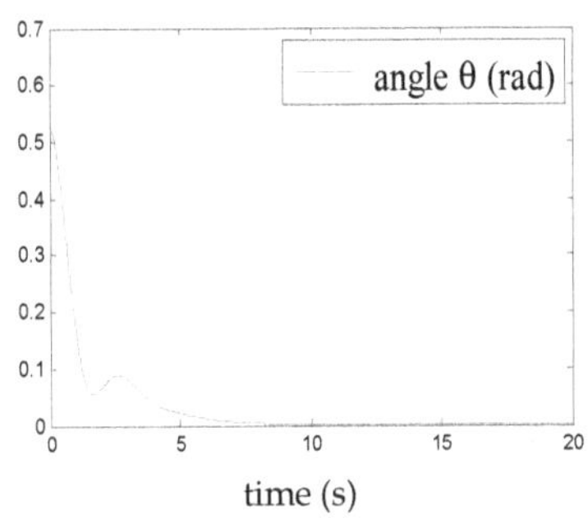

Figure 5. Convergence of the pitch angle.

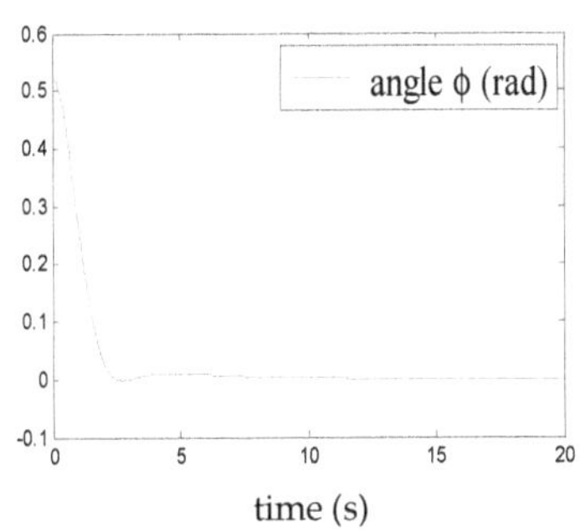

Figure 6. Convergence of the roll angle.

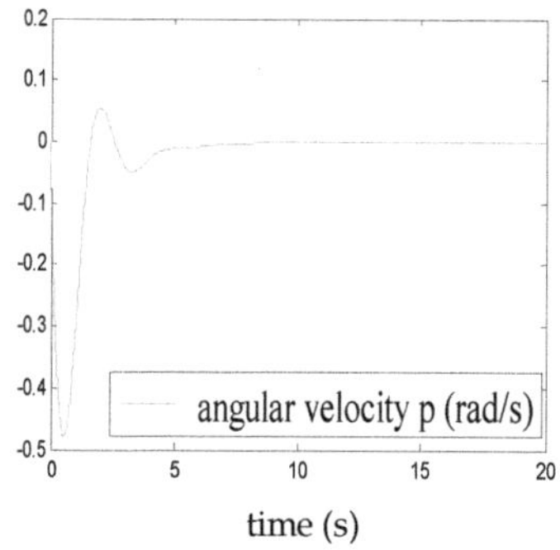

Figure 7. Convergence of p.

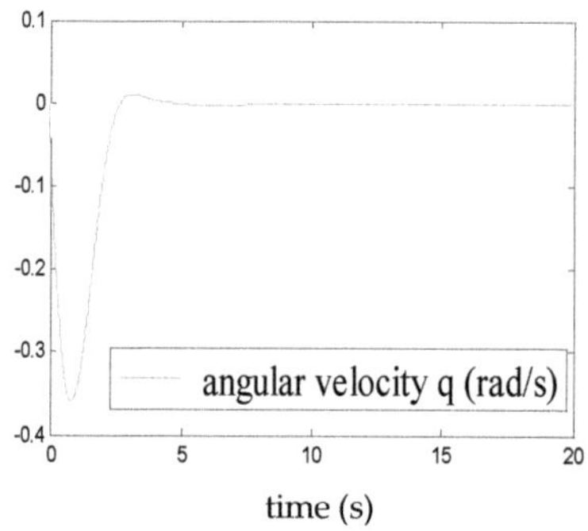

Figure 8. Convergence of q.

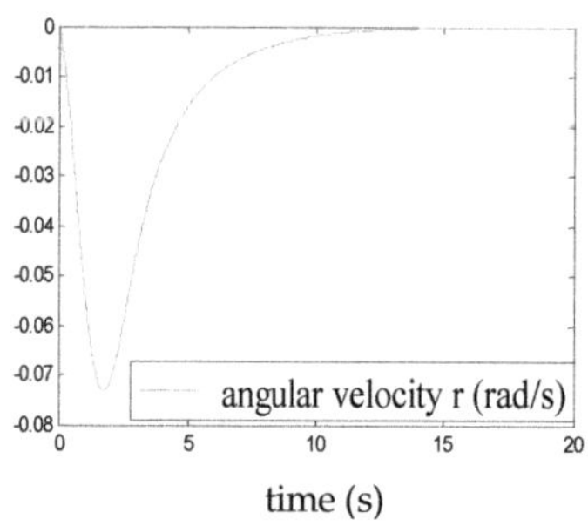

Figure 9. Convergence of r.

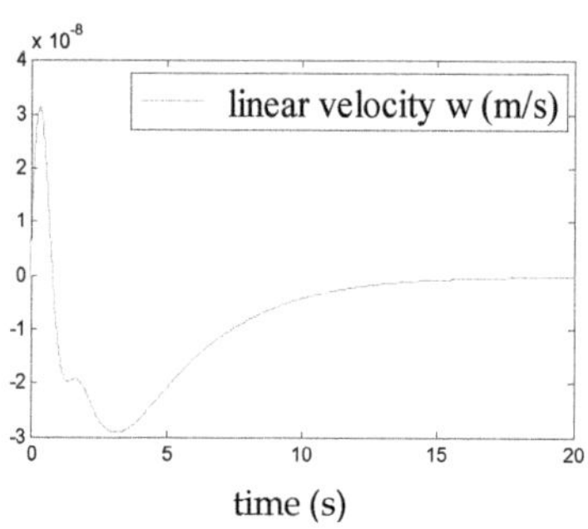

Figure 10. Convergence of w.

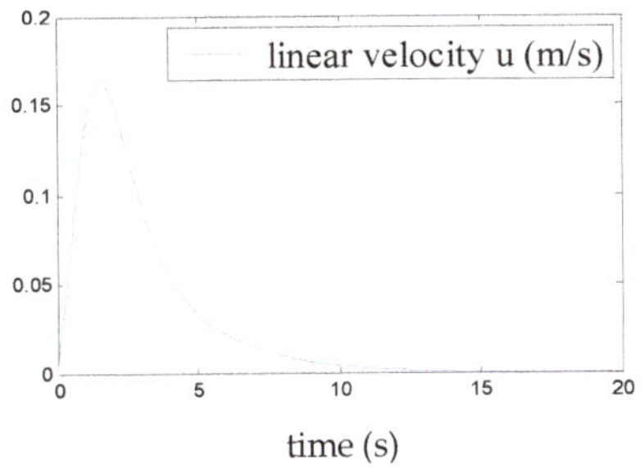

Figure 11. Convergence of u.

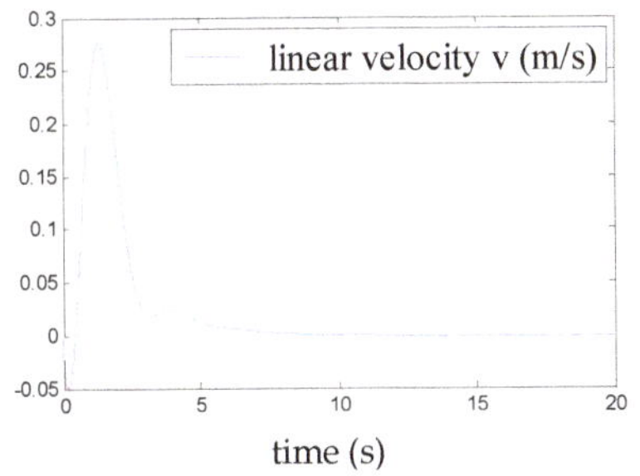

Figure 12. Convergence of v.

This demonstrates the ability of the proposed control to stabilise the airship around a point of equilibrium, even under the effect of a moderate squall of wind. The time evolution of the weighting functions is depicted in Figure 13. It can be clearly seen that the weighting functions h_i $(1 \leq i \leq 8)$ are between 0 and 1 and thus satisfy the convex sum property.

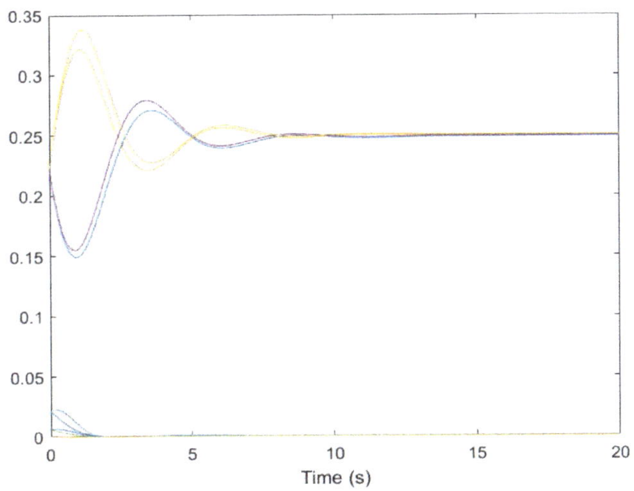

Figure 13. The weighting function. The different colors represent the different weighting functions h_i $(1 \leq i \leq 8)$.

The robustness of the proposed control is also tested on the complete nonlinear model (28).

We note, according to Figures 14–17, that even in the presence of a disturbance (presence of the wind) of maximum amplitude the 0.1, the airship remains stable. This demonstrates the performance of the elaborated control.

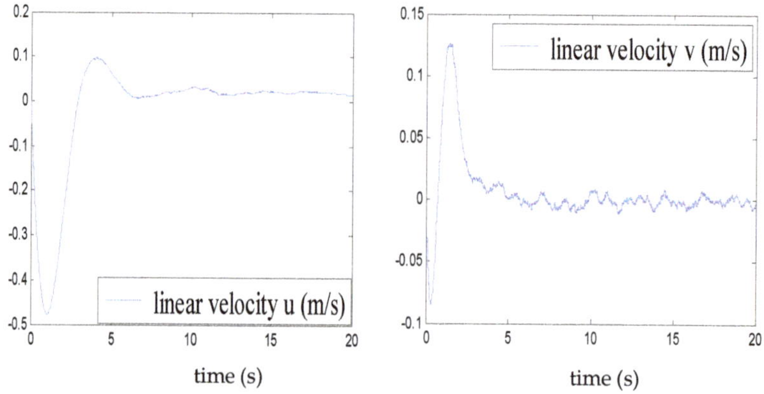

Figure 14. Velocities u and v in presence of disturbance.

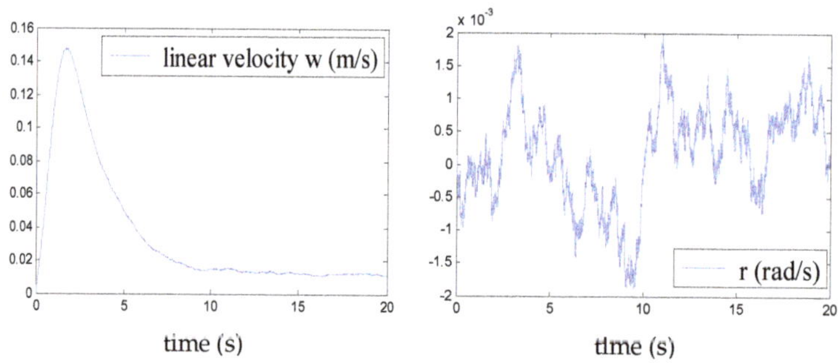

Figure 15. Velocities w and r in presence of disturbance.

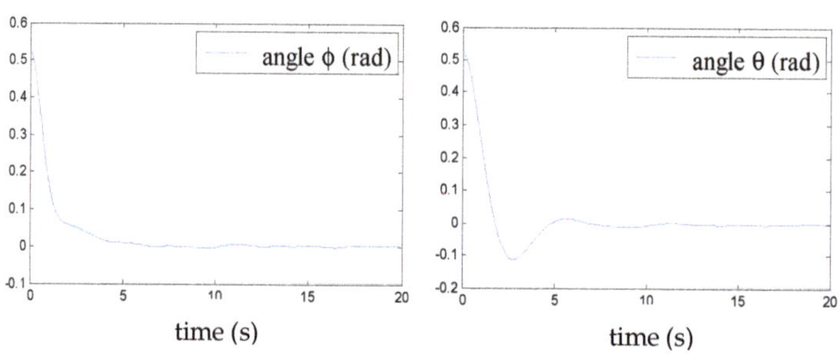

Figure 16. Roll angle ϕ and pitch angle θ in the presence of disturbance.

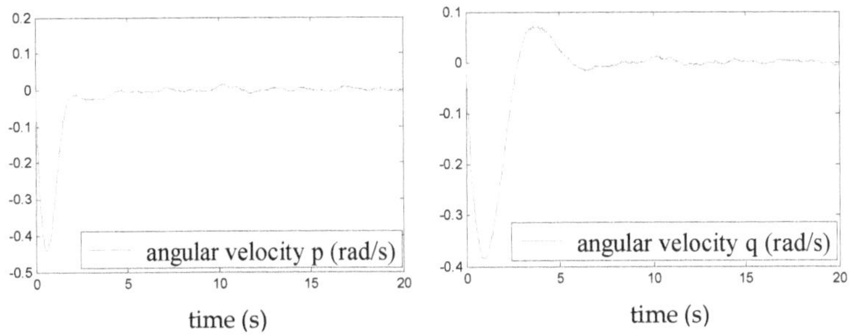

Figure 17. Angular velocities p and q in the presence of disturbance.

5. Conclusions

In the first part of this paper, we presented a dynamic model of an unconventional airship. The latter has a flying wing shape and is over-actuated.

The stabilisation of this flying object at a point of equilibrium above a loading area was the highlight of this work.

Specifically, we have presented a controller design technique for uncertain nonlinear systems with external disturbances described by the Quasi-LPV model. The main results concern the two theorems developed to ensure a sufficient condition in terms of LMI for the problem of tracking uncertain systems, expressed in Quasi-LPV form. The results of the numerical simulations are satisfactory and prove the robustness of our approach.

Author Contributions: Conceptualization, S.C. and N.A.; Methodology, S.C. and N.A.; software, S.C.; validation, S.C. and N.A.; investigation, S.C. and N.A.; writing—original draft preparation, S.C. and N.A.; writing—review and editing, S.C. and N.A.; supervision, N.A.; project administration, S.C. and N.A. All authors have read and agreed to the published version of the manuscript.

Funding: This research received no external funding.

Institutional Review Board Statement: Not applicable.

Informed Consent Statement: Not applicable.

Data Availability Statement: The data presented in this study are available on request from the corresponding author.

Conflicts of Interest: The authors declare no conflict of interest.

Appendix A

Appendix A.1

The sub matrixes A_i were determined as follows:

$$A_1 = \begin{pmatrix} 0 & z_3^{max} & 0 & 0 & 0 & c_1 z_4^{max} & 0 & 0 \\ -z_3^{max} & 0 & 0 & c_2 z_1^{max} & 0 & 0 & 0 & 0 \\ 0 & -z_2^{max} & 0 & 0 & 0 & 0 & 0 & 0 \\ 0 & 0 & 0 & 0 & 1 & 0 & 0 & 0 \\ 0 & 0 & 0 & a_1 z_1^{max} & 0 & 0 & a_2 z_2^{max} + a_3 z_3^{max} & 0 \\ 0 & 0 & 0 & 0 & 0 & 0 & 1 & 0 \\ 0 & 0 & 0 & 0 & a_6 z_2^{max} + a_4 z_3^{max} & a_5 z_4^{max} & 0 & a_6 z_3^{max} \\ 0 & 0 & 0 & a_7 z_1^{max} & 0 & 0 & a_8 z_2^{max} + a_9 z_3^{max} & 0 \end{pmatrix}$$

The weighting functions were:

$$
\begin{array}{lll}
h_1(z) = F_4^0 \cdot F_3^0 \cdot F_2^0 \cdot F_1^0 & h_6(z) = F_4^0 \cdot F_3^0 \cdot F_2^1 \cdot F_1^1 & h_{11}(z) = F_4^1 \cdot F_3^1 \cdot F_2^0 \cdot F_1^0 \\
h_2(z) = F_4^0 \cdot F_3^0 \cdot F_2^0 \cdot F_1^1 & h_7(z) = F_4^0 \cdot F_3^1 \cdot F_2^0 \cdot F_1^1 & h_{12}(z) = F_4^0 \cdot F_3^1 \cdot F_2^1 \cdot F_1^1 \\
h_3(z) = F_4^0 \cdot F_3^0 \cdot F_2^1 \cdot F_1^0 & h_8(z) = F_4^1 \cdot F_3^0 \cdot F_2^0 \cdot F_1^1 & h_{13}(z) = F_4^1 \cdot F_3^0 \cdot F_2^1 \cdot F_1^1 \\
h_4(z) = F_4^0 \cdot F_3^1 \cdot F_2^0 \cdot F_1^0 & h_9(z) = F_4^0 \cdot F_3^1 \cdot F_2^1 \cdot F_1^0 & h_{14}(z) = F_4^1 \cdot F_3^1 \cdot F_2^0 \cdot F_1^1 \\
h_5(z) = F_4^1 \cdot F_3^0 \cdot F_2^0 \cdot F_1^0 & h_{10}(z) = F_4^1 \cdot F_3^0 \cdot F_2^1 \cdot F_1^0 & h_{15}(z) = F_4^1 \cdot F_3^1 \cdot F_2^1 \cdot F_1^0 \\
h_{16}(z) = F_4^1 \cdot F_3^1 \cdot F_2^1 \cdot F_1^1 & &
\end{array}
$$

where: $F_1^0 = \frac{z_1 - z_1^{min}}{z_1^{max} - z_1^{min}}$; $F_1^1 = \frac{z_1^{max} - z_1}{z_1^{max} - z_1^{min}}$; $F_2^0 = \frac{z_2 - z_2^{min}}{z_2^{max} - z_2^{min}}$; $F_2^1 = \frac{z_2^{max} - z_2}{z_2^{max} - z_2^{min}}$; $F_3^0 = \frac{z_3 - z_3^{min}}{z_3^{max} - z_3^{min}}$; $F_3^1 = \frac{z_3^{max} - z_3}{z_3^{max} - z_3^{min}}$; $F_4^0 = \frac{z_4 - z_4^{min}}{z_4^{max} - z_4^{min}}$; $F_4^1 = \frac{z_4^{max} - z_4}{z_4^{max} - z_4^{min}}$.

Appendix A.2

We determined the terms of uncertainties, as shown in the following example:

$$
\Delta a_5 = \Delta\left(-\frac{B_u z_G}{M_{55}}\right) = \frac{\Delta(B_u z_G) \cdot M_{55} + B_u z_G \Delta M_{55}}{M^2{}_{55}} = \frac{M_{55}(z_G \Delta B_u + B_u \Delta z_G) + B_u z_G \Delta M_{55}}{M^2{}_{55}}
$$

$$
\Delta a_6 = \Delta\left(\frac{M_{46}}{M_{55}}\right) = \frac{\Delta M_{46} \cdot M_{55} + M_{46} \Delta M_{55}}{{M_{55}}^2}
$$

References

1. Liao, L.; Pasternak, I. A review of airship structural research and developments. *Prog. Aerosp. Sci.* **2009**, *45*, 83–96. [CrossRef]
2. Li, Y.; Nahon, M.; Sharf, I. Airship Dynamics modeling: A literature review. *Prog. Aerosp. Sci.* **2011**, *47*, 217–239. [CrossRef]
3. Jex, H.; Gelhausen, P. Control response measurements of the Skyship 500 Airship. In Proceedings of the 6th AIAA Conference Lighter than Air Technology, Norfolk, VA, USA, 26 June–28 July 1985; pp. 130–141.
4. Hygounenc, E.; Jung, I.K.; Soueres, P.; Lacroix, S. The autonomous blimp project of LAAS-CNRS: Achievements in flight control and terrain mapping. *Int. J. Robot. Res.* **2004**, *23*, 473–511. [CrossRef]
5. El Omari, K.; Schall, E.; Koobus, B.; Dervieux, A. Inviscid flow calculation around flexible airship. In Proceedings of the Mathematical Symposium Garcia de Galdeano, Jaca, Spain, 15–17 September 2003; Volume 31, pp. 535–544.
6. Bennaceur, S.; Azouz, N. Contribution of the added masses in the dynamic modelling of flexible airships. *Nonlinear Dyn.* **2012**, *67*, 215–226. [CrossRef]
7. Yang, Y.; Wu, J.; Zheng, W. Positioning Control for an Autonomous Airship. *J. Aircr.* **2016**, *53*, 1638–1646. [CrossRef]
8. Moutinho, A.; Azinheira, J.; Paiva, E.; Bueno, S.S. Airship robust path-tracking: A tutorial on airship modelling and gain-scheduling control design. *Control Eng. Pract.* **2016**, *50*, 22–36. [CrossRef]
9. Onat, C.; Kucukdemiral, I.; Sivrioglu, S.; Yuksek, I. LPV Model Based Gain-scheduling Controller for a Full Vehicle Active Suspension System. *J. Vib. Control* **2007**, *13*, 1629. [CrossRef]
10. Takagi, T.; Sugeno, M. Fuzzy identification of systems and its applications to modeling and control. *IEEE Trans. Syst. Man Cybern.* **1985**, *15*, 116–132. [CrossRef]
11. Tanaka, K.; Ikeda, T.; Wang, H.O. Fuzzy regulators and fuzzy observers: Relaxed stability conditions and LMI-based designs. *IEEE Trans. Fuzzy Syst.* **1998**, *6*, 250–265. [CrossRef]
12. Tanaka, K.; Wang, H.O. *Fuzzy Control Systems Design and Analysis: A Linear Matrix Inequality Approach*; John Wiley and Sons: Hoboken, NJ, USA, 2001.
13. Guerra, T.M.; Kruszewski, A.; Vermeiren, L.; Tirmant, H. Conditions of output stabilization for nonlinear models in the Takagi-Sugeno's form. *Fuzzy Sets Syst.* **2006**, *157*, 1248–1259. [CrossRef]
14. Tuan, H.D.; Apkarian, P.; Narikiyo, T.; Yamamoto, Y. Parameterized linear matrix inequality techniques in fuzzy control system design. *IEEE Trans. Fuzzy Syst.* **2001**, *9*, 324–332. [CrossRef]
15. Sala, A.; Arinõ, C. Asymptotically necessary and sufficient conditions for stability and performance in fuzzy control: Applications of Polya's theorem. *Fuzzy Sets Syst.* **2007**, *158*, 2671–2686. [CrossRef]
16. Hayat, A.; Shang, P. A quadratic Lyapunov function for Saint-Venant equations with arbitrary friction and space-varying slope. *Automatica* **2019**, *100*, 52–60.
17. Tanaka, K.; Hori, T.; Wang, H.O. A multiple Lyapunov function approach to stabilization of fuzzy control systems. *IEEE Trans. Fuzzy Syst.* **2003**, *11*, 582–589. [CrossRef]
18. Cao, Y.Y.; Frank, P.M. Analysis and synthesis of nonlinear time-delay systems via fuzzy control approach. *IEEE Trans. Fuzzy Syst.* **2000**, *8*, 200–211.
19. Ibrir, S.; Botez, R. Robust stabilization of uncertain aircraft active systems. *J. Vib. Control.* **2005**, *11*, 187–200. [CrossRef]
20. Goldstein, H. *Classical Mechanics*, 3rd ed.; Pearson, Ed.; Pearson: London, UK, 2001.
21. Shabana, A. *Dynamics of Multibody Systems*, 5th ed.; Cambridge University Press: Cambridge, UK, 2020.

2. Chaabani, S.; Azouz, N. Estimation of the Virtual Masses of a Large Unconventional Airship Based on Purely Analytical Method to Aid in the Preliminary Design. *Aircr. Eng. Aerosp. Technol.* **2021**, *94*, 531–540. [CrossRef]
3. Azouz, N.; Khamlia, M.; Lerbet, J.; Abichou, A. Stabilization of an Unconventional Large Airship when Hovering. *Appl. Sci.* **2021**, *11*, 3551. [CrossRef]
4. Petersen, I.R.; Hollot, C.V. A Riccati Equation Approach to the Stabilization of Uncertain Linear Systems. *Automatica* **1986**, *22*, 397–412. [CrossRef]
5. Boyd, S.; Ghaoui, L.; Feron, E.; Balakrishnan, V. *Linear Matrix Inequalities in Systems and Control Theory*; SIAM: Philadelphia, PA, USA, 1994.

aerospace

MDPI

Article

Effect of Lifting Gas Diffusion on the Station-Keeping Performance of a Near-Space Aerostat

Jun Li, Linyu Ling, Jun Liao *, Zheng Chen and Shibin Luo

School of Aeronautics and Astronautics, Central South University, Changsha 410083, China; lijun215@csu.edu.cn (J.L.); 15797890218@163.com (L.L.); chenzheng@csu.edu.cn (Z.C.); luoshibin@csu.edu.cn (S.L.)
* Correspondence: liaojun@csu.edu.cn

Abstract: During the long-endurance flight of a near-space aerostat, the characteristics of lifting gas diffusion have a great influence on the flight altitude adjustment and station-keeping performance. Thus, in this study, a lifting gas diffusion model and a dynamic model that consider thermal effects, which had not been studied in similar models before, were developed. The dynamic model and thermal model were validated by historic flight data, and the calculated lifting gas diffusion results were compared with the experimental data of other researchers. The variations in the flight endurance, flight altitude, lifting gas diffusion rate, and diffusion coefficient of a near-space aerostat were analyzed. The effects of the ratio of porosity to tortuosity and envelope radiation properties on the mass of the lifting gas and flight altitude were considered in detail. To analyze the effect mechanism of the ratio of porosity to tortuosity and the envelope radiation properties, the envelope and gas temperature, as well as the gas pressure, were studied. The results show that the lifting gas diffusion rate and diffusion coefficient are very sensitive to the change in the ratio of porosity to tortuosity and envelope temperature. The results obtained from the analysis of the lifting gas diffusion can lay a solid foundation for improving the flight performance of near-space aerostats and for providing improved design considerations for aerostats.

Keywords: lifting gas diffusion; near-space aerostat; theoretical study; thermal effect; station-keeping performance

1. Introduction

In recent years, more and more scientific research institutions and scholars have focused on utilizing lighter-than-air aerial flight vehicles as pseudo-satellites operating for extended durations of time in near space to achieve relevant military and civilian tasks previously accomplished using traditional spacecraft [1]. A near-space aerostat is an ideal high-altitude platform for Earth observation, homeland security, data and communications relay, and the relay of communications that need the regional station-keeping ability to adapt to multiple environments [2]. Whether in national defense or economic activities, a near-space aerostat can be used as a supplement to "space-based" vehicles.

For a near-space aerostat, its station-keeping performance is its main technical advantage [3]. To address station-keeping performance requirements, many scholars and researchers have carried out theoretical studies and simulations [4]. Ramesh et al. [5] evaluated the stationary performance of a dual-balloon system and a balloon–stratosail system and compared the best ranges of wind conditions for tether systems. The results showed that the dual-balloon system is more suitable for passive control in a wind field with quasi-zero wind layer and that a balloon–stratosail system can carry out appropriate wind resistance control. Van Wynsberghe et al. [6] presented an innovative design for a stratospheric platform that can achieve long-endurance horizontal station-keeping using electrohydrodynamic thrusters. Another innovation of this paper was its presentation of a wireless energy supplement scheme. Yang et al. [7] proposed a new conceptual design method and fuzzy adaptive backstepping control approach to adapt to inaccurate and non-real-time high-altitude wind field data. Based on the phenomenon of a quasi-zero wind layer at the bottom of the stratosphere, Du et al. [8] proposed the flight altitude adjustment method based on charging and discharging air in an aerostat ballonet to realize regional stationary flight or fixed-point air parking.

These studies provided a theoretical base for designing and controlling a near-space aerostat. These studies were mainly focused on flight performance, station-keeping endurance, and float altitude control. However, discussions of the effect of lifting gas diffusion on the station-keeping performance are rare. We all know that lifting gas diffusion is one of key elements in the design of envelope materials. Hall and Yavrouian [9] analyzed the effect of lifting gas loss on the internal gas pressure of a Venus super-pressure balloon (see in Figure 1). As a typical super-pressure lighter-than-air vehicle, the balloon's net buoyance and maximum flight altitude will gradually reduce with the loss of lifting gas. The loss of flight altitude adjustment capability means that there will be a reduction in maneuverability. During flight, the supplementation of lifting gas is almost impossible for high-altitude aerostats. Therefore, the lifting gas diffusion directly determines the success or failure of long-endurance missions carried out by near-space aerostats. The thermal performance and leakage of lifting gas have a direct impact on the flight altitude control of aerostats. The question of how to fully understand the temperature, pressure, and diffusion of lifting gas is one of the most important issues needing to be addressed in the process of the real-time flight control of an aerostat.

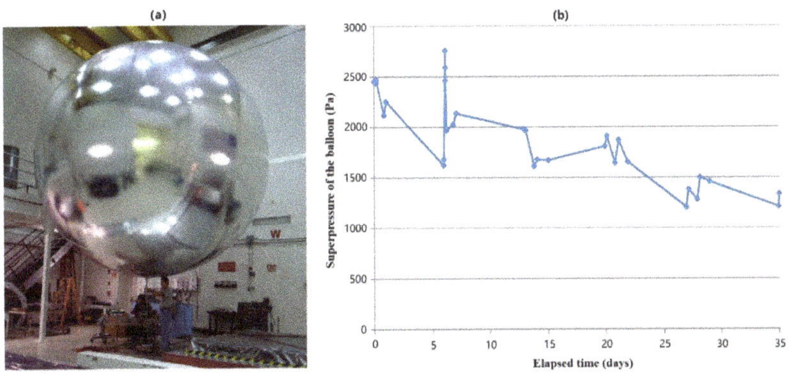

Figure 1. (**a**) A 5.5 m diameter prototype balloon undergoing testing; (**b**) 35-day buoyancy test data [9].

Inspired by studies of gas diffusion in the microelliptical groove gas diffusion layer, porous gas diffusion layer, and fibrous materials [10] under the influence of lifting gas loss, the gas diffusion characteristics of the envelope were studied here. The envelope is one of the major structural parts of a near-space aerostat. It is used to maintain the aerodynamic shape of an aerostat by filling it with lifting gas. The envelope material of a high-altitude aerostat is composed of a laminated membrane [11]. An envelope is made of multi-layer materials, including wearable, ultraviolet, lifting gas barrier, UV protection, anti-aging, and load-bearing layers, as shown in Figure 2 [12].

To date, the study of lifting gas diffusion caused by composite damage and external load has been largely focused on the mechanical properties of carbon-fiber-reinforced composites and the mode of crack propagation [13]. In one study, the lifting gas diffusion of orthogonal carbon-fiber-reinforced composite laminates under biaxial loading was measured [14]. The experimental results showed that the lifting gas diffusion coefficient on the surface of the laminates has a clear relationship with the load. In order to analyze the influence of thermodynamic properties and material matrix cracks, a diffusion model was developed obtain the diffusion coefficients of the composite laminates. It was found that the smaller the ply angle of the material matrix was (the reference values were 45°, 60°, 75° and 90°), the larger leak conductance at the crack intersection would be [15]. Based on the Lennard–Jones interactions, Thornton et al. [16] proposed a transport model of gas in various media to predict the minimum pore size needed for Knudsen diffusion. Yao et al. [17] studied the lifting gas leakage mechanism of damaged flexible composite materials and analyzed the influence of different types of damage on lifting gas leakage; this was of significance for the study of the damage-induced lifting gas leakage of envelope materials. The structure of an envelope material is different from that of carbon-fiber-reinforced composite laminates, and the mechanism of lifting gas diffusion caused by damage is different [18]. To study the lifting gas diffusion of an envelope material, it is necessary to establish a corresponding theoretical model according to its structural characteristics and use new experimental methods to obtain more accurate results.

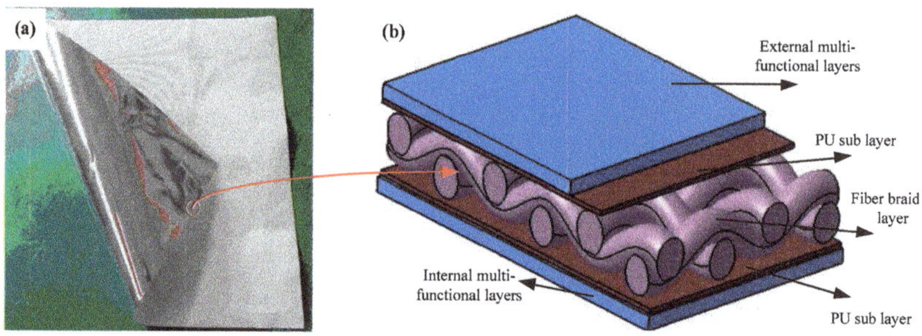

Figure 2. (**a**) Photograph of envelope; (**b**) cross-section of the aerostat envelope [12].

In view of the lightweight envelope material used and the harsh high-altitude thermal environment, the envelope of an aerostat is required to be processed very thinly, and the multi-functional composite layer has different mechanical properties that will lead to delamination and the damage of the envelope after being subjected to alternating thermal stress. Therefore, the study of the lift gas diffusion characteristics of flexible envelope materials is important for improving the station-keeping endurance and safety of near-space aerostats [19]. Based on a literature search, it can be seen that the pressure difference between the internal lifting gas and the external atmosphere can directly affect lifting gas diffusion at high altitudes. The gas pressure difference is directly related to the thermal performance of the lifting gas and the envelope [20]. Many experts and scholars have studied the thermal performance of aerostats [21]. Xiong et al. [22] proposed a simplified thermal model suitable for engineering to assess the steady equilibrium temperature in aerostats at float altitude. In addition to the temperature characteristics of the lifting gas and envelope, the thermal performance of the electronics within the system also needs to be studied to optimize the used thermal control methods.

From the abovementioned literature review, it can be seen that most studies to date have mainly focused on several key issues, such as the flight performance, station-keeping endurance, and thermal performance of near-space aerostats and the gas diffusion of flexible composite materials. However, investigations of the effects of the lifting gas diffusion on the station-keeping performance of a near-space aerostat have been rare. During the long-range flight of a near-space aerostat, it is necessary to understand the lifting gas diffusion characteristics. In this study of a naturally shaped super-pressure balloon made of laminated composite materials (as shown in Figure 2b), a lifting gas diffusion model and a dynamic model that considered thermal effects were developed. A numerical model was used to analyze the lifting gas diffusion and to study the effects of the envelope properties on the station-keeping endurance of the aerostat in detail. The contributions of this paper are summarized as follows: (1) A lifting gas diffusion model that considers thermal effects was established. The temperatures of the envelope and lifting gas were taken into consideration in this paper for the first time. The influence of the envelope thickness variation was considered in the theoretical model to make the lifting gas diffusion model more precise. (2) The effects of the main air tightness parameters of the envelope and the envelope radiation properties on the lifting gas diffusion coefficient were analyzed. The relationship between gas diffusion and flight duration was analyzed. The relationship between lifting gas diffusion and flight endurance and the mechanism of the variation of the gas diffusion coefficient were studied. The results obtained from the analysis of the lifting gas diffusion can contribute to improving the flight performance and station-keeping endurance of near-space aerostats and to providing improved design considerations for aerostats.

2. Theoretical Model

2.1. Lifting Gas Diffusion Model

In near-space aerostats, the lifting gas diffuses out by passing through the envelope material. Figure 3 represents the mass transfer mechanism of lifting gas through an aerostat envelope.

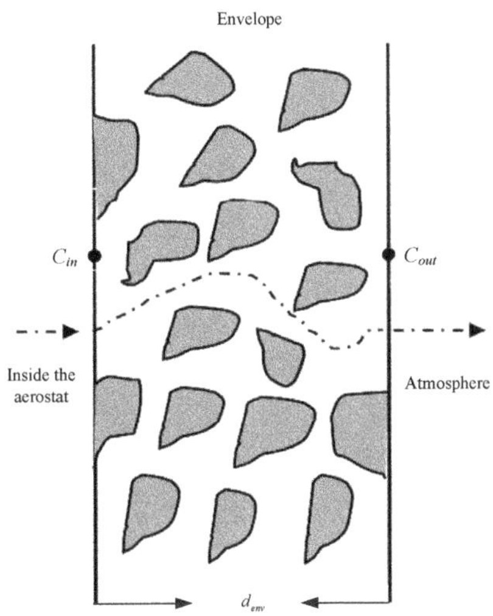

Figure 3. The mass transfer mechanism of lifting gas passing through an envelope [23].

Considering Figure 3, which demonstrates a case with a small envelope thicknesses, the gross mass of the diffusion lifting gas can be calculated as:

$$m_{diffusion} = M_{gas} \cdot n_{diffusion} \tag{1}$$

where M_{gas} is the lifting gas molecular weight in kg/kmol.

$n_{diffusion}$ is the amount of lifting gas that diffuses. The differential amount of lifting gas that diffuses can be calculated as:

$$dn_{diffusion} = N_{He} \cdot A_b \cdot dt \tag{2}$$

where N_{He} is the lifting gas flux and A_b is the effective cross-sectional area of heat and mass transfer.

$$N_{He} = D_{real} \cdot \frac{P_{gas}}{P_{air}} \cdot \frac{(C_{in} - C_{out})}{d_{env}} \tag{3}$$

where C_{in} and C_{out} are the lifting gas and external atmosphere concentrations inside and outside of the aerostat envelope, $\text{kmol}/(\text{m}^2 \cdot \text{s})$, respectively. C_{in} and C_{out} can be calculated as:

$$C_{in} = \rho_{gas} / M_{gas} \tag{4}$$

$$C_{out} = \rho_{air} / M_{air} \tag{5}$$

where P_{gas} and P_{air} are the pressures of the lifting gas and external air, respectively; ρ_{gas} is the lifting gas density; ρ_{air} is the density of external air; M_{gas} is the molecular weights of lifting gas; M_{air} is the molecular weights of external air; and d_{env} is the thickness of the near-space aerostat envelope.

The transfer velocity is critically affected by the real effective diffusivity. D_{real} is the real effective diffusivity, and it can be calculated as [24]:

$$\frac{1}{D_{real}} = \frac{1}{D_{gas\_air\_eff}} + \frac{1}{D_{K\_gas\_eff}} \tag{6}$$

The effective diffusion coefficient can be calculated by considering the molecular and Knudsen diffusions [25]. $D_{gas\_air\_eff}$ and $D_{K\_gas\_eff}$ are the effective molecular diffusivity and effective Knudsen diffusivity, respectively.

Due to the envelope being made of porous material, $D_{gas\_air\_eff}$ can be calculated by considering the gas diffusion through the porous solid material.

$$D_{gas\_air\_eff} = \frac{\varepsilon_{env}}{\tau_{ss}} \cdot D_{gas\_air} \tag{7}$$

where ε_{env} is the porosity of the envelope material and τ_{ss} is the tortuosity. Tortuosity is used in an attempt to account for the longer distance traversed in the pores. Tortuosities usually range between two and six, averaging about three [26].

D_{gas_air} is the molecular diffusivity, which can be calculated from the Chapman–Enskog equation [27]:

$$D_{gas\_air} = \frac{\alpha_1 \cdot T_{env}^{3/2}}{P_{gas} \cdot r_{col}^2 \cdot \Omega_{gas\_air}} \cdot \left(\frac{M_{gas} + M_{air}}{M_{gas} \cdot M_{air}} \right)^{1/2} \tag{8}$$

where α_1 = 1.8583 × 10 ^{7} [28], T_{env} is the envelope temperature, P_{gas} is the pressure in atmospheres, r_{col} is the collision diameter, and r_{gas} and r_{air} are the collision diameters of the lifting gas and ambient air, respectively.

$$r_{col} = \frac{1}{2}\left(r_{gas} + r_{air} \right) \tag{9}$$

The dimensionless quantity Ω_{gas_air} is more complex but usually of an order of one [26]. The detailed calculation is affected by the interaction between the two species [29].

The effective Knudsen diffusivity $D_{K\_gas\_eff}$ can be calculated as:

$$D_{K\_gas\_eff} = \frac{\varepsilon_{env}}{\tau_{ss}} \cdot D_{K\_gas} \tag{10}$$

For a porous envelope material, the Knudsen diffusion is the main influencing factor [30].

$$D_{K\_gas} = \frac{2}{3} \cdot r_P \cdot \sqrt{\frac{8 \cdot R_u \cdot T_{env}}{\pi \cdot M_{gas}}} \tag{11}$$

where R_u is the universal gas constant (8314.47 $\mathrm{Pa \cdot m^3 / (kmol \cdot K)}$) [31] and r_P is the pore radius, which is in the order of 10^{-6} m.

Based on Equations (7) and (10), we can obtain the porosity of the envelope material and tortuosity, as the air tightness parameters of the envelope material directly affect the effective diffusion coefficient. Analyzing the effective diffusion coefficient in the mass transfer equation as a function of the ratio of porosity to tortuosity, we can obtain the intermediate variable used to indicate the air tightness of materials.

$$ratio = \varepsilon_{env}/\tau_{ss} \tag{12}$$

2.2. Balloon Geometry

The expansion of the balloon as a function of altitude can safely be assumed to be governed by the ideal gas law. According to the ideal gas law, the state of an amount of gas can be determined by its pressure, volume, and temperature using the equation:

$$V_b = \begin{cases} V_{gas} & V_{gas} \leq V_{\max} \\ V_{\max} & V_{gas} > V_{\max} \end{cases} \tag{13}$$

where V_b is the volume of the balloon in the current flight environment,V_{gas} is the volume of lifting gas after free expansion in the current flight environment, and $V_{\max}$ is the maximum design volume of the balloon.

$$V_{gas} = \frac{m_{gas} \cdot R_{gas} \cdot T_{gas}}{P_{gas}} \tag{14}$$

where m_{gas} is the mass of the lifting gas, T_{gas} is the thermodynamic temperature of the lifting gas, R_{gas} is the specific gas constant of the lifting gas, $R_{He} = 2078.5\ \mathrm{J/(kg \cdot K)}$ for helium, and P_{gas} is the pressure of the lifting gas.

$$P_{gas} = \begin{cases} P_{air} & V_{He} \leq V_{\max} \\ \frac{m_{gas} \cdot R_{gas} \cdot T_{gas}}{V_b} & V_{He} > V_{\max} \end{cases} \tag{15}$$

The differential pressure between the lifting gas and external air, ΔP, can be written as:

$$\Delta P = P_{gas} - P_{air} \tag{16}$$

The balloon envelope thickness (here assuming that the thickness of the skin is uniform and that the thickness is the same everywhere) can be calculated with envelope mass, m_{env}, and envelope material, ρ_{env}.

$$d_{env} = \frac{m_{env}}{A_b \cdot \rho_{env}} \tag{17}$$

The balloon diameter (top view) can be calculated according to the volume of the balloon. For a high-altitude balloon, β_{dia} is a coefficient representing the effect of the balloon volume and shape change on the equivalent diameter of the balloon [32].

$$d_{top} = \beta_{dia} \cdot V_b^{1/3} \tag{18}$$

2.3. Dynamic Model

The vertical acceleration of a near-space aerostat is mainly affected by the pressure and temperature of the outside atmosphere. In the vertical direction, the main forces on the aerostat are buoyancy and drag, which will affect the speed change. The following equation demonstrates the influence of these forces on the climbing rate:

$$a_z = \frac{d^2 z}{dt^2} = F_z/m_{tot} \tag{19}$$

where m_{tot} is the total mass of the aerostat system and F_z is the resultant force on the aerostat in the vertical direction, which can be written as:

$$\begin{gathered} F_z = B - m_b \cdot g - D_b \cdot \left| \frac{v_{z\_rel}}{v_b} \right| \\ m_{total} = m_b + m_{add} \end{gathered} \tag{20}$$

where m_{add} is the air mass displaced by the motion of the balloon, named the virtual mass, which can be calculated using the virtual mass coefficient c_{add}. The virtual mass coefficient c_{add} can range from 0.25 to 0.5 and is specified as 0.5 here [33]. A precise calculation of the virtual mass can be obtained from references [34,35].

$$m_{add} = c_{add} \cdot \rho_{air}(z) \cdot V_b(z) \tag{21}$$

where B is the total buoyancy of an aerostat.

$$B = V_b(z) \cdot \left(\rho_{air}(z) - \rho_{gas}\right) \cdot g \tag{22}$$

$V_b(z)$ is the aerostat volume.

$$\rho_{gas}(z) = \frac{m_{gas}}{V_b(z)} \tag{23}$$

F_D is the aerodynamic drag, which can be expressed as:

$$F_D = \frac{1}{2} \cdot \rho_{air}(z) \cdot v_b^2 \cdot C_d \cdot A_{eff} \tag{24}$$

A_{eff} is the effective area that used to calculate the aerodynamic drag.

$$A_{eff} = \pi \cdot d_{top}^2 / 4 \tag{25}$$

The influence of the balloon shape is mainly considered in the calculation of the aerodynamic drag coefficient, C_d.

$$C_d = \frac{4}{3} \cdot \frac{g \cdot d_{top}}{v_{z\_rel}^2} \cdot \left(1 - \frac{M_{gas} \cdot T_{air}}{M_{air} \cdot T_{gas}} \cdot \left(1 + \frac{m_b}{m_{gas}} + \frac{m_{total}}{m_{gas}} \cdot \frac{a_z}{g}\right)\right) \tag{26}$$

2.4. Thermal Model

The thermal environment of a near-space aerostat is complicated and consists of direct solar radiation, internal and external infrared radiation, scattered radiation, and internal and external convection [36], as depicted in Figure 4.

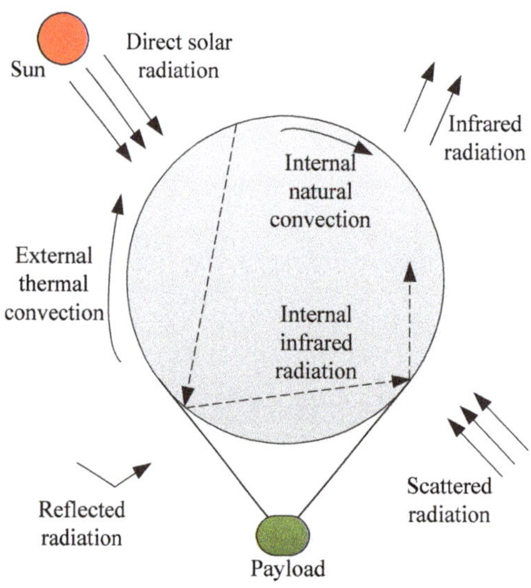

Figure 4. Thermal balance of a near-space aerostat.

In this paper, the Earth standard atmosphere was used to analyze the thermal characteristics of the aerostat [37].

$$T_{atm} = \begin{cases} 288.15 - 6.5 \cdot h,\ K & 0 \leq h \leq 11\text{km} \\ 216.15,\ K & 11 \leq h \leq 20\text{km} \\ 216.15 + (h - 20),\ K & 20 \leq h \leq 32\text{km} \end{cases} \tag{27}$$

The atmosphere dynamic viscosity can be expressed by Sutherland's formula:

$$u_a = u_0 \cdot \left(\frac{T_0 + C_a}{T_a + C_a} \right) \cdot \left(\frac{T_a}{T_0} \right)^{3/2} \tag{28}$$

The atmospheric pressure at any altitude can be calculated using the following formula:

$$P_a = P_0 \cdot \left(1 - \frac{L \cdot h}{T_0} \right)^{M_{air} \cdot g/(R \cdot L)} \tag{29}$$

The change in the lifting gas temperature in the aerostat is mainly related to the convective heat transfer between the gas and the envelope, as well as the volume change of lifting gas. The lifting gas temperature rate can be obtained through:

$$\frac{dT_{gas}}{dt} = \frac{q_{in,gas}}{M_{gas} \cdot c_v} + (\gamma - 1) \cdot T_{gas} \cdot \left(\frac{1}{m_{gas}} \cdot \frac{dm_{gas}}{dt} - \frac{1}{V_{gas}} \cdot \frac{dV_{gas}}{dt} \right) \tag{30}$$

where $\gamma = c_p / c_v$ and $\gamma - 1 = R_{gas} / c_v$. c_v is the specific heat at a constant volume of the lifting gas, R_{gas} is the specific gas constant, and $q_{in,gas}$ is the internal thermal convective heat load [38].

The mass change rate can be expressed as:

$$\frac{dm_{gas}}{dt} = \Delta m_{valve} + \Delta m_{diffusion} \tag{31}$$

The mass change of the lifting gas mainly includes two parts: discharge through the valve, Δm_{valve}, and gas diffusion through the envelope, $\Delta m_{diffusion}$. The gas diffusion through the envelope can be calculated by the lifting gas diffusion model established in Section 2.1:

$$\Delta m_{valve} = -A_{valve} \cdot c_{disch} \cdot \sqrt{2 \cdot \Delta P_{valve} \cdot \rho_{gas}} \tag{32}$$

$$\Delta m_{diffusion} = -M_{gas} \cdot D_{real} \cdot \frac{P_{gas}}{P_{air}} \cdot \frac{(C_{in} - C_{out})}{d_{env}} \cdot A_b \tag{33}$$

where A_{value} is the outlet area of the valve, c_{disch} is the discharge coefficient, and ΔP_{valve} is the differential pressure across the area interface.

The envelope temperature differential equation can be expressed as:

$$m_{env} \cdot c_{env} \cdot \frac{dT_{env}}{dt} = q_{in,\ env} \tag{34}$$

The heat flux terms $q_{in,env}$ in Equation (34) are given as:

$$q_{in,\ env} = q_D + q_S + q_{albedo} + q_{IR} + q_{conv} \tag{35}$$

where q_D is the absorbed direct sunlight heat. The absorbed direct solar radiation includes the heat absorbed by the outer surface of the envelope and the heat absorbed by the inner surface through the envelope [39]. The effective absorption of the envelope will be improved due to the multiple types of reflection taking place.

$$q_D = \alpha_{env} \cdot A_{proj} \cdot I_{sun} \cdot \tau_{atm} \cdot \left(1 + \frac{\tau_{env}}{1 - \gamma_{env}} \right) \tag{36}$$

where α_{env} is the absorption coefficient of the envelope material for solar radiation, A_{proj} is the projected area of the balloon, I_{sun} is the direct solar irradiance, τ_{atm} is the transmission coefficient of the atmosphere for sunlight, and τ_{env} and γ_{env} are the transmittance and reflectance of the envelope material, respectively [20].

q_S the absorbed scattered radiation and can be expressed as:

$$q_S = \alpha_{env} \cdot \kappa_S \cdot A_b \cdot I_{sun} \cdot \tau_{atm} \cdot (1 + \tau_{env} \cdot (1 + \gamma_{env})) \tag{37}$$

where κ_S is the atmospheric scattering empirical coefficient.

q_{albedo} is the ground albedo radiation, and its intensity is mainly affected by the direct solar radiation intensity and the average albedo of the ground.

$$q_{albedo} = \alpha_{env} \cdot A_b \cdot V_F \cdot I_{albedo} \cdot (1 + \tau_{env} \cdot (1 + \gamma_{env})) \tag{38}$$

where $I_{albedo} = \varepsilon_G \cdot I_{sun} \cdot \sin(\alpha_{ele})$, ε_G is the ground albedo, α_{ele} is the sun elevation angle, and V_F is the angle coefficient between the balloon surface and the Earth's surface, which is related to the flight altitude [40].

q_{IR} is the absorbed infrared radiation heat, which mainly includes ground infrared radiation, $q_{IR,E}$; infrared radiation from the sky, $q_{IR,sky}$; and infrared radiation self-glow from the interior, $q_{IR,env}$.

$$q_{IR} = q_{IR,E} + q_{IR,sky} + q_{IR,env} \tag{39}$$

$$\begin{aligned} q_{IR,E} &= \alpha_{IR} \cdot A_{surf} \cdot I_{IR,E} \cdot V_F \cdot (1 + \tau_{IR}/(1 - r_{IR})) \\ q_{IR,sky} &= \alpha_{IR} \cdot A_{surf} \cdot I_{IR,sky} \cdot (1 - V_F) \cdot (1 + \tau_{IR}/(1 - r_{IR})) \\ q_{IR,env} &= \alpha_{env} \cdot \varepsilon \cdot A_{surf} \cdot T_{env}^4 \cdot (1 + \alpha_{IR}/(1 - r_{IR}) + \tau_{IR}/(1 - r_{IR})) \end{aligned} \tag{40}$$

$I_{IR,sky} = \varepsilon_g \cdot \sigma \cdot T_{sky}^4 \cdot \tau_{atm\_IR}$, and T_{sky} is the effective temperature of the sky [41].

q_{conv} is the convective heat load on the envelope.

$$q_{conv} = h_{ex} \cdot A_b \cdot (T_{env} - T_{atm}) + h_{in} \cdot A_b \cdot (T_{env} - T_{gas}) \tag{41}$$

The external thermal convection coefficient (which is mainly composed of natural convection, $h_{free\text{-}ex}$, and the forced convection heat transfer coefficient, $h_{forced\text{-}ex}$) is given as [42]:

$$h_{ex} = \left(h_{free\text{-}ex}^3 + h_{forced\text{-}ex}^3\right)^{1/3} \tag{42}$$

3. Model Validation

3.1. Verification of the Dynamic Model

Based on the thermal and dynamic models established in this paper, a computer program was developed. The high-altitude experimental data obtained by Yang et al. [33] were used to verify the validity and accuracy of the dynamic model developed in this study. The same balloon parameters, launch times, and launch sites were used in our comparative analysis. These critical parameters are listed in Table 1. During ascent, the differential pressure between the lifting gas and external air was approximately equal to 0 until the maximum design altitude was reached. By comparing the results of the numerical simulation and the experimental data (as shown in Figure 5), it can be seen that the maximum relative error of the flight altitude was less than 5%. There were some differences in flight altitude during ascent, mainly due to the influence of the aerodynamic drag coefficient on the shape change. The variation in balloon flight altitude with the flight time distribution remained almost identical. Hence, the dynamic model developed in this paper can be applied to indicate flight status with reasonable accuracy.

After reaching the design altitude, the pressure and temperature of the internal lifting gas of the aerostat rose rapidly. Then, there were some valve-switching operations that have not been described in the literature; these operations affected the accuracy of temperature prediction based on Equation (30). As can be seen in Figure 6, the prediction result and reference values corresponded with deviations below 9%, which means that the present thermal model can be used to predict the temperature characteristics of aerostats.

Table 1. Design parameters of the near-space aerostat used for comparison [33].

Parameters	Value
Design altitude (km)	20
Total mass (kg)	5700
Helium gas mass (kg)	900
Envelope mass (kg)	1200
Maximum volume (m^3)	65,000
Absorption coefficient, α_{env}	0.3
Transmission coefficient, τ_{env}	0.5
Launch time	1 July at 07:00
Launch site	110° E, 30° N

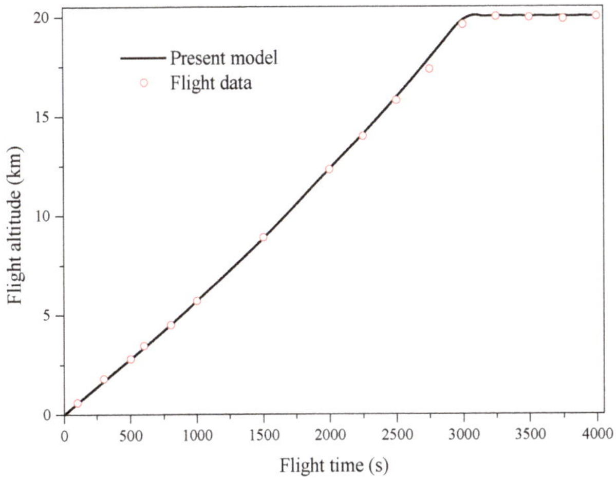

Figure 5. Comparison between the predicted data and the experimental data (flight altitude).

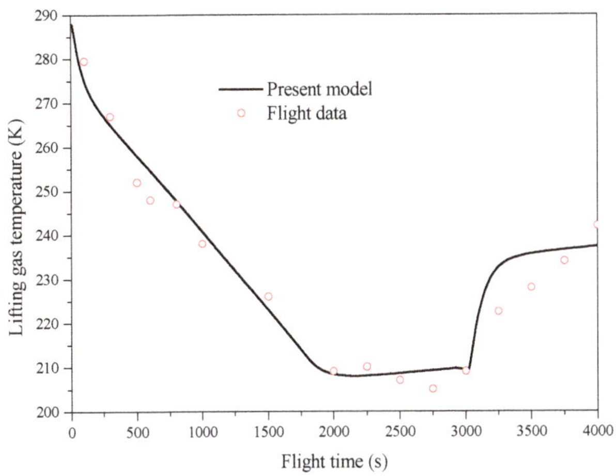

Figure 6. Comparison between the predicted data and the experimental data (lifting gas temperature).

3.2. Verification of Lifting Gas Diffusion Model

In order to understand the mechanism of the helium leakage of aerostat envelope materials, Wu [43] carried out theoretical and experimental studies. The results of these studies were used to verify the accuracy of our mass transfer model. The design parameters used for analyzing the envelope diffusion coefficient are listed in Table 2. During the flight, the temperature of the balloon envelope generally varied from 216 to 336 K. Therefore, in [43], the diffusion coefficient of an envelope under a range of different envelope temperatures was analyzed. As shown in Figure 7, the diffusion coefficients of envelopes given in the literature and predicted in this paper increased with increases in envelope temperature. In addition, the deviations between the comparison data and the predicted results were less than 6%, indicating that the mass transfer model established in this paper can be used to estimate the lifting gas leakage and flight endurance in the preliminary design of an aerostat.

Table 2. Design parameters used for analyzing an envelope diffusion coefficient.

Parameters	Value
Envelope thickness (m)	2.7×10^{-4}
Pressures of the lifting gas (mPa)	0.1
Porosity of the envelope material	0.2%
Tortuosity (kg/(m s^2))	2
Collision diameters of lifting gas	2.551
Collision diameters of ambient air	3.711
Pore radius (m)	0.2×10^{-6}

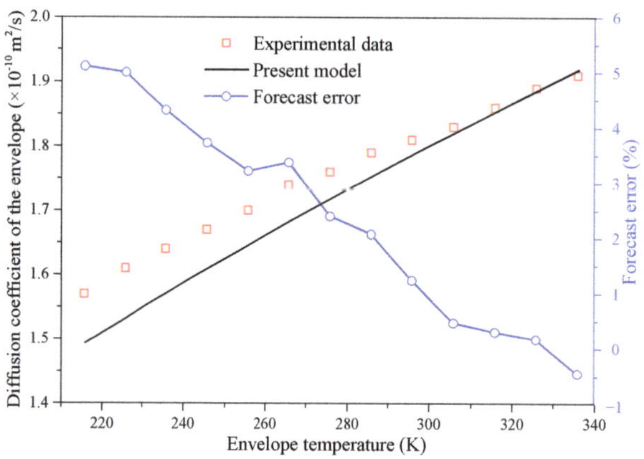

Figure 7. Comparisons of the diffusion coefficient of the envelope [43].

4. Discussion

4.1. Simulation Conditions

The theoretical model established in this paper was used to predict the gas diffusion characteristics and flight performance of a near-space super-pressure balloon. The main design parameters of the balloon are listed in Table 3.

4.2. Effects of $\varepsilon_{env}/\tau_{ss}$

Based on the theory established in this paper, it can be stated that the mass of the lifting gas inside a near-space aerostat directly affects the volume and pressure of the lifting gas. Additionally, the total buoyancy of a near-space aerostat varies with the volume of the aerostat at a given altitude. Furthermore, the atmospheric pressure decreases with increases in the altitude, as does the density of the atmosphere. Based on this analysis, it can be summarized that a change in the mass of the lifting gas will affect the flight altitude of an aerostat. In order to meet the needs of long-term monitoring and long-distance communication, the near-space aerostat—a brand-new concept intended to meet the requirements of long-endurance station-keeping performance—was developed. This means that the flight altitude of an aerostat needs to be guaranteed above a certain limit. Therefore, the effects of the air tightness parameters of the envelope material and the ratio of porosity to tortuosity on the mass of lifting gas inside a near-space aerostat should be studied in detail. For this purpose, five simulations were conducted to study the ratio of porosity to tortuosity ($\varepsilon_{env}/\tau_{ss} = 0.001, 0.002, 0.004, 0.008$, and 0.012).

Table 3. Main parameters of the aerostat used for the simulation.

Parameters	Value
Design altitude (km)	20
Failure altitude (km)	15
Total mass (kg)	145
Initial helium gas mass (kg)	20
Payload mass (kg)	67.2
Envelope mass (kg)	57.7
Maximum volume (m^3)	1625
Maximum overpressure (Pa)	600
Launch date	1 January
Launch site	Changsha, Hunan

Figure 8 presents the variations in the mass of lifting gas inside a near-space aerostat with the $\varepsilon_{env}/\tau_{ss}$ and the ratio of porosity to tortuosity at different flight times. In order to meet the needs of long-term monitoring and long-distance communication, a minimum flight altitude of 15 km was used in this study. In order to achieve station-keeping at this altitude, a sufficient level of lifting gas inside the near-space aerostat had to be maintained. It was found that 16 kg of lifting gas is required for an aerostat to fly at the altitude of 15 km. Under the same flight endurance, the mass of lifting gas inside a near-space aerostat was found to increase with decreases in the ratio of porosity to tortuosity when all aerostats are flying above the lowest altitude. The calculations showed that the amount of lifting gas diffusion taking place increases as the ratio of porosity to tortuosity increases. Additionally, the airtightness of the envelope material tends to worsen over time.

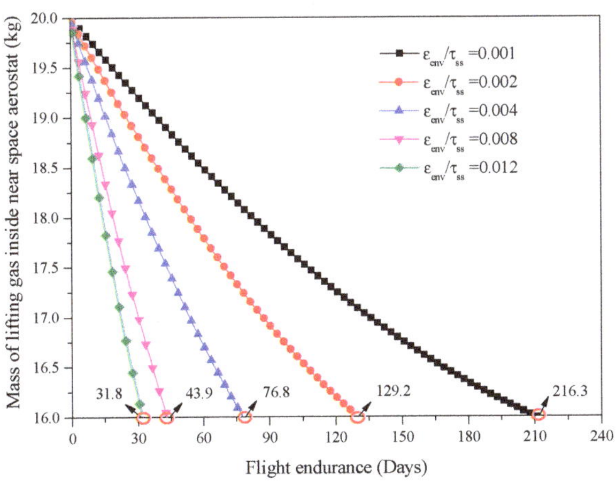

Figure 8. Effects of $\varepsilon_{env}/\tau_{ss}$ on the mass of lifting gas inside a near-space aerostat.

Furthermore, it is obvious that flight endurance will increase with decreases in $\varepsilon_{env}/\tau_{ss}$. The flight endurance corresponding to different $\varepsilon_{env}/\tau_{ss}$ values was found to be 31.8, 43.9, 76.8, 129.2, and 216.3 days. It can be concluded that flight endurance at a higher ratio of porosity to tortuosity, $\varepsilon_{env}/\tau_{ss}$, dramatically decreases, whereas it is fairly stable for lower ratios of $\varepsilon_{env}/\tau_{ss}$. This indicates that an optimal ratio of porosity to tortuosity may exist for the selection of envelope materials.

The effects of $\varepsilon_{env}/\tau_{ss}$ on the flight altitude of near-space aerostats are shown in Figure 9. It was found that when the ratio of porosity to tortuosity decreases, the flight altitude increases. Under the same $\varepsilon_{env}/\tau_{ss}$, the flight altitude of a near-space aerostat decreases with the flight time when all aerostats are flying above the lowest altitude. Additionally, the rate of height change gradually increases, especially at smaller $\varepsilon_{env}/\tau_{ss}$.

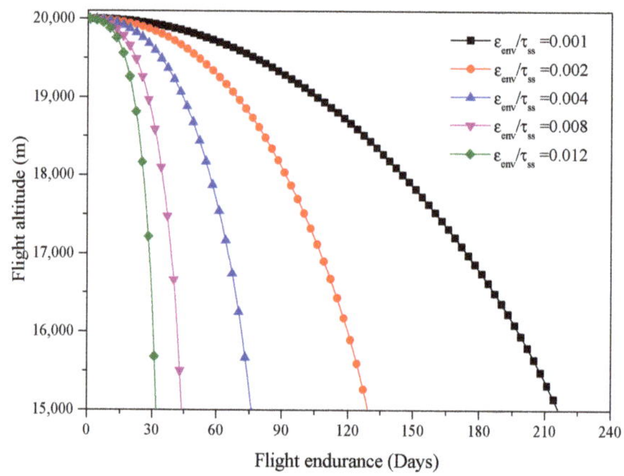

Figure 9. Effects of $\varepsilon_{env}/\tau_{ss}$ on the flight altitude of near-space aerostats.

In order to evaluate the effects of $\varepsilon_{env}/\tau_{ss}$ on the diffusion coefficient of an envelope, a series of values for the ratio of porosity to tortuosity were considered. The variation in the diffusion coefficient with $\varepsilon_{env}/\tau_{ss}$ is shown in Figure 10. The results showed that with increases in the flight endurance, the diffusion coefficient decreases. The diffusion coefficient was found to decrease with $\varepsilon_{env}/\tau_{ss}$ due to an increase in the molecular diffusivity. If $\varepsilon_{env}/\tau_{ss}$ becomes larger, the diffusion coefficient decreases faster.

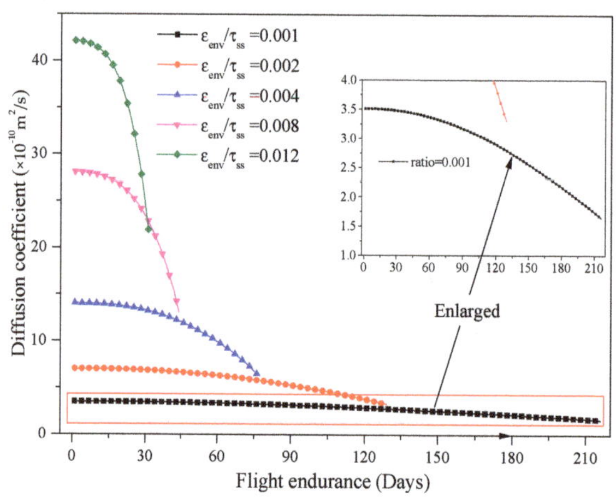

Figure 10. Effects of $\varepsilon_{env}/\tau_{ss}$ on the diffusion coefficient of the envelope.

Furthermore, the ratio of $\varepsilon_{env}/\tau_{ss}$ was found to have a significant influence on the envelope and gas temperature, as well as the gas pressure, as shown in Figure 11. With the increase in the flight endurance, the maximum temperatures of the envelope and lifting gas per day were found to first rapidly and then slowly increase before finally decreasing. These results showed that the maximum temperature of the envelope and lifting gas per day are mainly affected by the flight date. The initial flight date was set to be 1 January. With the increase in the flight endurance, the heat caused by direct sunlight was found to gradually increase until the summer solstice. On the same flight date, the higher the $\varepsilon_{env}/\tau_{ss}$ is, the lower the maximum temperatures of the envelope and lifting gas were shown to be, although the difference was small, as shown in Figure 11a,b. This temperature difference is mainly caused by the leakage of lifting gas. The maximum gas pressure per day was found to rapidly increase with the increase in flight endurance, as shown in Figure 11c, because of the decrease in the flight altitude. As shown in Figure 11d, the maximum gas pressure difference ($P_{gas} - P_{air}$) per day was shown to first decrease and then increase with the increase in flight endurance. The main factor influencing the descending section was found to be the leakage of lifting gas, which is caused by the initial high diffusion coefficient of the envelope based on Equation (3). The main factor influencing the rising section was found to be the increase in the lifting gas temperature.

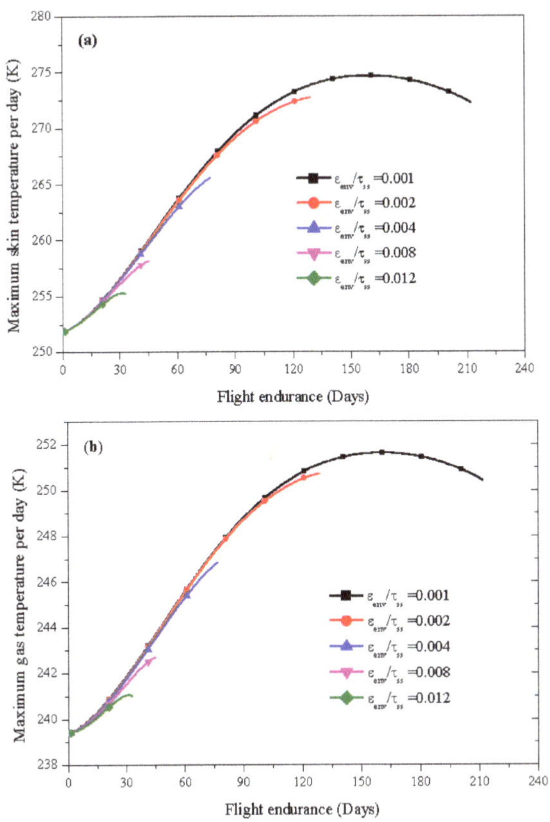

Figure 11. *Cont.*

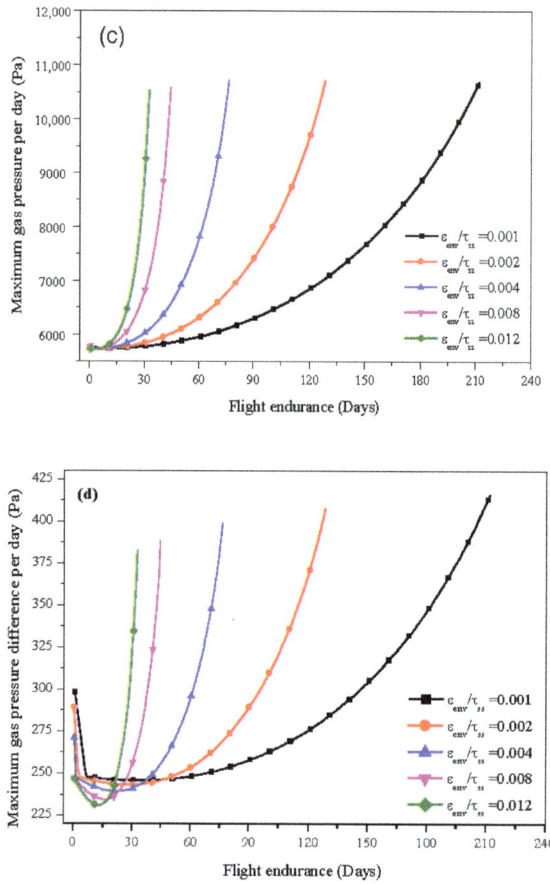

Figure 11. Effects of $\varepsilon_{env}/\tau_{ss}$ on the (**a**) maximum skin temperature per day, (**b**) maximum gas temperature per day, (**c**) maximum gas pressure per day, and (**d**) maximum gas pressure difference per day.

4.3. Effect of Envelope Radiation Properties

As shown in Figure 12, the effects of the envelope temperature on the lifting gas diffusion rate, envelope thickness, and diffusion coefficient of the near-space aerostat were studied. Under normal conditions, the lowest temperature of a near-space aerostat envelope would be about 210 K, while the highest temperature at noon would be 310 K. Therefore, the lifting gas diffusion rate, envelope thickness, and diffusion coefficient in this temperature variation range were analyzed in this paper. It was found that when envelope temperature rose from 210 to 310 K, the lifting gas diffusion rate increased from 1.33 to 2.32 g/s and the diffusion coefficient increased from 2.75×10^{-10} to 4.79×10^{-10} m/s^{-2}. The envelope thickness of the near-space aerostat decreased with increases in the envelope temperature. The reason for this is that the effective diffusion coefficient, which was calculated considering the Knudsen and molecular diffusions, was found to be positively related to the envelope temperature. This result implies that the envelope temperature can affect the lifting gas diffusion rate by changing the effective diffusion coefficient.

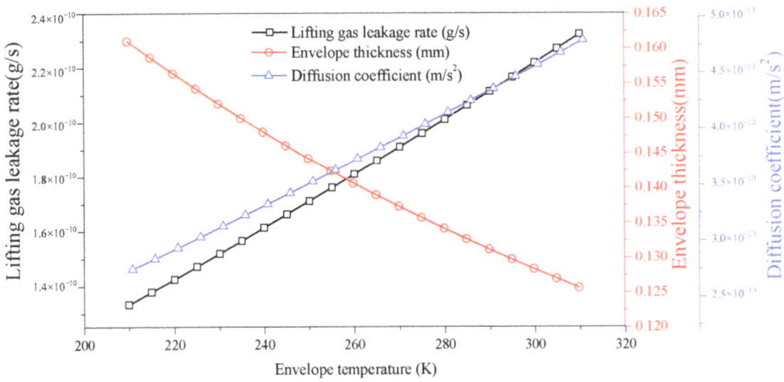

Figure 12. The lifting gas diffusion rate, envelope thickness, and diffusion coefficient.

Figure 13 illustrates the temperatures of the lifting gas and envelope and the lifting gas diffusion rate at the summer and winter solstices. It can be seen that there was an obvious peak value of temperature change in the day, which is consistent with the theoretical expectation. The temperatures of the envelope could achieve their highest values, 273.3 and 249.8 K at noon during the summer and winter solstices, respectively. The highest temperatures of the lifting gas were found to be 244.2 and 233.3 K, which were lower than those of the envelope. Furthermore, it is obvious that the temperature differences of the lifting gas and inner air presented the same expressional tendencies. The maximum lifting gas diffusion rate on the day of the summer solstice is obviously larger than that observed on the day of the winter solstice. This result implies that the mass of lifting gas diffused on the day of the summer solstice will be obviously larger than that on the day of the winter solstice.

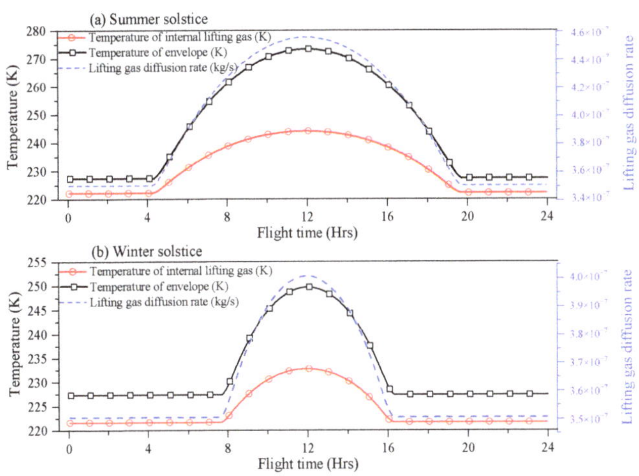

Figure 13. The temperatures of the lifting gas and envelope and the lifting gas diffusion rate at the (**a**,**b**).

Examples of how the flight altitude of the near-space aerostat and diffusion coefficient were shown to be influenced by the envelope absorptivity are shown in Figures 14 and 15. Envelope absorptivities of 0.02, 0.05, 0.1, 0.2, and 0.4 were used, as these represent the absorptivity range of most envelope materials. Under the same level of flight endurance, the flight altitude of the near-space aerostat was shown to increases with decreases in the envelope absorptivity when the aerostat was flying above the lowest altitude. It can be seen in Figure 15 that over the whole flight period, the diffusion coefficient first slowly and then sharply declined as the flight time increased. The results showed that the trends of the flight altitude of the simulated and measured diffusion coefficients were consistent. These results indicate that a higher envelope absorptivity will result in a higher temperature of the lifting gas and envelope, which will consequently increase the mass of lifting gas diffused. Based on Equation (8), the diffusion coefficient is a function of $T_{env}^{3/2}/P_{gas}$. As can be seen from Figure 16, the maximum skin temperature and maximum gas pressure per day were found to gradually increase along with the flight endurance, but the rate of change of rising speed was different, leading to a nonlinear downward trend.

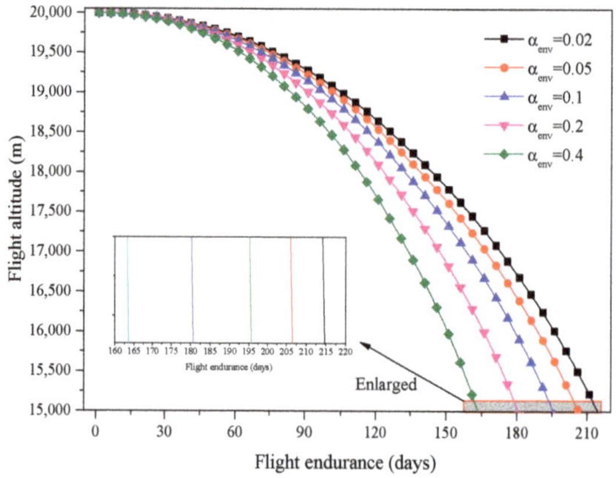

Figure 14. Effects of envelope absorptivity on the flight altitude.

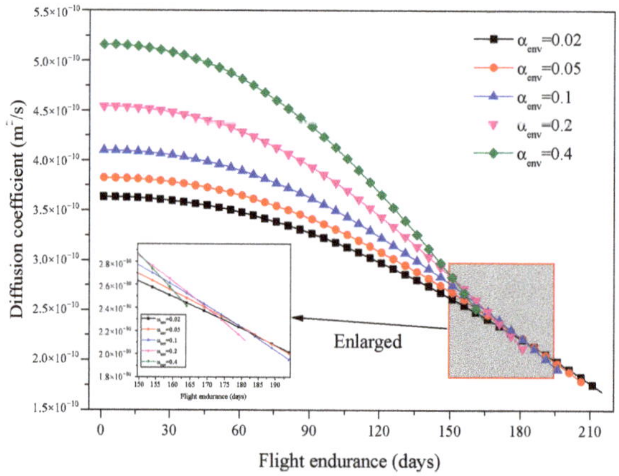

Figure 15. Effects of envelope absorptivity on the diffusion coefficient.

The effects of envelope absorptivity on the envelope and gas temperature, as well as the gas pressure, were studied in order to analyze the mechanism of the effect of absorptivity on the diffusion properties, as shown in Figure 16. With the increase in the flight endurance, the maximum temperatures of the envelope and lifting gas per day firstly increased and then decreased, as shown in Figure 16a,b. On the same flight date, the envelope temperature with a higher absorptivity was significantly higher than that with a lower absorptivity. The maximum gas pressure and maximum gas pressure difference per day increased with the increase in envelope absorptivity, as shown in Figure 16c,d.

As shown in Figure 17, five envelope emissivities were considered here to investigate their impact on the flight altitude of a near-space aerostat. Several common envelope materials were considered for this parametric study. It can be seen that the flight altitude of the aerostat decreased as the flight time increased. The flight endurances of the near-space aerostat under these situations remained fairly steady for 190 days. The results of this research demonstrate that the envelope emissivity has some effect on the flight altitude. Figure 18 presents the effects of the envelope emissivity on the diffusion coefficient. It can be easily seen that the diffusion coefficient was reduced by almost 50% as the flight time increased. The effects of the envelope emissivity on the envelope and gas temperature, as well as the gas pressure, were studied in order to analyze the mechanism of the effect of the envelope emissivity on the diffusion properties, as shown in Figure 19. With the increase in the flight endurance, the maximum temperatures of the envelope and lifting gas per day firstly increased and then decreased, as shown in Figure 19a,b. On the same flight date, the temperature of the envelope with a lower emissivity was significantly higher than that of the envelope with a higher emissivity. The maximum gas pressure and maximum gas pressure difference per day decreased with increases in the envelope emissivity, as shown in Figure 19c,d.

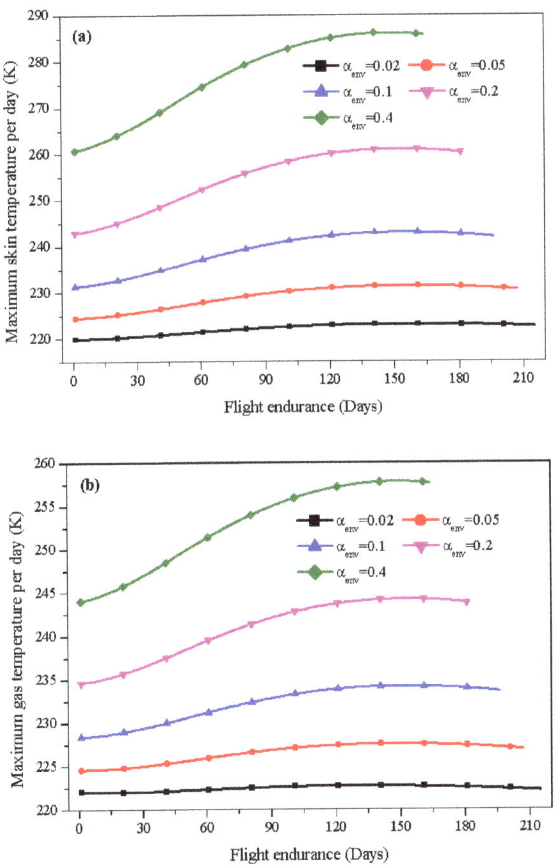

Figure 16. *Cont.*

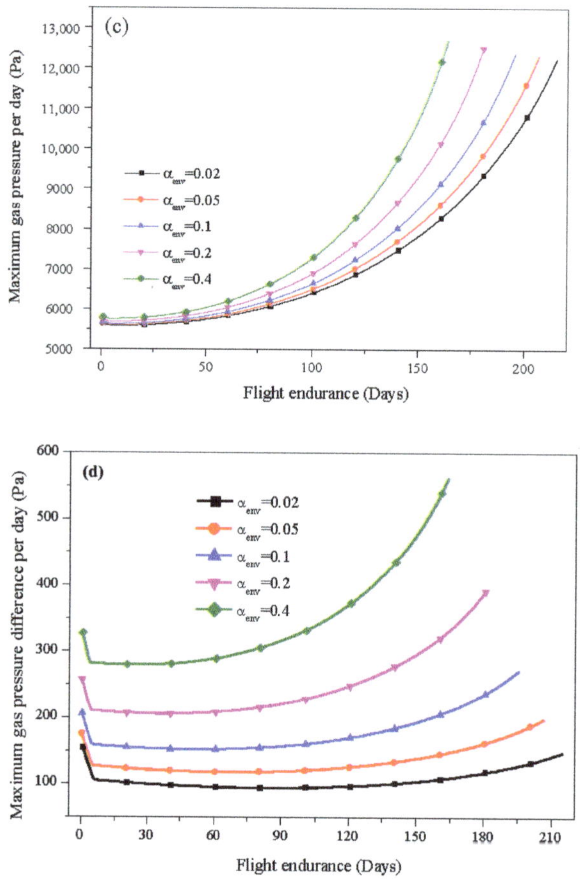

Figure 16. Effects of envelope absorptivity on the (**a**) maximum skin temperature per day, (**b**) maximum gas temperature per day, (**c**) maximum gas pressure per day, and (**d**) maximum gas pressure difference per day.

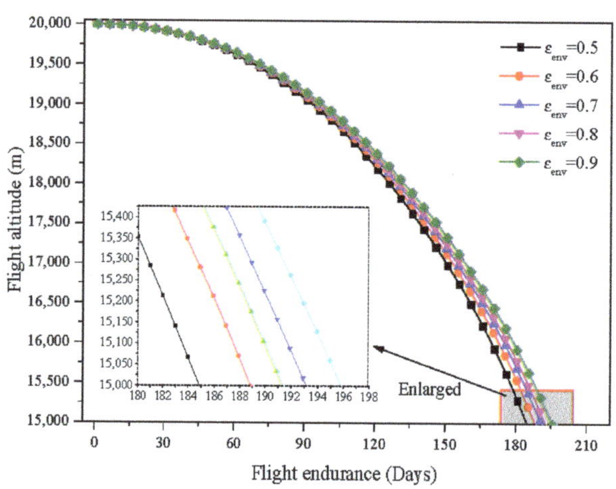

Figure 17. Effects of envelope emissivity on flight altitude.

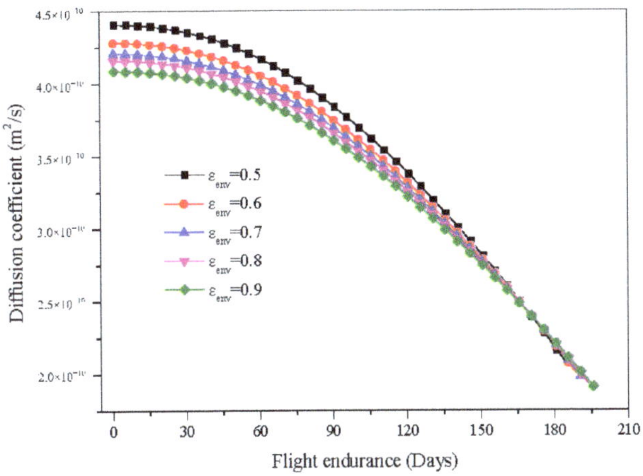

Figure 18. Effects of envelope emissivity on diffusion coefficient.

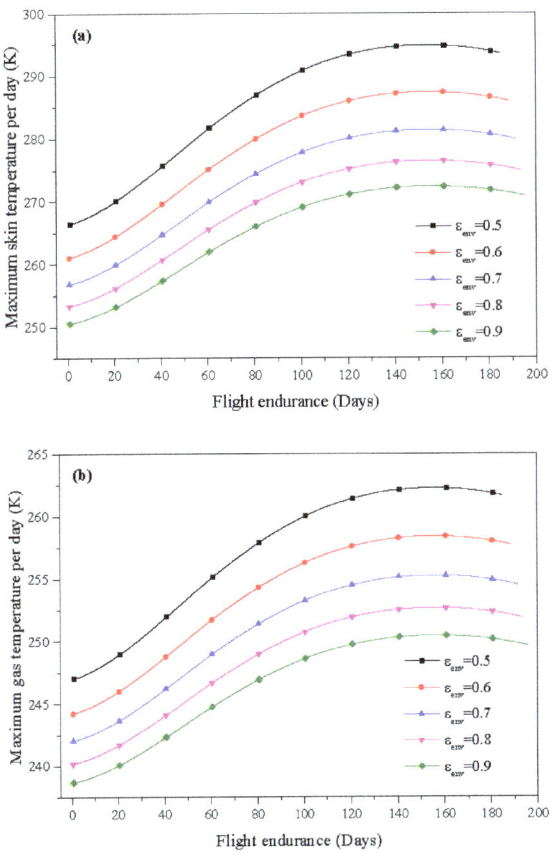

Figure 19. *Cont.*

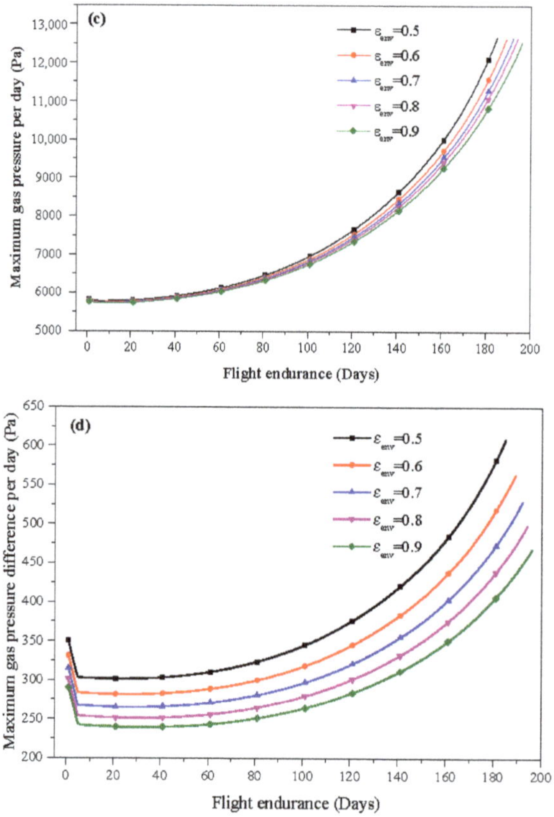

Figure 19. Effects of envelope emissivity on the (**a**) maximum skin temperature per day, (**b**) maximum gas temperature per day, (**c**) maximum gas pressure per day, and (**d**) maximum gas pressure difference per day.

5. Conclusions

Based on the lifting gas diffusion model and dynamic model with thermal effect established in this paper, the lifting gas diffusion behavior and the effects of the envelope properties on the flight performance of the aerostat were numerically investigated and validated. The main results were as follows:

(1) Based on the verification of results, it was found that the lifting gas diffusion model proposed in this paper, which considers the thermal effect for a near-space aerostat during a long-endurance flight, can be utilized to study lifting gas permeability and flight performance.

(2) The ratio of porosity to tortuosity was found to have a significant influence on the gas diffusion coefficient, directly leading to a sharp decline in flight endurance with increases in the ratio of porosity to tortuosity. In the preliminary design of an aerostat, it is helpful to choose an envelope material with an optimal ratio of porosity to tortuosity in order to improve the aerostat's flight performance.

(3) During high-altitude flight, the lifting gas diffusion rate and diffusion coefficient are very sensitive to changes in envelope temperature. Compared to the envelope infrared emissivity, the envelope absorptivity was found to have a stronger influence on the lifting gas diffusion and the thermal performance of near-space aerostats. A higher envelope absorptivity would result in higher temperatures of the lifting gas and envelope, which would consequently increase the mass of lifting gas diffused.

Author Contributions: Conceptualization, J.L. (Jun Li) and L.L.; methodology, Z.C. and L.L.; software, J.L. (Jun Liao); validation, J.L. (Jun Liao), J.L. (Jun Li) and L.L.; investigation, S.L.; data curation, L.L.; writing—original draft preparation, L.L. and Z.C.; writing—review and editing, J.L. (Jun Liao) and J.L. (Jun Li); visualization, J.L. (Jun Li); supervision, S.L. All authors have read and agreed to the published version of the manuscript.

Funding: This work was supported by the Key R&D Projects of Hunan Province (2021GK2011) and the Natural Science Foundation of Hunan Province No. 2019JJ50773, as well as National Key Research and Development Program of China (2016YFB0500801). This research was also supported by the Innovation-Driven Project of Central South University (No. 2018CX024).

Institutional Review Board Statement: Not applicable.

Informed Consent Statement: Not applicable.

Data Availability Statement: Not applicable.

Conflicts of Interest: The authors declare no conflict of interest.

References

1. Bellemare, M.G.; Candido, S.; Castro, P.S.; Gong, J.; Machado, M.C.; Moitra, S.; Ponda, S.S.; Wang, Z. Autonomous navigation of stratospheric balloons using reinforcement learning. *Nature* **2020**, *588*, 77–82. [CrossRef] [PubMed]
2. Gangadhar, A.; Manikandan, M.; Rajaram, D.; Mavris, D. Conceptual design and feasibility study of winged hybrid airship. *Aerospace* **2022**, *9*, 8. [CrossRef]
3. Gemignani, M.; Marcuccio, S. Dynamic characterization of a high-altitude balloon during a flight campaign for the detection of ISM radio background in the stratosphere. *Aerospace* **2021**, *8*, 21. [CrossRef]
4. Du, H.; Lv, M.; Li, J.; Zhu, W.; Zhang, L.; Wu, Y. Station-keeping performance analysis for high altitude balloon with altitude control system. *Aerosp. Sci. Technol.* **2019**, *92*, 644–652. [CrossRef]
5. Ramesh, S.S.; Ma, J.; Lim, K.M.; Lee, H.P.; Khoo, B.C. Numerical evaluation of station-keeping strategies for stratospheric balloons. *Aerosp. Sci. Technol.* **2018**, *80*, 288–300. [CrossRef]
6. van Wynsberghe, E.; Turak, A. Station-keeping of a high-altitude balloon with electric propulsion and wireless power transmission: A concept study. *Acta Astronaut.* **2016**, *128*, 616–627. [CrossRef]
7. Yang, Y.; Wu, J.; Zheng, W. Station-keeping control for a stratospheric airship platform via fuzzy adaptive backstepping approach. *Adv. Space Res.* **2013**, *51*, 1157–1167. [CrossRef]
8. Du, H.; Li, J.; Zhu, W.; Qu, Z.; Zhang, L.; Lv, M. Flight performance simulation and station-keeping endurance analysis for stratospheric super-pressure balloon in real wind field. *Aerosp. Sci. Technol.* **2019**, *86*, 1–10. [CrossRef]
9. Hall, J.L.; Yavrouian, A. Pinhole effects on Venus superpressure balloon lifetime. In Proceedings of the AIAA Balloon Systems (BAL) Conference, Daytona Beach, FL, USA, 25–28 March 2013.
10. Yin, B.; Xu, S.; Yang, S.; Dong, F. Influence of microelliptical groove gas diffusion layer (GDL) on transport behavior of proton exchange membrane fuel cell (PEMFC). *Int. J. Heat Mass Transf.* **2021**, *180*, 121793. [CrossRef]
11. Shi, T.; Chen, W.; Gao, C.; Hu, J.; Zhao, B.; Wang, M. Biaxial strength determination of woven fabric composite for airship structural envelope based on novel specimens. *Compos. Struct.* **2018**, *184*, 1126–1136. [CrossRef]
12. Kang, W.; Suh, Y.; Lee, I. Mechanical property characterization of film-fabric laminate for stratospheric airship envelope. *Compos. Struct.* **2006**, *75*, 151–155. [CrossRef]
13. García-Salaberri, P.A.; Hwang, G.; Vera, M.; Weber, A.Z.; Gostickd, J.T. Effective diffusivity in partially-saturated carbon-fiber gas diffusion layers: Effect of through-plane saturation distribution. *Int. J. Heat Mass Transf.* **2015**, *86*, 319–333. [CrossRef]
14. Kumazawa, H.; Aoki, T.; Susuki, I. Influence of stacking sequence on leakage characteristics through CFRP composite laminates. *Compos. Sci. Technol.* **2006**, *66*, 2107–2115. [CrossRef]
15. Yokozeki, T.; Ogasawara, T.; Ishikawa, T. Evaluation of gas leakage through composite laminates with multilayer matrix cracks: Cracking angle effects. *Compos. Sci. Technol.* **2006**, *66*, 2815–2824. [CrossRef]
16. Thornton, A.W.; Hilder, T.; Hill, A.J.; Hill, J.M. Predicting gas diffusion regime within pores of different size, shape and composition. *J. Membr. Sci.* **2009**, *336*, 101–108. [CrossRef]
17. Yao, X.F.; Lei, Y.M.; Xiong, C.; Wang, X.Q.; Wang, Y.Q. Experimental study of helium leakage parameters in flexible composite. *J. Appl. Polym. Sci.* **2010**, *116*, 3562–3568. [CrossRef]
18. Yin, Y.; Qu, Z.G.; Zhang, J.F. Multiple diffusion mechanisms of shale gas in nanoporous organic matter predicted by the local diffusivity lattice Boltzmann model. *Int. J. Heat Mass Transf.* **2019**, *143*, 118571. [CrossRef]
19. Kayhan, Ö.; Hastaoglu, M.A. Modeling of stratospheric balloon using transport phenomena and gas compress–release system. *J. Thermophys. Heat Transf.* **2014**, *28*, 534–541. [CrossRef]
20. Zhu, W.; Xu, Y.; Du, H.; Li, J. Thermal performance of high-altitude solar powered scientific balloon. *Renew. Energy* **2019**, *135*, 1078–1096. [CrossRef]

21. Wu, J.; Fang, X.; Wang, Z.; Hou, Z.; Ma, Z.; Zhang, H.; Dai, Q.; Xu, Y. Thermal modeling of stratospheric airships. *Prog. Aerosp Sci.* **2015**, *75*, 26–37. [CrossRef]
22. Xiong, J.; Bai, J.B.; Chen, L. Simplified analytical model for predicting the temperature of balloon on high-altitude. *Int. J. Therm Sci.* **2014**, *76*, 82–89. [CrossRef]
23. Kayhan, Ö.; Hastaoglu, M.A. Control of high altitude telecommunication balloons via transport phenomena, material and fuel cell. In Proceedings of the 6th International Conference on Recent Advances in Space Technologies IEEE, Istanbul, Turkey, 12–14 June 2013.
24. Li, Y.; Kalantari-Dahaghi, A.; Zolfaghari, A.; Dong, P.; Negahban, S.; Zhou, D. A new model for the transport of gaseous hydrocarbon in shale nanopores coupling real gas effect, adsorption, and multiphase pore fluid occupancies. *Int. J. Heat Mas Transf.* **2020**, *148*, 119026. [CrossRef]
25. Yuan, J.; Sundén, B. On mechanisms and models of multi-component gas diffusion in porous structures of fuel cell electrodes. *Int J. Heat Mass Transf.* **2014**, *69*, 358–374. [CrossRef]
26. Cussler, E.L. *Diffusion: Mass Transfer in Fluid Systems*; Cambridge University Press: Cambridge, UK, 2009; pp. 191–206.
27. Pan, L.; Xu, K. Two-stage fourth-order gas-kinetic scheme for three-dimensional Euler and Navier-Stokes solutions. *Int. J. Comput Fluid Dyn.* **2018**, *10*, 395–411. [CrossRef]
28. Hastaoglu, M.A. Transient modelling of a packed tower: Mass and heat transfer with reaction. *Fuel* **1995**, *74*, 1624–1631. [CrossRef
29. Sadagopan, A.; Camci, C. Viscous flow and performance issues in a 6:1 supersonic mixed-flow compressor with a tandem diffuse *Aerosp. Sci. Technol.* **2019**, *88*, 9–21. [CrossRef]
30. Cuenya, B.R. Synthesis and catalytic properties of metal nanoparticles: Size, shape, support, composition, and oxidation state effects. *Thin Solid Films* **2010**, *518*, 3127–3150. [CrossRef]
31. Müller, D.; Warneke, T.; Rixen, T.; Müller, M.; Mujahid, A.; Bange, H.W.; Notholt, J. Fate of peat-derived carbon and associated CO2 and CO emissions from two Southeast Asian estuaries. *Biogeosci. Discuss.* **2016**, *12*, 8299–8340.
32. Farley, R.E. BalloonAscent-3-D simulation tool for the ascent and float of high-altitude balloons. In Proceedings of the AIAA 5th Aviation, Technology, Integration, and Operations Conference, Arlington, VA, USA, 26–28 September 2005.
33. Yang, X.; Zhang, W.; Hou, Z. Improved Thermal and Vertical Trajectory Model for Performance Prediction of Stratospheric Balloons. *J. Aerosp. Eng.* **2015**, *28*, 04014075. [CrossRef]
34. Korotkin, A.I. *Added Masses of Ship Structures*; Springer Science & Business Media: Berlin, Germany, 2008; pp. 1–111.
35. Said, C.; Naoufel, A. Estimation of the virtual masses of a large unconventional airship based on purely analytical method to aid in the preliminary design. *Aircr. Eng. Aerosp. Technol.* **2022**, *94*, 531–540.
36. Zhang, Y.; Liu, D. Influences of initial launch conditions on flight performance of high altitude balloon ascending process. *Adv Space Res.* **2015**, *56*, 605–618. [CrossRef]
37. Saleh, S.; He, W. New design simulation for a high-altitude dual-balloon system to extend lifetime and improve floating performance. *Chin. J. Aeronaut.* **2018**, *31*, 247–256. [CrossRef]
38. Liu, Q.; Wu, Z.; Zhu, M.; Xu, W.Q. A comprehensive numerical model investigating the thermal-dynamic performance of scientific balloon. *Adv. Space Res.* **2014**, *53*, 325–338. [CrossRef]
39. Liu, Y.; Xu, Z.; Du, H.; Lv, M. Multidisciplinary optimization of thermal insulation layer for stratospheric airship with a solar array. *Aerospace* **2022**, *9*, 83. [CrossRef]
40. Saleh, S.; He, W. Ascending performance analysis for high altitude zero pressure balloon. *Adv. Space Res.* **2017**, *59*, 2158–2172 [CrossRef]
41. Dai, Q.; Fang, X.; Zhao, Y.; Xing, D. An empirical model for estimating the atmospheric transmittance of upward infrared radiation at different altitudes. *Adv. Space Res.* **2016**, *58*, 2453–2459. [CrossRef]
42. Bergman, T.L.; Incropera, F.P.; Dewitt, D.P.; Lavine, A.S. *Fundamentals of Heat and Mass Transfer*; John Wiley & Sons: New York, NY USA, 2011; pp. 378–419.
43. Qing, W. Theoretical and Experimental Study of Helium Leakage Mechanism of Aerostat Envelope Materials. Master's Thesis Tsinghua University, Beijing, China, 2010.

aerospace

MDPI

Article

Layout Analysis and Optimization of Airships with Thrust-Based Stability Augmentation

Carlo E. D. Riboldi *,† and Alberto Rolando †

Dipartimento di Scienze e Tecnologie Aerospaziali, Politecnico di Milano, Via La Masa 34, 20156 Milan, Italy; alberto.rolando@polimi.it

* Correspondence: carlo.riboldi@polimi.it

† These authors contributed equally to this work.

Abstract: Despite offering often significant advantages with respect to other flying machines, especially in terms of flight endurance, airships are typically harder to control. Technological solutions borrowed from the realm of shipbuilding, such as bow thrusters, have been largely experimented with to the extent of increasing maneuverability. More recently, also thrust vectoring has appeared as an effective solution to ameliorate maneuverability. However, with an increasing interest for high-altitude airships (HAAs) and autonomous flight and the ensuing need to reduce weight and lifting performance, design simplicity is a desirable goal. Besides saving weight, it would reduce complexity and increase time between overhauls, in turn enabling longer missions. In this perspective, an airship layout based on a set of non-tilting thrusters, optimally placed to be employed for both propulsion and attitude control, appears particularly interesting. If sufficiently effective, such configurations would reduce the need for control surfaces on aerodynamic empennages and the corresponding actuators. Clearly, from an airship design perspective, the adoption of many smaller thrusters instead of a few larger ones allows a potentially significant departure from more classical airship layouts. Where on one side attractive, this solution unlocks a number of design variables—for instance, the number of thrusters, as well as their positioning in the general layout, mutual tilt angles, etc.—to be set according simultaneously to propulsion and attitude control goals. In this paper, we explore the effect of a set of configuration parameters defining three-thrusters and four-thrusters layout, trying to capture their impact on an aggregated measure of control performance. To this aim, at first a stability augmentation system (SAS) is designed so as to stabilize the airship making use of thrusters instead of aerodynamic surfaces. Then a non-linear model of the airship is employed to test the airship in a set of virtual simulation scenarios. The analysis is carried out in a parameterized fashion, changing the values of configuration parameters pertaining to the thrusters layout so as to understand their respective effects. In a later stage, the choice of the optimal design values (i.e., the optimal layout) related to the thrusters is demanded to an optimizer. The paper is concluded by showing the results on a complete numerical test case, drawing conclusions on the relevance of certain design parameters on the considered performance, and commenting the features of an optimal configuration.

Keywords: flight dynamics; airship; optimal layout; thruster layout; high-altitude airship; HAA; unmanned; lighter-than-Air; LTA; stability augmentation system; SAS; thrust-based stabilization; airship design; control design

1. Introduction

In recent times, airships have regained the interest of researchers and engineers, as potentially advantageous solutions for carrying out different types of missions. These include primarily high-altitude pseudo-satellites (HAPS). The design of high-altitude airships (HAAs) has been in the focus of several research endeavors [1–8], which due to the relative immaturity of design techniques specifically targeting this mission, mostly consider preliminary design. They consequently focus on understanding the general features of an airship for such mission, such as weight, installed power, volume, shape, area of solar cells, etc. Despite reaching in some instances the prototype stage (as is the case for the HiSentinel series and Hale-D in the United States or YuanMeng in China), existing HAAs are typically proofs of concept, featuring a relatively standard overall layout, compared to older airships.

The latter is also generally the case for the most recent design studies and applications in the field of low altitude flights, where airships are gaining industrial interest as alternatives to unmanned multi-copters [9–13].

Over the last decade, the application of novel propulsion concepts in aeronautics has been focusing particularly on the inclusion of an electric component in the powertrain [14–17], and related enabling technologies [18]. On airships, the adoption of electric motors, to drive the propellers instead of piston-powered units, allows benefiting from more compact and simpler thrusters. Conjugated with solar cells, such thrusters may be fed for long periods of time especially when considering HAPS missions, at altitudes where the solar incident radiation is higher. As highlighted in some studies [3,4], corresponding HAAs may take on missions spanning several weeks at once. Therefore, the limit to even higher flight times would be attributed to other issues, including overhauling of the payload and of the airship plants (as practically observed in the case of the HiSentinel campaign). The latter would be decreased in the most effective way by pushing on design simplicity, i.e., removing movable parts and the corresponding actuators as much as possible.

One such design simplification may be obtained by suppressing movable control surfaces, exploiting the adoption of more electric thrusters for control. Of course, due to the generally marginal stability characteristics of airships, the artificial stability augmentation typically on-board would be demanded to the thrusters, which would be employed simultaneously to push the aircraft in forward flight (or station-keeping against a constant wind for HAAs in HAPS missions), maneuvers, and to stabilize the system in presence of disturbances.

A very effective, solution in this sense comes from thrust vectoring, which has been employed also on some of the few airships currently flying regularly (e.g., Zeppelin LZ N07-101).

However, to the aim of increasing simplicity, a smart placement (i.e., layout) of the thrusters onboard would be possibly a more promising solution. On the other hand, the placement of thrusters would be also a more delicate task, since it should answer the need for several propulsion and control goals at once, as pointed out. Furthermore, the assignment of a configuration requires the definition of a relatively large set of geometric parameters, including the three-dimensional positioning of the thrusters with respect to the hull (as well as with respect to the buoyancy center and center of gravity), the tilt angle of the thrusters, and clearly their number. This makes the analysis and the selection of an optimal configuration a potentially challenging problem [8,19].

In the present paper, we try to explore this design problem. Taking inspiration from existing airship layouts, a baseline airship featuring a cruciform, tailback configuration with movable surfaces, and a single pushing thruster, is selected. The baseline layout is altered, making computations assuming to remove the existing thruster and replacing it with three or four thrusters in multiple layouts. In particular, free design parameters include the three-dimensional positioning of all thrusters with respect to the buoyancy center of the hull, as well as the fixed vertical tilt angle of the thrusters with respect to the longitudinal axis. In order to stabilize the system, a stability augmentation system (SAS) is designed and implemented, taking as inputs rates usually measurable on flying vehicles, and producing as outputs stabilizing values of the thrust settings.

In order to synthetically capture the performance of a certain layout and allow mutual comparisons, a cost function sensible to both the departure from a trimmed condition and the energy required by the actuators to control the airship is defined. A non-linear model of the airship is introduced, which is originally implemented and subject to the mentioned purpose-designed control laws, integrating forward-in-time its behavior starting from perturbations of trimmed, horizontal flight conditions. The action of the SAS is such to restore trimmed flight, with a performance which is the direct result of a choice of the layout.

A parameter analysis is carried out first to preliminarily map the cost function just introduced and to assess its sensitivity to the assumed design parameters (i.e., geometric quantities defining the layout of the thrusters configurations).

Finally, an optimization algorithm is deployed to automatically find the optimal configuration, minimizing the selected cost function. A comparison of results in significant scenarios is proposed, trying to understand the dependence of the optimal solution on trim airspeed as well as geometric features of the layout. Finally, exploiting the nature of the proposed cost function, an assessment of the relative advantage of a thruster layout with respect to another is attempted.

2. Dynamic Model of the Airship

In order to ease the comparison between different layouts of the thrusters, the formulation for the dynamics of the airship has been written in an agnostic form, according to a vectorial formalism. Considering the dynamic equilibrium for displacement and rotational motion, the generalized vector form is assumed, yielding the following:

$$\boldsymbol{M_P}\dot{\boldsymbol{w}}_{\boldsymbol{P}} + \boldsymbol{w}_{\boldsymbol{P}_{\times}}\boldsymbol{M_P}\boldsymbol{w_P} = \boldsymbol{s_P} \tag{1}$$

where generalized mass matrix $\boldsymbol{M_P}$ is a function of the mass m, static moment $\boldsymbol{S_P}$, and moment of inertia $\boldsymbol{J_P}$, which are arranged as follows.

$$\boldsymbol{M_P} = \begin{bmatrix} m\boldsymbol{I} & \boldsymbol{S_P}^T \\ \boldsymbol{S_P} & \boldsymbol{J_P} \end{bmatrix}. \tag{2}$$

In Equation (1) the generalized velocity vector is $\boldsymbol{w_P} = \{\boldsymbol{v_P}, \boldsymbol{\omega_{\mathcal{B}}}\}^T$, i.e., listing, respectively, the velocity of the generic point $\boldsymbol{P}$ with respect to an inertial observer, and the rotational speed of the airship body $\mathcal{B}$ with respect to the same observer. The southwest product operator $\times$ applied to $\boldsymbol{w_P}$ yields the following.

$$\boldsymbol{w}_{\boldsymbol{P}_{\times}} = \begin{bmatrix} \boldsymbol{\omega}_{\mathcal{B}_\times} & \boldsymbol{0} \\ \boldsymbol{v}_{\boldsymbol{P}_\times} & \boldsymbol{\omega}_{\mathcal{B}_\times} \end{bmatrix}. \tag{3}$$

The generalized forcing term in Equation (1) is defined as $\boldsymbol{s_P} = \{\boldsymbol{f}, \boldsymbol{m_P}\}^T$, where the two components represent, respectively, the resultants of the force and moment (the latter with respect to the measuring point $\boldsymbol{P}$) due to aerodynamics ($\boldsymbol{s}_{\boldsymbol{P}}^{a}$), buoyancy ($\boldsymbol{s}_{\boldsymbol{P}}^{b}$), gravity ($\boldsymbol{s}_{\boldsymbol{P}}^{g}$), and propulsion ($\boldsymbol{s}_{\boldsymbol{P}}^{t}$).

In order to make Equation (1) explicit, a body reference system is chosen (indicated with $(\cdot)^{\mathcal{B}}$), centered in the center of buoyancy $\boldsymbol{CB}$, with the first axis pointing to the nose of the airship, the third to the bottom, and consequently the second to the right. The two systems are shown in a sketch in Figure 1. With respect to an inertial reference on the ground (indicated with $(\cdot)^{\mathcal{I}}$), it is possible to obtain the body reference from the ground one via a rotation tensor $\boldsymbol{R}_{\mathcal{I}\to\mathcal{B}} = \boldsymbol{R}_{\mathcal{I}\to\mathcal{B}}(\varphi, \vartheta, \psi)$, where φ, ϑ, and ψ are angular parameters, physically interpreted as attitude angles of roll, pitch, and yaw, respectively [20].

The generalized mass in Equation (2) will be localized in the center of buoyancy, according to the choice $\boldsymbol{P} \equiv \boldsymbol{CB}$. It is noteworthy that the static moment $\boldsymbol{S_{CB}}$ is not identically null (unlike in typical winged aircraft dynamics applications), since the reference point is not in the center of gravity CG. In body reference, the components of the speed vector take the scalar definitions $\boldsymbol{v}_{\boldsymbol{CB}}^{\mathcal{B}} = \{U, V, W\}^T$, whereas the components of the rotational speed $\boldsymbol{\omega}_{\mathcal{B}}^{\mathcal{B}} = \{p, q, r\}^T$. Consequently, the inertial part (i.e., the left hand side) of dynamic equilibrium in Equation (1) will produce six scalar expressions, functions of inertia components in the body frame, and the six scalar kinematic variables just introduced (and their first-order time derivatives).

By components in the body frame $\mathcal{B}$, the forcing term s_{CB} on the right hand side of Equation (1) can be written as described next.

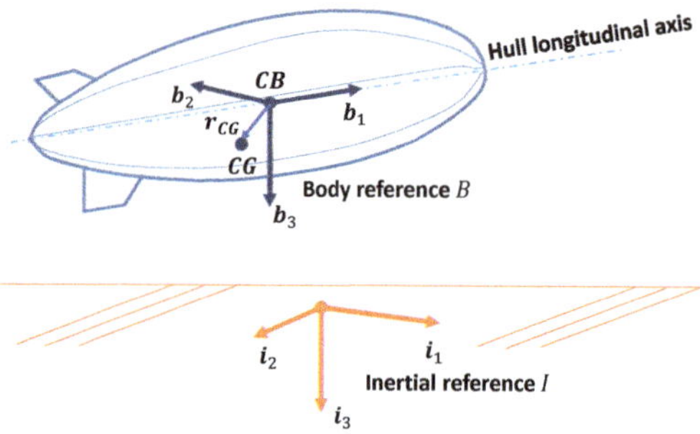

Figure 1. Inertial reference on ground, and body reference.

2.1. Aerodynamics

An aerodynamic model allowing a balance of accuracy and simplicity deemed suitable for the present study, centered on flight dynamics, has been described in [21–23]. The contribution s^a_{CB} can be split into two components, an active term similar to most winged flying vehicles $s^{a,b}_{CB}$, and a reaction term due to the displacement of surrounding air by action of the hull, called $s^{a,m}_{CB}$. By components in the body frame $\mathcal{B}$, the active term can be written according to the aircraft dynamics nomenclature [20], where the dependence of each component is modeled according to a first-order expansion with respect to the kinematic variables. Taking into account the mutual balance between coefficients for an airship with a standard hull and tail configuration and proportion, it is possible to reduce the number of parameters in the expansion, thus yielding for $s^{a,b^{\mathcal{B}}}_{CB}$ [22]:

$$s^{a,b^{\mathcal{B}}}_{CB} = \begin{Bmatrix} X \\ Y \\ Z \\ L_{CB} \\ M_{CB} \\ N_{CB} \end{Bmatrix} = \begin{bmatrix} X_U & 0 & X_W & 0 & X_q & 0 \\ 0 & Y_V & 0 & Y_p & 0 & Y_r \\ Z_U & 0 & Z_W & 0 & Z_q & 0 \\ 0 & L_{CB_V} & 0 & L_{CB_p} & 0 & L_{CB_r} \\ M_{CB_U} & 0 & M_{CB_W} & 0 & M_{CB_q} & 0 \\ 0 & N_{CB_V} & 0 & N_{CB_p} & 0 & N_{CB_r} \end{bmatrix} \begin{Bmatrix} U \\ V \\ W \\ p \\ q \\ r \end{Bmatrix} + \boldsymbol{V}\boldsymbol{u}. \quad (4)$$

The scalar sensitivities appearing in the first matrix on the right hand side of Equation (4) can be modeled via the Munk–DeLaurier theory [21,24], making use of the potential flow theory for the hull part, as well as of standard incompressible lifting surface aerodynamics for the empennages [20]. The corresponding modeling is a function of the specific geometry of the hull and fins. For the former, a detailed analytic description is required, usually attainable at least for axis-symmetric, regular shapes of the envelope. For tail lifting surfaces, methods for obtaining lumped force and moment coefficients for an assigned platform, span, and aerodynamic profile, originally developed for aircraft, still hold for the airship case. Corrections for the interaction between the envelope and tail surfaces allow an accurate mix of teh contributions of the two most aerodynamically relevant components of the airship [22].

In Equation (4), $\boldsymbol{u}$ is the array of controls, and matrix $\boldsymbol{V}$ modulates the effect of controls on the aerodynamic force. For standard airship steering and stability augmentation, based on the deflection of aerodynamic surfaces on the tail, this component will be a sparse non-zero matrix, with coefficients modeled according to standard methods for deflectable surfaces, inherited from the winged aircraft case [20]. In this work, where thrust-based control is adopted and there are no deflectable (i.e., control) parts on the tail surfaces, the definition of the control term as $\boldsymbol{u} = \left\{\delta_{T_1}, \ldots, \delta_{T_{N_t}}\right\}^T$ applies, where each scalar δ_{T_i} represents the thrust setting of the corresponding thruster, and N_t is the number of thrusters.

Since the effect of this control is on propulsion force, in Equation (4) we have $\boldsymbol{V} = \boldsymbol{0}$. The control vector, and the corresponding control action, will be modeled as part of the propulsion force (see later Section 2.3).

The reaction component $s_{CB}^{a,m}$ of the aerodynamic forcing terms can be modeled according to a model making use of Munk's shape-specific coefficients k_1, k_2, k_3, and k', and is proportional to the time rate of the generalized velocity $\dot{w}_{CB}$ [22]. In body components, for an axis-symmetric body of volume $\mathcal{V}$, this forcing term yields the following:

$$s_{CB}^{a,m^{\mathcal{B}}} = -\rho \begin{bmatrix} k_1\mathcal{V} & & & & & \\ & k_2\mathcal{V} & & & & \\ & & k_3\mathcal{V} & & & \\ & & & 0 & & \\ & & & & k'J_{CB_{yy}} & \\ & & & & & k'J_{CB_{zz}} \end{bmatrix} \begin{Bmatrix} \dot{U} \\ \dot{V} \\ \dot{W} \\ \dot{p} \\ \dot{q} \\ \dot{r} \end{Bmatrix}, \tag{5}$$

where ρ is the density of air, $J_{CB_{yy}}$ and $J_{CB_{zz}}$ are diagonal components of $J_{CB}^{\mathcal{B}}$, expressed in the body reference centered in CB. Since the term in Equation (5) is a function of the rate $\dot{w}_{CB}$, it is typically moved to the left hand side of Equation (1) and treated as an additional mass term in dynamic equilibrium.

2.2. Buoyancy and Gravity

The term $s_{CB}^{b^{\mathcal{B}}}$ is composed of a non-null force term, and an identically null moment, since buoyancy acts by definition in the center of buoyancy $\boldsymbol{CB}$. An expression of buoyancy force is particularly straightforward in the ground inertial reference, where it bears a single non-null component pointing up from the ground.

Similarly, gravity force bears an expression which is structurally close to that of buoyancy (i.e., a single component normal to the ground). However, since $\boldsymbol{CB}$ and $\boldsymbol{CG}$ are not in the same location, gravity exerts a non-null moment with respect to $\boldsymbol{CB}$.

By components in the body reference, buoyancy, and gravity yield, we have the following:

$$s_{CB}^{b^{\mathcal{B}}} = \begin{Bmatrix} R_{\mathcal{I}\to\mathcal{B}}^{\mathcal{I}|\mathcal{B}^T} \begin{Bmatrix} 0 \\ 0 \\ -\rho g \mathcal{V} \end{Bmatrix} \\ \begin{Bmatrix} 0 \\ 0 \\ 0 \end{Bmatrix} \end{Bmatrix}, \quad s_{CB}^{g^{\mathcal{B}}} = \begin{Bmatrix} R_{\mathcal{I}\to\mathcal{B}}^{\mathcal{I}|\mathcal{B}^T} \begin{Bmatrix} 0 \\ 0 \\ mg \end{Bmatrix} \\ r_{CG}^{\mathcal{B}} \times R_{\mathcal{I}\to\mathcal{B}}^{\mathcal{I}|\mathcal{B}^T} \begin{Bmatrix} 0 \\ 0 \\ mg \end{Bmatrix} \end{Bmatrix} \tag{6}$$

where $R_{\mathcal{I}\to\mathcal{B}}^{\mathcal{I}|\mathcal{B}^T}$ is the transpose of the matrix representation in either $\mathcal{I}$ or $\mathcal{B}$ of the rotation tensor $R_{\mathcal{I}\to\mathcal{B}}$, $r_{CG}^{\mathcal{B}}$ is the position vector of CG from CB in body components, and g is the modulus of gravitational acceleration.

It can be pointed out that gravity and buoyancy bring the additional variables wrapped in array $e_{321}^{\mathcal{B}} = \{\varphi, \vartheta, \psi\}^T$ as unknowns into the expression of dynamic equilibrium. This increase in the number of scalar unknowns in the formulation of dynamic equilibrium is balanced by invoking the kinematic equations relating $\dot{e}_{321}^{\mathcal{B}}$ to the body components of the rotational rate $\omega_{\mathcal{B}}^{\mathcal{B}} = \{p, q, r\}^T$, yielding the following:

$$\dot{e}_{321}^{\mathcal{B}} = S_{321}^{\mathcal{B}^{-1}} \omega_{\mathcal{B}}^{\mathcal{B}}, \tag{7}$$

where matrix $S_{321}^{\mathcal{B}} = S_{321}^{\mathcal{B}}(\varphi, \vartheta, \psi)$. This allows obtaining a balanced 9-by-9 system of first-order, non-linear differential equations, representing the dynamic equilibrium of the system.

2.3. Thrust

Total thrust force and moment around $\boldsymbol{CB}$ are obtained from the sum of the contributions of each of the thrusters. The thrust vector $\boldsymbol{T}_i$ pertaining to the i-th thruster can be oriented with respect to the body reference through a swing angle σ_i and a tilt angle λ_i, according to Figure 2.

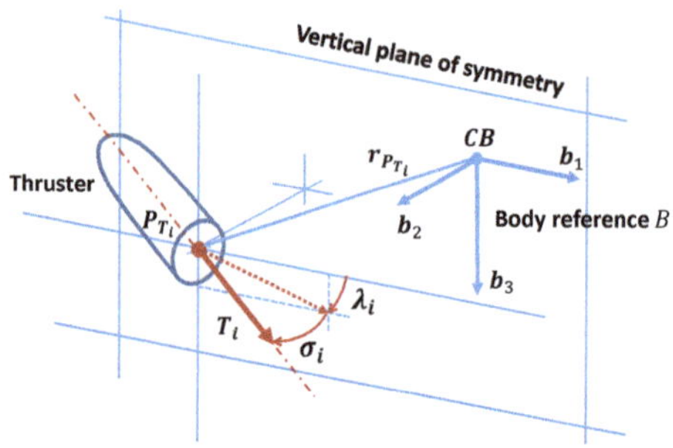

Figure 2. Definition of thruster position and orientation angles with respect to body reference. Swing angle σ_i and tilt angle λ_i are positive as shown with respect to reference $\mathcal{B}$.

Therefore, the total force and moment in body reference components are as follows:

$$\boldsymbol{s}_{CB}^{t\mathcal{B}} = \sum_{i=1}^{N_t} T_i \left\{ \begin{array}{c} \left\{ \begin{array}{c} \cos\sigma_i \cos\lambda_i \\ \sin\sigma_i \\ \cos\sigma_i \sin\lambda_i \end{array} \right\} \\ \boldsymbol{r}_{P_{T_i}}^{\mathcal{B}} \times \left\{ \begin{array}{c} \cos\sigma_i \cos\lambda_i \\ \sin\sigma_i \\ \cos\sigma_i \sin\lambda_i \end{array} \right\} \end{array} \right\}, \tag{8}$$

wherein $\boldsymbol{r}_{P_{T_i}}^{\mathcal{B}}$ is the representation in body components of the position vector pointing from $\boldsymbol{CB}$ to the point of application of $\boldsymbol{T}_i$, named $\boldsymbol{P}_{T_i}$. Intensity T_i is expressed as a function of the control variable δ_{T_i} as follows:

$$T_i = \tilde{T}_i \breve{K}_i(\delta_{T_i})\delta_{T_i}, \tag{9}$$

where $\tilde{T}_i$ is the nominal value of thrust from the i-th thruster. The control δ_{T_i} modulates the nominal value through the shape function $\breve{K}_i(\delta_{T_i})$. The latter allows the reproduction of possible regime-dependent efficiency effects, or non-linear features related to the specific thruster technology. Furthermore, according to the technology implemented in the thruster, a different range of δ_{T_i} values can be considered. For instance, for a piston engine, this control would be limited to a positive value, whereas for electric motors or ion thrusters it may be also negative, thus producing an inversion of the thrust force, exploiting a well-known advantage of these types of thrusters with respect to more standard piston-powered ones.

It should be noted that the formulation just introduced through Equation (8) bends itself to the inclusion of thrust vectoring, which would require properly subordinating σ_i and λ_i to a control logic, instead of assigning them as geometrical parameters, as is of interest in this work.

3. Thrusters Layout and Thrust-Based Control

As stated in Section 1, in this paper the effect on the dynamic response of the airship sorted by the value assumed by a set of geometrical parameters is of interest. A shortlist of configurations of potential interest, for which quantitative design analyses will be performed, is formed mainly based on two drivers:

1. Number of thrusters (N_t). Where on the one hand it may be interesting to increase this number to achieve better controllability and investigate arbitrary configurations, in practice a higher number of thrusters would proportionately increase the chance of failure, and consequently system downtime. Together with a greater complexity of the control system which should be faced when dealing with an increasing number of control variables (i.e., thrust settings), a greater number of thrusters would likely increase also the cost of manufacture. Therefore, layouts with a low number of thrusters are preferred in this work.
2. Positioning of thrusters. Even considering the equilibrium flight conditions most often encountered, i.e. straight horizontal flight at different airspeeds (which also includes station-keeping in constant stratospheric wind for a HAA), airships of standard back-tailed configuration are usually marginally dynamically stable or slightly unstable. In particular, recurring to a modal analysis in a linearized framework, carried out around a trimmed condition in straight, horizontal, steady flight, lateral-directional motion of airships is typically dynamically unstable (due to sideslip subsidence), and longitudinal motion may feature a low-damped pendulum oscillatory mode [22]. For conditioning the dynamics of the system and easing piloting, artificial stability can be obtained in forward flight (not in hover) by implementing a SAS acting on the tail control surfaces. When deleting the (movable) control surfaces, while leaving the tail as a purely passive (stabilizing) assembly, such artificial stability augmentation is delegated to the thrusters. Clearly, this puts constraints on the placement and orientation of the thrusters. For instance, two-thruster configurations would not be able to stabilize the system in terms of both longitudinal and lateral-directional dynamics. Furthermore, when all thrusters are aligned with the longitudinal body axis, no direct roll control could be achieved, despite retaining control abilities through the intrinsic yaw-roll aerodynamic coupling of the airship. Therefore, basic layouts for quantitative investigations need to be preliminarily checked in terms of their ability in principle to perform a control action around all airship body axes.

According to these two drivers, three thruster layouts have been considered in the present work—namely (A), (B), and (C)—which appear as compromises between simplicity and control authority. They are described in the sketches in Figure 3.

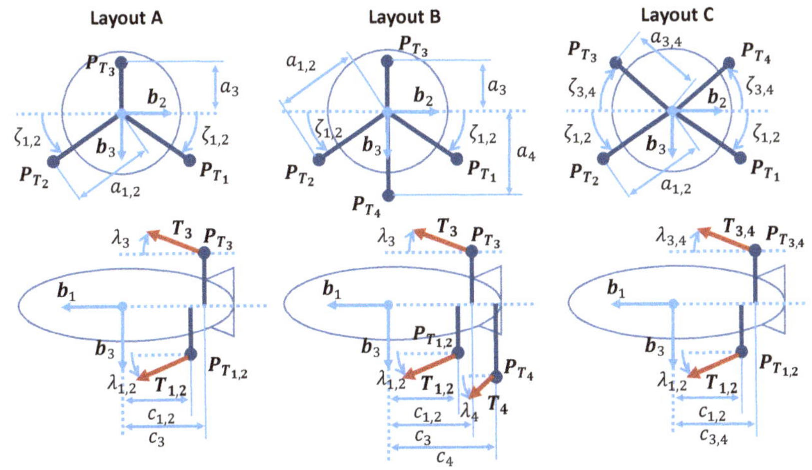

Figure 3. Three-thruster layout configurations and geometrical parameters.

From a geometrical standpoint, all configurations share the same arrangement of the baseline airship less the thrusters—hull, tail empennages, and payload—and differ only for the number and positioning of the thrusters. From the viewpoint of inertia, the number and positioning of the thrusters in the layout imply a different positioning of the CG, and changing values of the overall mass, static, and inertia moment, i.e., virtually all components in Equation (2) shall be affected.

A description of the configurations presented in Figure 3 will be proposed next.

3.1. Description of Thruster Layouts

3.1.1. Features of Layout (A)

Layout (A) features a set of three thrusters, where #1 and #2 share the vertical body coordinate and are symmetrically placed with respect to the vertical plane of symmetry of the airship. Thruster #3 is located on the plane of symmetry.

Thrusters #1 and #2 feature a line of thrust passing below CB, whereas that of #3 passes above that point. This allows thrusters in this configuration to control the motion around the pitch axis. The laterally symmetric positioning of #1 and #2 allows yaw control. In order to achieve direct roll control, however, a non-null tilt $\lambda_1 = \lambda_2 = \lambda_{1,2} \neq 0$ of the couple #1 and #2 needs to be implemented. In this manner, a differential use of #1 and #2 thrusters would produce a torque around the roll axis. For instance, starting from a trimmed condition, and a small positive $\lambda_{1,2} > 0$, i.e., thrusters pointing slightly downwards with respect to the longitudinal body axis, a small increase on #1 thrust and a corresponding decrease on #2 would introduce a negative rolling moment, which would not be achieved if the two tilt angles of the thrusters were null. Clearly, it can be observed that such differential action would produce also a yawing moment.

A down-tilt of #1 and #2 thrusters may be compensated by an up-tilt of #3, i.e., $\lambda_3 < 0$, to allow avoiding an increase in body vertical speed V as an effect of a homogeneous (i.e., not differential) action on #1 and #2—as in trimmed forward flight.

In principle, from a control standpoint, an opposite mutual setting of the thruster tilt may be hypothesized as well (namely $\lambda_{1,2} < 0$ and $\lambda_3 > 0$), but considering the positioning of the thrusters towards the tailcone of the hull, that choice would point the thrusters towards the hull, which due to the aerodynamic flow around the latter would bear a counter-intuitive and uncommon configuration.

In this layout, of particular interest are the longitudinal positions of the thrusters, $c_{1,2}$ and c_3, as well as the dihedral angle $\zeta_{1,2}$ of #1 and #2 with respect to the horizontal. Moreover, the lengths $a_{1,2}$ and a_3 of the radii of the respective thrusters from the longitudinal body axis are also of interest.

Broadly speaking, configuration (A) can be expected to be the cheapest and simpler to design, as well as the least prone to propulsion fault thanks to the lowest possible number of thrusters, but also the one providing the least control effectiveness, making use of the shortest possible number of control degrees of freedom for stabilization around all body axes.

3.1.2. Features of Layout (B)

The second layout is intended to explore the baseline '+'-shaped cruciform configuration. Besides an increased number of thrusters with respect to (A), in layout (B) control in the longitudinal plane (i.e., around pitch axis) is delegated to the combined use of thrusters #3 and #4, whereas lateral-directional control is obtained through the differential use of thrusters #1 and #2. As observed for configuration (A), a down-tilt of #1 and #2 is needed to obtain direct roll control, to be compensated by an up-tilt of #3 (on account of a down-tilt of #4, according to the shape of the tailcone).

Quantities of interest in this configuration are the longitudinal positions $c_{1,2}$, which is the same for #1 and #2 for lateral symmetry, as well as c_3 and c_4. The position of thrusters #3 and #4 is on the vertical plane of symmetry, with a radius of a_3 and a_4 respectively from the longitudinal axis, whereas the lateral thrusters #1 and #2 can be at a dihedral angle

$\zeta_{1,2}$ from the horizontal body plane, which is a relaxation from a strictly '+'-shaped layout, and at a radius of $a_{1,2}$ from the longitudinal axis.

Configuration (B) may be likely more expensive than (A), due to the additional thruster, but artificial stabilization should be in principle easier to design, thanks to a possible decoupling between the longitudinal and lateral-directional action. It would be also possibly more effective, i.e., such to achieve a faster convergence following perturbation through a more limited action of the controls.

3.1.3. Features of Layout (C)

The third layout has been envisaged to explore the features of a '×'-shaped configuration of four thrusters. The couple of #1 and #2 thrusters is attached to the lower part of the airship, to the right and left respectively. Similarly, #3 and #4 are located on the top part of the airship, symmetrically with respect to the vertical plane of symmetry.

The geometry is defined by the longitudinal distances $c_{1,2}$ and $c_{3,4}$ of the corresponding thrusters from $\boldsymbol{CB}$, by the radii $a_{1,2}$ and $a_{3,4}$ from the longitudinal axis, and by the dihedral angles $\zeta_{1,2}$ and $\zeta_{3,4}$, which measure the angular distance respectively of the lower (#1 and #2) and upper (#3 and #4) thrusters from the horizontal body plane.

Similarly to the other cases, direct roll control can be implemented by setting proper values of $\lambda_{1,2}$ and $\lambda_{3,4}$.

From a control standpoint, artificial stabilization around all body axes can be effectively achieved in this configuration. An increase in the thrust of (#1, #2) together with a decrease on (#3, #4) shall provide positive pitching moment and vice-versa. A decrease in (#1, #4) setting and a simultaneous increase in (#2, #3) shall provide a positive yawing moment, and vice-versa. Finally, for a down-tilted (#1, #2) and an up-tilted (#3, #4) couple, a positive rolling moment could be obtained in principle by only increasing (#1, #3) setting, and a negative rolling moment by increasing (#2, #4). However, a more effective roll control can be achieved by additionally decreasing (#2, #4) for a positive rolling moment, and (#1, #3) for a negative one.

3.2. *Thrust-Based Artificial Stability Augmentation for the Considered Layouts*

A stability augmentation system (SAS) should counteract rotations around the three body axes, which in turn requires feeding back the rates $\boldsymbol{\omega}_B^B = \{p, q, r\}^T$, typically available from electronic inertial measurements unit (IMU) sensors. Considering as a reference a trimmed straight, unaccelerated flight condition, the reference values of such rates would be null (i.e., $p_r, q_r, r_r = 0$, with $(\cdot)_r$ standing for reference), and the corresponding feedback rates should be driven to zero by the controller. Furthermore, to more effectively counter the attitude drift in steady state following a perturbation, the integral of the three signals should be targeted, thus configuring a proportional-integral control law structure.

Considering a control system for a reference steady, straight flight conditions, two SAS sub-systems, for longitudinal and lateral-directional dynamics, can be synthesized in a decoupled fashion. This is the result of a decoupling of dynamics achieved in this type of flight [22,23], similar to the case of winged aircraft. The fed-back signals for longitudinal dynamics will be $\boldsymbol{y}_L = \{q, \int q\}^T$, whereas for lateral-directional dynamics $\boldsymbol{y}_D = \{p, r, \int p, \int r\}^T$. Concerning the control array, as outlined in Section 3.1, the choice of the layout produces different requirements on the contributions of scalar variables listed in it, which in turn are reflected by the identity of non-zero gains, as well as in mutual relationships among non-zero ones. This will be shown in detail in this section. However, it should be observed that in principle all δ_{T_i} controls may contribute to both longitudinal and lateral-directional control. The corresponding control arrays $\boldsymbol{u}_L = \left\{\delta_{T_{1_L}}, \ldots, \delta_{T_{N_{tL}}}\right\}^T$ and $\boldsymbol{u}_D = \left\{\delta_{T_{1_D}}, \ldots, \delta_{T_{N_{tD}}}\right\}^T$ will be obtained through a control law as follows.

$$\begin{cases} \boldsymbol{u}_L = \boldsymbol{K}_L \boldsymbol{y}_L \\ \boldsymbol{u}_D = \boldsymbol{K}_D \boldsymbol{y}_D \end{cases}. \tag{10}$$

The complete control input signals to the airship dynamic system will be the composition of a trimming reference value $\boldsymbol{u}_r$, the stabilizing SAS contribution of Equation (10), and the pilot's control $\boldsymbol{u}_p$, yielding the following.

$$\boldsymbol{u} = \boldsymbol{u}_r + (\boldsymbol{u}_L + \boldsymbol{u}_D) + \boldsymbol{u}_p. \tag{11}$$

For layout (A), the number of thrusters is $N_t = 3$, yielding for the control arrays, either for longitudinal and lateral-directional SAS, $\boldsymbol{u}_L = \left\{\delta_{T_{1_L}}, \delta_{T_{2_L}}, \delta_{T_{3_L}}\right\}^T$ and $\boldsymbol{u}_D = \left\{\delta_{T_{1_D}}, \delta_{T_{2_D}}, \delta_{T_{3_D}}\right\}^T$. Considering the geometry of the configuration, and the hypothesis of a positive (i.e., down-) tilt $\lambda_{1,2} > 0$ and a negative (i.e., up-) tilt $\lambda_3 < 0$, the following structure of the SAS gain matrices has been hypothesized:

$$\boldsymbol{K}_L = \begin{bmatrix} -k^P_{1,2_q} & -k^I_{1,2_q} \\ -k^P_{1,2_q} & -k^I_{1,2_q} \\ k^P_{3_q} & k^I_{3_q} \end{bmatrix}, \quad \boldsymbol{K}_D = \begin{bmatrix} -k^P_{1,2_p} & k^P_{1,2_r} & -k^I_{1,2_p} & k^I_{1,2_r} \\ k^P_{1,2_p} & -k^P_{1,2_r} & k^I_{1,2_p} & -k^I_{1,2_r} \\ 0 & 0 & 0 & 0 \end{bmatrix}, \tag{12}$$

where $k^P_{(\cdot)}$ and $k^I_{(\cdot)}$ are proportional and integral gains, respectively, and all $k^{(\cdot)}_{(\cdot)}$ are positive values so that their sign is explicit in the expressions of Equation (12) (as well as in Equations (13) and (14) to follow). In particular, the sign takes into account the positive verse of body rotation rates. It can be remarked that there is no contribution of #3 thruster on lateral-directional stabilization. The number of scalar gains to assign is eight, according to this architecture. A sketch of the corresponding logical scheme is shown in Figure 4.

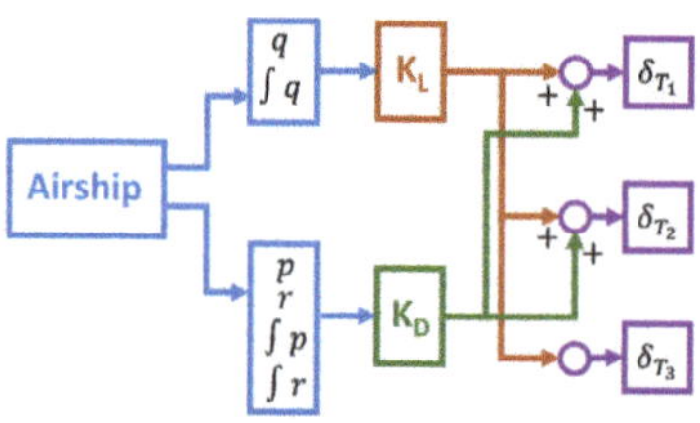

Figure 4. Stability augmentation logic for layout (A).

Moving on to layout (B), the two control arrays are here augmented by a fourth control ($N_t = 4$), yielding $\boldsymbol{u}_L = \left\{\delta_{T_{1_L}}, \delta_{T_{2_L}}, \delta_{T_{3_L}}, \delta_{T_{4_L}}\right\}^T$ and $\boldsymbol{u}_D = \left\{\delta_{T_{1_D}}, \delta_{T_{2_D}}, \delta_{T_{3_D}} \delta_{T_{4_D}}\right\}^T$. According to the hypothesis on the down-tilt of the lateral thrusters, $\lambda_{1,2} > 0$, the following structure can be hypothesized for the SAS gain matrices.

$$\boldsymbol{K}_L = \begin{bmatrix} 0 & 0 \\ 0 & 0 \\ k^P_{3_q} & k^I_{3_q} \\ -k^P_{4_q} & -k^I_{4_q} \end{bmatrix}, \quad \boldsymbol{K}_D = \begin{bmatrix} -k^P_{1,2_p} & k^P_{1,2_r} & -k^I_{1,2_p} & k^I_{1,2_r} \\ k^P_{1,2_p} & -k^P_{1,2_r} & k^I_{1,2_p} & -k^P_{1,2_r} \\ 0 & 0 & 0 & 0 \\ 0 & 0 & 0 & 0 \end{bmatrix}. \tag{13}$$

As can be noticed from Equation (13), here, the longitudinal and lateral-directional control components are fully decoupled, delegating lateral-directional control to thrusters #1 and #2, and longitudinal control to #3 and #4 (as reported in Figure 5). This is not the only possible choice, since especially thrusters #1 and #2 might contribute to longitudinal stabilization (i.e., producing a pitching moment), with an effectiveness which is proportional to the dihedral angle $\zeta_{1,2}$, all things being equal. However, it was chosen to analyze the chance offered by layout (B) to decouple the two stabilization systems, a control architecture that allows retaining the number of gains at eight (hence the same control design complexity as configuration (A)), despite an increase in the number of thrusters (and corresponding control variables).

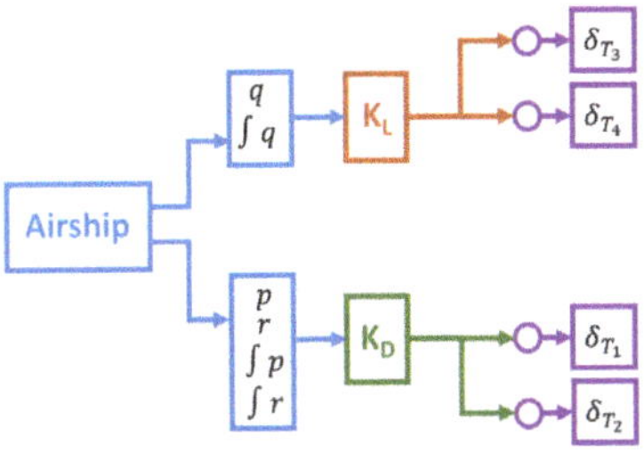

Figure 5. Stability augmentation logic for layout (B).

The last considered layout (C) shares the size of the control arrays with the previous one ($N_t = 4$). Similar to layout (A), also in this case, it is not possible to decouple the control action on longitudinal dynamics from that on lateral-directional dynamics. Therefore, due to the increased number of thrusters with respect to (A), the number of gain coefficients to assign will be greater. Under the hypothesis of a down-tilt of the lower thrusters ($\lambda_{1,2} > 0$) and an up-tilt of the upper ones ($\lambda_{3,4} < 0$), the following structure of the gain matrices can be envisaged.

$$\boldsymbol{K}_L = \begin{bmatrix} -k^P_{1,2_q} & -k^I_{1,2_q} \\ -k^P_{1,2_q} & -k^I_{1,2_q} \\ k^P_{3,4_q} & k^I_{3,4_q} \\ k^P_{3,4_q} & k^I_{3,4_q} \end{bmatrix}, \quad \boldsymbol{K}_D = \begin{bmatrix} -k^P_{1,3_p} & k^P_{1,4_r} & -k^I_{1,3_p} & k^I_{1,4_r} \\ k^P_{2,4_p} & -k^P_{2,3_r} & k^I_{2,4_p} & -k^I_{2,3_r} \\ -k^P_{1,3_p} & -k^P_{2,3_r} & -k^I_{1,3_p} & -k^I_{2,3_r} \\ k^P_{2,4_p} & k^P_{1,4_r} & k^I_{2,4_p} & k^I_{1,4_r} \end{bmatrix}. \tag{14}$$

The number of gains to be assigned is raised to twelve in this configuration, which is shown in the sketch of Figure 6.

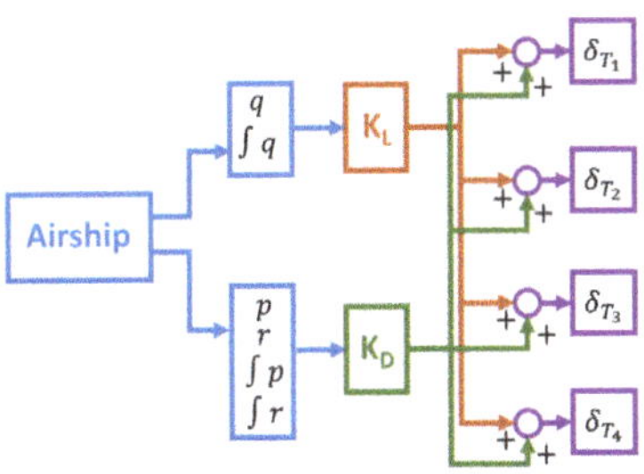

Figure 6. Stability augmentation logic for layout (C).

The ability of the SAS systems proposed for each configuration to stabilize the airship, when perturbing a horizontal, steady, forward flight condition at different airspeeds spanning the operational envelope of a testbed, has been assessed by using linear and non-linear methods, namely by eigenanalyses on a closed-loop linearized representation of the system obtained in a trimmed, steady forward flight condition, as well as through time-marching simulations on the non-linear model described in Section 2, controlled by the proposed SAS and subject to perturbations of the initial condition. Example results will be shown in the results section. At this level, it can be anticipated that all systems can be easily designed to be capable of effectively stabilizing the system in the considered testing scenario. More comments on this will be added when describing the results of the analyses carried out in the design analysis and optimization phase introduced in the next section.

4. Measure of Performance for a Layout and Corresponding Optimization

As stated in the introductory Section 1, the goals of the present paper are not limited to the introduction of a thrust-based control scheme, but also include an investigation of the potential of each considered layout in ameliorating dynamic performance, trying to understand what is the optimal way to loft an airship based on a certain thruster configuration, and where possible, to compare outcoming layouts, also featuring different numbers of thrusters, to each other.

With this ambition, a comprehensive measure of performance was sought, to the aim of setting up an optimal problem where an optimal design solution—in terms of a geometry such to optimize performance—would be automatically found by an optimization algorithm. Since the comparison is centered on dynamic performance, and in particular on stabilization, a natural measure of performance comes in the form of an integral norm of the quadratic deviation of both the states and controls from the equilibrium condition. Considering for instance a steady, horizontal trimmed flight condition, when a perturbation is induced, the reaction of a (closed-loop) stable system would be that of incurring in a deviation of the state and control signals from the respective reference values, which would then be damped, taking the system back to the starting equilibrium condition. In this scenario, usually, the slacker the control dynamics, the farther the initial departure of the state from the reference, and the more limited the control action. Conversely, the more reactive the controller, the shorter the departure of the system state from the reference, but the more intense the movement of the control signals. In both cases, for opposite reasons, the quadratic norm would increase. A minimal (i.e., optimal) value of this cost function would be the result of the best compromise between control cost (i.e., the energy required for control action) and dynamic performance (i.e., in a broad sense the dynamic stability of the system, reflected in damping, settling time, etc. of the system state).

The minimization of this cost function is the base of consolidated control law synthesis methods (e.g., linear-quadratic regulators, LQR [25]), which are typically employed to produce fully coupled MIMO (multiple-input, multiple-output) gain matrices in a linear (or linearized) full-state feedback framework. However, in this study, the quadratic cost function just introduced is not used for control synthesis, but for a parametric analysis on the effect of geometrical layout variables, thus making a different use of it. Actually, in order to make comparisons among layouts easier, the construction of the cost function is somewhat more articulated. Considering the system in Equation (1) with the kinematic equations Equation (7), the states of the system can be collected in $\boldsymbol{x} = \{U, V, W, p, q, r, \varphi, \vartheta, \psi\}^T$, whereas the array of controls would be $\boldsymbol{u} = \left\{\delta_{T_1}, \ldots, \delta_{T_{N_t}}\right\}$. Considering a single time-marching simulation of time extension $[0, T_f]$, it is possible to assemble the following quantity.

$$
\begin{aligned}
F_k &= F_{k_x} + F_{k_u} \\
&= \left[\frac{1}{T_f}\int_0^{T_f} (\boldsymbol{x}(t) - \boldsymbol{x}_r)^T \boldsymbol{W}_x^T \boldsymbol{Q} \boldsymbol{W}_x (\boldsymbol{x}(t) - \boldsymbol{x}_r) \mathrm{d}t\right] \frac{1}{\sum_{j=1}^{9} Q_j} \\
&+ \left[\frac{1}{T_f}\int_0^{T_f} (\boldsymbol{u}(t) - \boldsymbol{u}_r)^T \boldsymbol{W}_u^T \boldsymbol{R} \boldsymbol{W}_u (\boldsymbol{u}(t) - \boldsymbol{u}_r) \mathrm{d}t\right] \frac{1}{\sum_{j=1}^{N_t} R_j}.
\end{aligned} \tag{15}
$$

In Equation (15), diagonal matrices $\boldsymbol{Q}$ and $\boldsymbol{R}$ can be used to select which states and controls will contribute to the cost function, and their mutual relevance. Diagonal matrices $\boldsymbol{W}_x$ and $\boldsymbol{W}_u$ instead are used to normalize the dimensional value of the array of states and controls, so as to make them more comparable in case of numerically very different values (such as, for instance, for control variables, varying here in the range $[-1, 1]$, and attitude angles or rotational rates, which may be in the order of 10^{-2} rad or rad/s, respectively, and hence very different).

In the present study, $\boldsymbol{Q}$ and $\boldsymbol{R}$ have been used as triggers to select the states and controls to appear in the cost function, and their coefficients take a value which is either 1 or 0. In particular, concerning states, speed and rotational rate components contribute equally to the cost function. Concerning control, all thrusters have been selected as contributors and then $\boldsymbol{R} = \boldsymbol{I}(N_t)$, i.e., an identity matrix of size equal to the number of thrusters. However, considering two different layouts, in particular (A) vs. (B) or (C), which feature a different number of thrusters, an increase in the cost function would arise by moving from a three-thruster configuration to a four-thruster one, due to the plain addition of the contribution pertaining to a further thruster. This would make quantitative comparisons among the outcome of three- and four-thrusters configurations more difficult. To mitigate this effect, in Equation (15), the two components F_{k_x} and F_{k_u} are normalized by the sum of the components of $\boldsymbol{Q}$ and $\boldsymbol{R}$, respectively.

Now, taking inspiration from certification procedures, several test cases are taken into account to assemble an aggregated performance value, richer in information than the outcome of a single simulation. In particular, different trim airspeeds in steady, horizontal forward flight are considered, and for each of them, a number of perturbations to the initial conditions. This produces a testing scenario where N_s simulations are carried out, yielding for the actual cost function the following expression.

$$J = \frac{1}{N_S} \sum_{k=1}^{N_S} (F_{k_x} + F_{k_u}). \tag{16}$$

The actual numerical settings for the analysis of the configurations, including the pool of simulations actually carried out, will be described in more detail in the following section devoted to numerical results. It can be anticipated that the proposed cost function in Equation (16) with the components in Equation (15) has been extensively tested, showing a good balance between sensitivity to the parameters appearing in the problem (i.e., geometrical parameters to assign the layout) and regularity with respect to their change. This will be shown in the results as well.

Based on the cost function just introduced, it is possible to formulate a corresponding optimal problem, where the function in Equation (16) is minimized by acting on a set of parameters.

It should be pointed out here that, since the focus of the analysis in this paper is on geometrical parameters defining the layout, these have been considered as optimization parameters. Geometrical parameters have a significant effect on dynamic performance, and understanding this very effect is part of the aim of the present work. This raises a methodological issue, however, since the best way to make use of an assigned (i.e., also geometrically sized) layout is in principle not decoupled from the tuning of the control gains. The latter, indeed, bears an effect on dynamic performance. However, in order to try to isolate the contribution to dynamic performance due to the geometrical sizing from that due to control gains, a two-step analysis procedure has been envisaged. In the first stage, for each layout, a set of control gains has been found, capable of stabilizing the system for three reference airspeeds, covering the performance span of the airship testbed. This rules out an airspeed-based gain scheduling, which is desirable also in terms of control simplicity and ease of design. Furthermore, in the same step, in order to make comparisons among different layouts fairer, such control gain matrices are all based on the same reference gain values. Considering gain matrices in Equations (12)–(14), this working hypothesis implies the following:

$$
\begin{aligned}
k^P_{(\cdot)_q} &= k^P_q, \quad k^I_{(\cdot)_q} = k^I_q, \\
k^P_{(\cdot)_p} &= k^P_p, \quad k^I_{(\cdot)_p} = k^I_p, \\
k^P_{(\cdot)_r} &= k^P_r, \quad k^I_{(\cdot)_r} = k^I_r,
\end{aligned}
\tag{17}
$$

where the six values appearing on the right hand side of Equation (17) are assigned and define all gain components in Equations (12)–(14). While it is not the only possible choice for easing comparisons, this has been deemed suitable a posteriori when comparing results produced in this research.

Once a set of reference gains $(k^P_q, k^P_p, k^P_r, k^I_q, k^I_p, k^I_r)$ has been selected such to stabilize all considered layouts (considering a baseline numerical sizing, i.e., an assigned set of thruster positioning, tilt, etc.) in all considered testing scenarios (trim speeds and perturbations), optimal analyses have been launched in a second step. Thanks to the generality of the designed cost function (Equations (15) and (16)), the optimal problem can be written as follows:

$$
\min_{z} J(z) \text{ s.t. } \boldsymbol{h}(z) \tag{18}
$$

and employed for different comparisons without any formal change to its definition. In Equation (18), the set of optimization parameters z depends on the layout, as detailed in the following Table 1. The problem in Equation (18) is constrained by a set $\boldsymbol{h}(z)$ of direct bounds and geometrical inequality constraints on the parameters, in consideration of their physical meaning, according to definitions in Figures 2 and 3.

In particular, bounds are defined in Table 1. Geometric inequality constraints instead express the need to keep the thrusters physically out of the surface of the envelope. The latter is mathematically assigned as an axial-symmetric surface, with a known radius r_h, function of the position along the longitudinal axis of the hull. The ensuing generic scalar form of the non-linear constraint just mentioned writes the following:

$$
a_i - (r_h(c_i) + d_h) > 0, \tag{19}
$$

which is replicated for each i-th thruster in the airship layout. In Equation (19), d_h is an assigned constant positive buffer distance.

Table 1. Sets of optimization parameters z and corresponding bounds according to geometrical definitions, for the three considered thruster layouts.

Layout ID	(A) z	(A) Bounds	(B) z	(B) Bounds	(C) z	(C) Bounds
	$a_{1,2}$	$a_{1,2_{lb}}, a_{1,2_{ub}}$	$a_{1,2}$	$a_{1,2_{lb}}, a_{1,2_{ub}}$	$a_{1,2}$	$a_{1,2_{lb}}, a_{1,2_{ub}}$
	a_3	$a_{3_{lb}}, a_{3_{ub}}$	a_3	$a_{3_{lb}}, a_{3_{ub}}$	$a_{3,4}$	$a_{3,4_{lb}}, a_{3,4_{ub}}$
	ζ	$0, \frac{\pi}{2}$	a_4	$a_{4_{lb}}, a_{4_{ub}}$	$\zeta_{1,2}$	$0, \frac{\pi}{2}$
	$c_{1,2}$	$c_{1,2_{lb}}, c_{1,2_{ub}}$	ζ	$0, \frac{\pi}{2}$	$\zeta_{3,4}$	$0, \frac{\pi}{2}$
	c_3	$c_{3_{lb}}, c_{3_{ub}}$	$c_{1,2}$	$c_{1,2_{lb}}, c_{1,2_{ub}}$	$c_{1,2}$	$c_{1,2_{lb}}, c_{1,2_{ub}}$
	$\lambda_{1,2}$	$0, \frac{\pi}{2}$	c_3	$c_{3_{lb}}, c_{3_{ub}}$	$c_{3,4}$	$c_{3,4_{lb}}, c_{3,4_{ub}}$
	λ_3	$-\frac{\pi}{2}, 0$	c_4	$c_{4_{lb}}, c_{4_{ub}}$	$\lambda_{1,2}$	$0, \frac{\pi}{2}$
			$\lambda_{1,2}$	$0, \frac{\pi}{2}$	$\lambda_{3,4}$	$-\frac{\pi}{2}, 0$
			λ_3	$-\frac{\pi}{2}, 0$		
			λ_4	$0, \frac{\pi}{2}$		

As observed, the choice of the bounds for λ in some cases reflect basic configuration choices, where the thruster alignment is not pointing towards the hull, to reduce the interference with the aerodynamic flow around it. Furthermore, the selection of the sign of the gains as in Equations (12)–(14) constrains the tilt orientation so that the resulting control action is coherent with the sign taken by the fed-back state variables. On the other hand, broad changes of the tilt angles are generally allowed, as shown.

More details concerning the implementation of the optimal problem will be provided in the results Section 5, but it can be anticipated here that the cost function introduced in Equation (16) has been tested in terms of regularity with respect to the elements of the array of optimization parameters $\boldsymbol{z}$ shown in Table 1, for each of the three considered layouts. The good regularity shown by the cost function has allowed to select a gradient-based method for the solution of the optimal problem in Equation (18), comprising bounds and non-linear constraining equation(s) as in Equation (19) (as many scalar equations as the number of thrusters in the considered configuration to be optimized, as explained).

5. Results

As a testbed for the methodologies introduced in the previous sections, the Lotte airship [22,23] has been selected as a baseline. This prototype airship has been extensively studied and characterized in previous research, and its main features are provided in Table 2 for reference. This unmanned airship is intended for low-altitude operations, and not as a HAA. However, the abundance of reliable and accurate data, seldom found in the literature for airships, makes it adequate for feeding realistic computations. Furthermore, the configuration is rather standard and not dissimilar from the expected one of airships to operate by design either in the higher atmosphere or closer to the ground—both interesting operative conditions, as stated in Section 1.

Table 2. Basic specifications of Lotte airship [22].

Parameter	Value
Volume $\mathcal{V}$ (m^3)	107.42
Length (m)	16.0
Max. diameter (m)	4.0
Tail span (m)	4.6
Mass m (kg)	136.8
Max. airspeed (m/s)	12

A virtual non-linear model of the airship has been assembled populating the formulation described in Section 2.

A major difference included with respect to the baseline is obviously in the thrusters. The real prototype features a single motor-propeller assembly in the tailcone. This has been taken out of the design, and substituted with three or four thrusters, for layout (A) and (B), (C) respectively. Each thruster has been hypothesized to be an electric motor-propeller assembly, and provide a nominal thrust of $\tilde{T}_i = 400$ N each, and a mass of 2 kg. Considering Equation (9), the values of the control variables δ_{T_i} have been limited between $[-1, 1]$, where negative values are achievable for electric motors. Furthermore, on account of the starkly lower expected efficiency of the propeller in reverse rotation conditions (i.e., for $\delta_{T_i} < 0$), the shape function has been coarsely set to $\tilde{K}_i = 0.5$ for $\delta_{T_i} < 0$, and $\tilde{K}_i = 1.0$ otherwise. The inertial data (mass and position) of the original single motor have been used to modify that of the original airship, in terms of m, $\boldsymbol{r}_{CG}$, $\boldsymbol{S}_{CB}$, $\boldsymbol{J}_{CB}$. The new thrusters have been included modifying the airship inertia correspondingly, modeling them as point masses, accurately placed according to their actual positions $\boldsymbol{r}_{P_{T_i}}$ in the considered layout. The overall mass change—taking out the original thruster and replacing it with others—is sufficiently small to provide a negligible change in the buoyancy ratio when leaving the volume of the envelope at its original value.

Another difference with respect to the baseline is in the tail. The tail assembly has been retained, but the surfaces have been considered to have only a stabilizing function, and no moving parts.

As anticipated in the formulation Section 2, the aerodynamic forcing terms have been characterized computing the stability derivatives in Equation (4) using analytic and semi-empirical methods [22,26], according to an accurate description of the specific geometry of the hull and fins.

5.1. Quality Assessment of the Stability Augmentation System

In order to select the values of the gains and generally check the suitability of the proposed control action, a testing scenario has been set up as follows.

Starting from the airspeed range of the original airship, the conditions of horizontal forward flight have been considered as reference at 4, 8, 12 m/s, at an altitude of 200 m.

For each of the three considered layouts, a set of geometrical values completely assigning the configuration has been tested. According to Table 1, eight parameters have been chosen for layouts (A) and (C), and ten parameters for (B). Geometrical settings have been experimented with broadly so as to test a specific set of gains on different geometries, in view of the next optimization phase, where geometrical parameters are free to change between wide bounds. Table 3 displays the values adopted for Figures 7 and 8 to follow. They do not correspond to the satisfaction of any optimality criterion and are presented as an example of a possible sub-optimal choice.

Table 3. Geometry settings for all considered configurations for SAS performance study.

Layout ID	(A) Param.	(A) Value	(B) Param.	(B) Value	(C) Param.	(C) Value
			$a_{1,2}$	2 m		
	$a_{1,2}$	2.2 m	a_3	2.1 m	$a_{1,2}$	2.0 m
	a_3	2.0 m	a_4	2.2 m	$a_{3,4}$	2.2 m
	ζ	45 deg	ζ	45 deg	$\zeta_{1,2}$	45 deg
	$c_{1,2}$	5.0 m	$c_{1,2}$	5.0 m	$\zeta_{3,4}$	45 deg
	c_3	6.0 m	c_3	3.0 m	$c_{1,2}$	5.0 m
	$\lambda_{1,2}$	5 deg	c_4	6.0 m	$c_{3,4}$	6.0 m
	λ_3	−20 deg	$\lambda_{1,2}$	5 deg	$\lambda_{1,2}$	20 deg
			λ_3	−20 deg	$\lambda_{3,4}$	−20 deg
			λ_4	10 dcg		

For each reference speed, the free (i.e., uncontrolled) models corresponding to the three layouts have been trimmed by solving Equation (1) in steady state ($\dot{\boldsymbol{w}}_{CB} = \boldsymbol{0}$) in symmetric, horizontal, forward flight. The trim solution comes in terms of the body components of airspeed U_r and W_r, as well as a pitch angle ϑ_r, all other components of the state array having been forcibly set to zero. In terms of controls, equilibrium values $\delta_{T_{i_r}}$ are computed in the solution of the trim problem.

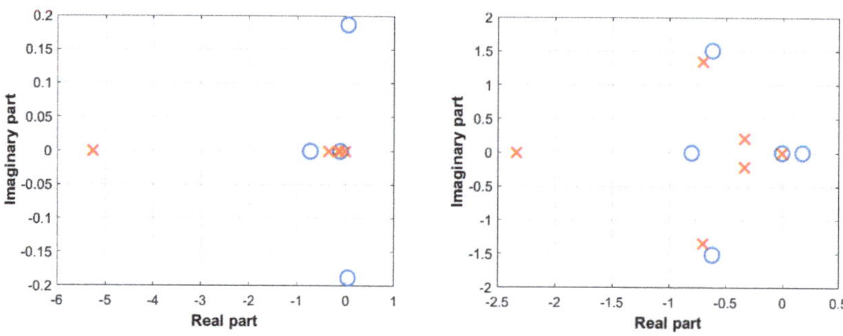

Figure 7. *Cont.*

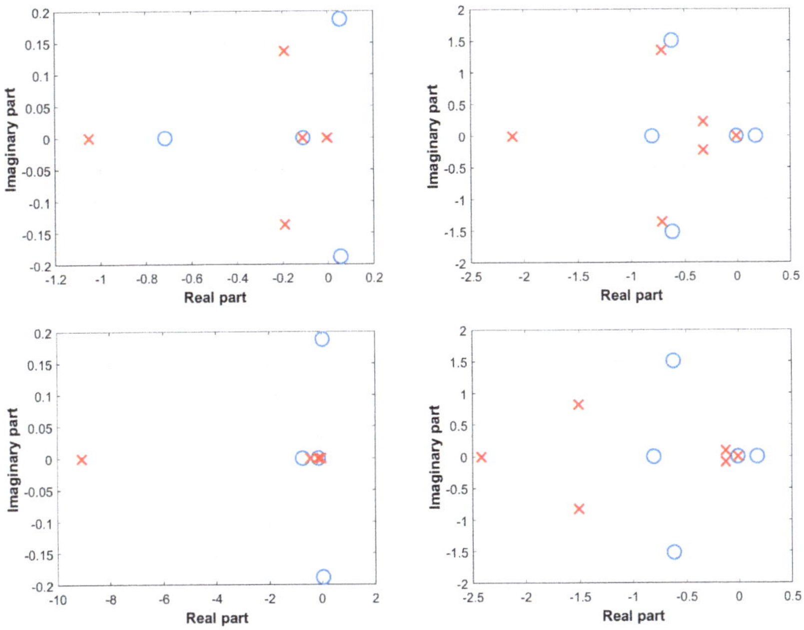

Figure 7. Maps of eigenvalues for free (blue circles) and artificially stabilized (red crosses) systems, trim speed 8 m/s. **Left**: longitudinal dynamics. **Right**: lateral-directional dynamics. **Top row**: layout (A). **Mid row**: layout (B). **Bottom row**: layout (C).

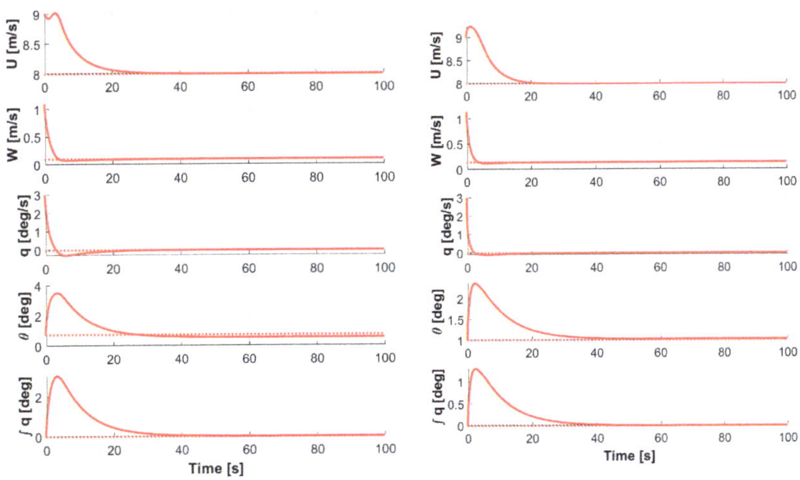

Figure 8. *Cont.*

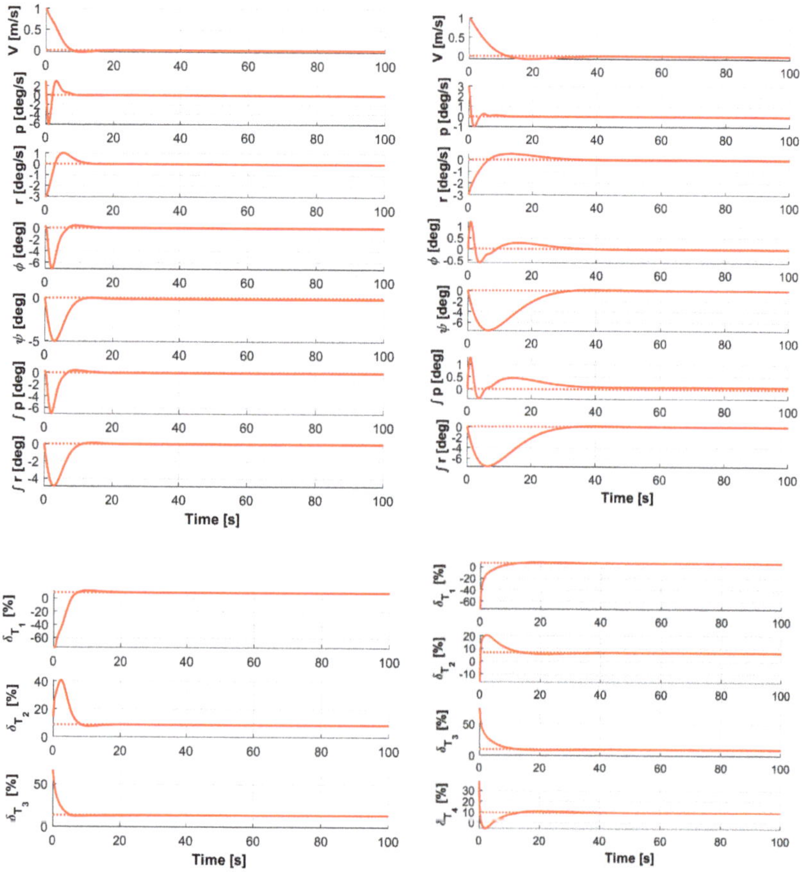

Figure 8. Time-marching simulation of the non-linear airship system subject to artificial stabilization. **Left**: layout (A). **Right**: layout (C). Forward flight at 8 m/s, with a perturbation in the initial condition of the states. **Top row**: longitudinal states. **Mid row**: lateral-directional states. **Bottom row**: controls. Dotted line: trim solution.

As previously stated, checking the control law has been performed both through a linear and a non-linear analysis. For the former, decoupled linearized models have been obtained for each of the three layouts and for each of the trim speeds (where the latter imply a change in the stability derivatives as well as trim values of the states), for longitudinal and lateral-directional dynamics. Similarly to winged aircraft, decoupling is achievable under the hypothesis of steady, horizontal, symmetric forward flight. The corresponding analytic models, taken from the literature and amended to account for integral states in the SAS feedback, are reported in Appendix A for completeness. The eigendynamics of the free system are compared to those for the artificially stabilized system, assessing the change in the positions of the eigenvalues.

As an example, in Figure 7, the free-response and stabilized response of the longitudinal and lateral-directional systems are shown for the three layouts and the corresponding SASs, for a speed of 8 m/s. Qualitatively, similar results are obtained for other considered trim speeds (not shown for brevity).

It can be observed how the eigenvalues of the controlled response are significantly damped both in terms of longitudinal and lateral-directional dynamics. Considering longitudinal dynamics, in the presented examples, the marginally unstable pendulum mode is stabilized and damped while retaining an oscillatory nature (layout (B)) or even damped to produce real stable eigenvalues (layouts (A)–(C)). A neutrally stable eigenvalue appears, associated to the integral state $\int q$ in the state matrix of the stabilized system (see matrices in Appendix A). Concerning lateral-directional motion, the unstable sideslip subsidence eigenvalue and neutrally stable pure-yaw eigenvalue (null mode) are substituted by a stable, well-damped oscillatory motion. Roll oscillation, already stable in all considered cases, is slightly further damped. The appearance of neutrally stable eigenvalues in the controlled case is again due to the integral states $\int p$ and $\int r$ in the augmented state of the SAS-controlled system (see Appendix A).

Further stability and performance analyses have been carried out on the non-linear system, by checking the response of the system to perturbations in the initial conditions or in the controls with respect to the trim condition, when performing time-marching simulations. A Runge-Kutta integration scheme was selected to run the integration on a time window of $T_f = 100$ s from the initial condition.

As an example, in Figure 8, the time histories of the longitudinal and lateral-directional states, and of the controls, are shown for layouts (A) and (C) (with the geometry proposed in Table 3), considering a simultaneous perturbation of the initial condition on the state as $\Delta U_0 = +1$ m/s, $\Delta V_0 = +1$ m/s, $\Delta W_0 = -1$ m/s, $\Delta p_0 = +3$ deg/s, $\Delta q_0 = +3$ deg/s, $\Delta r_0 = -3$ deg/s, where subscript $(\cdot)_0$ defines the initial condition of the simulation.

From time-marching simulations it is possible to better assess the actual deviation from the reference of the states, and correspondingly the control effort—both contributing to the merit function in Equation (15). The gains selected by means of the eigenvalue analysis could be thus further checked, deeming the action of the controller realistic (especially not too intense) and the state oscillation acceptable. To the aim of further avoiding unrealistically intense action of the controller, the SAS contribution has been subjected to a saturation to a threshold value which is 25% lower than the actual top value of the thrust control variables (i.e., 0.75 instead of 1.0). This should simulate leaving some spare control authority for pilot action or for more intense control activity, in case of perturbations more intense than those actually simulated.

Comparing the two layouts (A) and (B) in Figure 8, the performance of the corresponding stabilization systems is clearly somewhat different, as expected. Actually, the meaning of controls (bottom row), and consequently their use, is different by design in the two cases. Looking at the states however, it can be noted that a good degree of stability is confirmed in both conditions, as forecast by the eigenvalue analysis (carried out on decoupled, linearized systems).

It can be further observed from Figure 8 that the states are driven to the trimmed steady-state value following perturbation, with a relatively limited action of the controllers.

As already pointed out, the presented structure of the SAS system has produced acceptable results without much effort in the tuning of the gains. This in turn has allowed to introduce the hypothesis in Equation (17) so that, irrespective of the layout, the gain coefficients are numerically based on a set of three values of proportional gains and three values of integral gains, collected in $(k_q^P, k_p^P, k_r^P, k_q^I, k_p^I, k_r^I)$. As explained, the latter hypothesis eases mutual comparisons among layouts, by allowing to exclude the intensity of the control action from the changing parameters in the parameter analysis and optimization, linking any differences in performance to changes in the airship sizing.

By performing analyses similar to those shown in Figures 7 and 8 on a set of six sets of initial conditions, for changing values of all geometrical parameters defining the layout (i.e., the position and attitudes of the thrusters), and across all three considered layouts (A), (B) and (C), it was checked that the response of the system did not diverge, keeping the gains $(k_q^P, k_p^P, k_r^P, k_q^I, k_p^I, k_r^I)$ fixed. The actual gain selection has been carried out deploying a trial and error procedure, where gains providing acceptable results on layout (A) have been tried on the other two considered layouts. Following this check, the set of gains has been consolidated, and kept frozen for the optimal analyses described next, which have been invariably run on the artificially stabilized system.

5.2. Optimal Analyses

A major ambition of the presented optimal methodology is that of finding the optimal positioning of the thrusters for a certain layout of the thrusters, and possibly ranking the performance of a layout compared to another in easing the stabilization task.

In order to first assess the suitability of the proposed cost function in Equation (16) before attempting the numerical solution of an optimal problem on it, such cost function has been computed for a set of $N_s = 18$ simulations, corresponding to as many sets of perturbed initial conditions, separately for the trim speeds of 4, 8, and 12 m/s, on all three layouts. For each layout, parameterized analyses have been carried out, running the simulations changing one of the geometrical parameters in Table 1 at a time, and computing the cost function to check its regularity. All controls, speed and rate components contribute equally to the cost function. As an example, in Figure 9 are shown some of the analyzed dependencies of the cost function, computed from $N_s = 18$ simulations corresponding to as many initial conditions, at 12 m/s for layout (B). In particular, starting from the geometry in Table 3 for layout (B), the plots in Figure 9 have been obtained changing $a_{1,2}$, c_4, $\zeta_{1,2}$, and λ_3.

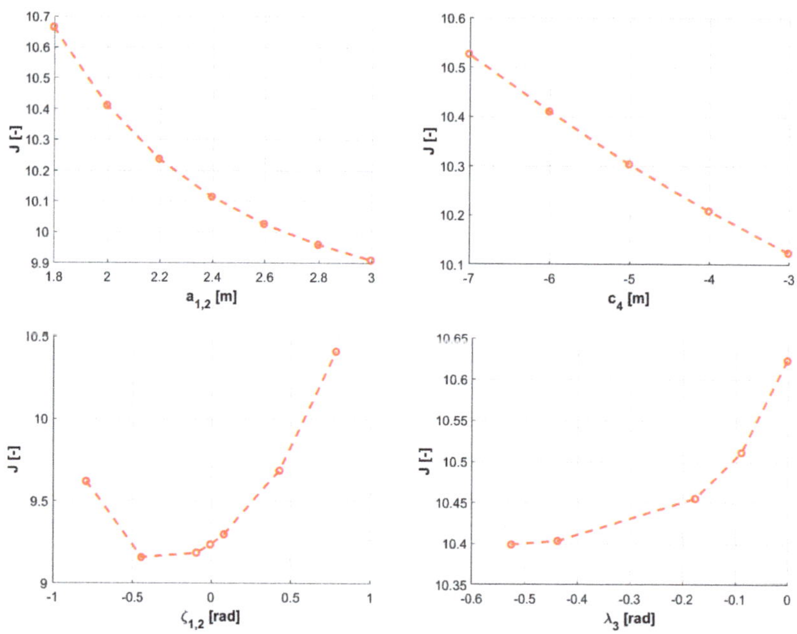

Figure 9. Example studies on the sensitivity of the cost function with respect to parameters $a_{1,2}$, c_4, $\zeta_{1,2}$, and λ_3 for layout (B). Trim speed 12 m/s. Computation of the cost function based on $N_s = 18$ simulations corresponding to as many perturbed initial conditions.

It should be remarked that the examples in Figure 9 portray the dependence of the cost function from a single parameter at a time, leaving all others to an assigned (unchanged) value. Therefore, such results are related to a scenario much more limited with respect to the one faced by an optimization algorithm, which changes *all* geometric quantities within an iteration. However, it can be observed from Figure 9 that the cost function appears very regular with respect to the changing parameters. With this notion, as anticipated in Section 4, a gradient-based method capable of dealing with non-linear constraints was deemed suitable for an optimal search.

Within an optimization loop, the optimizer sets the parameters in Table 1 (corresponding to the considered layout), adjusts the inertia of the airship correspondingly, finds a trim solution, then performs N_s time-marching simulations of the non-linear system subject to artificial stabilization, and finally computes the cost function in Equation (16). The optimization algorithm tunes the geometrical parameters in search for the minimum of the cost function, and compliant with the bounds in Table 1 and constraint(s) in Equation (19).

The output of an optimization is a set of geometrical parameters (optimal solution) corresponding to the optimum of the cost function, as well as the corresponding value of the cost function.

For a more thorough exploration of the dependence of the optimum from the working condition of the airship, for each considered layout, an optimization was first run for each of the three considered reference trim speeds. The bounds in Table 1 have been set specifically to the following:

- $a_{1,2_{lb}} = a_{3_{lb}} = a_{3,4_{lb}} = a_{4_{lb}} = 0.5$ m, $a_{1,2_{ub}} = a_{3_{ub}} = a_{3,4_{ub}} = a_{4_{ub}} = 2.5$ m
- $c_{1,2_{lb}} = c_{3_{lb}} = c_{3,4_{lb}} = c_{4_{lb}} = 2.0$ m, $c_{1,2_{ub}} = c_{3_{ub}} = c_{3,4_{ub}} = c_{4_{ub}} = 8.0$ m

and distance d_h in Equation (19) is set to $d_h = 0.8$ m, meaning a minimum clearance of the thrusters from the hull's surface set to that value.

The behavior of the optimal solution as a function of the trim airspeed (4, 8 and 12 m/s) is shown for the three layouts in Figures 10–12. The plots of the optimal geometric variables are reported side-by-side with those of the trim solutions, for the same airspeeds and geometrical settings.

In the case of layout (A) (Figure 10), it can be observed that except for a reduction at higher airspeeds of the arm of the lower thrusters ($a_{1,2}$), the optimal solution pushes the radial position of the thrusters to the boundary value. The same happens for the longitudinal position of the top thruster (c_3), which is placed as far back to the tail as possible. Conversely, the longitudinal position of the two bottom thrusters ($c_{1,2}$) as well as the dihedral angle $\zeta_{1,2}$ are significantly changed as functions of the airspeed. This is in accordance with the behavior of the trim solution (right plot). Interestingly, the optimal behavior of the tilt angle $\lambda_{1,2}$ of the lower thrusters versus the airspeed is such to keep them basically aligned with the longitudinal body axis. Conversely, the top thruster is tilted by a significant angle $\lambda_3 \approx -70$ deg. This configuration of the tilt angles can be matched with the control trim values, where it can be noticed that δ_{T_1} and δ_{T_2} are set to generally much higher values than δ_{T_3}. Qualitatively, it appears that the optimal solution for layout (A) features two side thrusters mainly responsible for propulsion, and a top one, put at a significant angle with respect to the airship longitudinal axis, providing pitch control and little thrust. Another remark concerning the null value of $\lambda_{1,2}$ is related to roll control. As previously stated, a non-null down-tilt of the side thrusters was envisaged for granting control authority around the roll axis. This does not appear to be required in an optimal perspective (i.e., when weighting the deviation of p and $\int p$ together with that of other state and control variables), delegating the control around roll axis to roll-yaw coupling, achieved by means of airship eigendynamics or through the combined action of the side and top thrusters.

Considering layout (B) (Figure 11), the longitudinal position of the thrusters ($c_{1,2}$, c_3 and c_4) is remarkably similar for all of them and far from the bounds, displaying only mild changes for different airspeeds. A similar behavior is found also in the radii of the lateral and top thrusters ($a_{1,2}$, a_3), whereas for the bottom thruster the value (a_4) is generally lower and more variable with the airspeed. Again, the side thrusters are set at an almost null tilt angle $\lambda_{1,2} \approx 0$ deg, and are associated to a very limited dihedral $\zeta_{1,2}$, i.e., a basically horizontal mounting for these thrusters. As remarked for layout (A), roll authority is delegated to the intrinsic roll-yaw coupling and the combined action of top, bottom and side thrusters. Similarly to layout (A), the top and bottom thrusters are associated with larger tilt angles λ_3 and λ_4, albeit more limited here than for layout (A), yet the pushing action appears more equally distributed among all thrusters at trim.

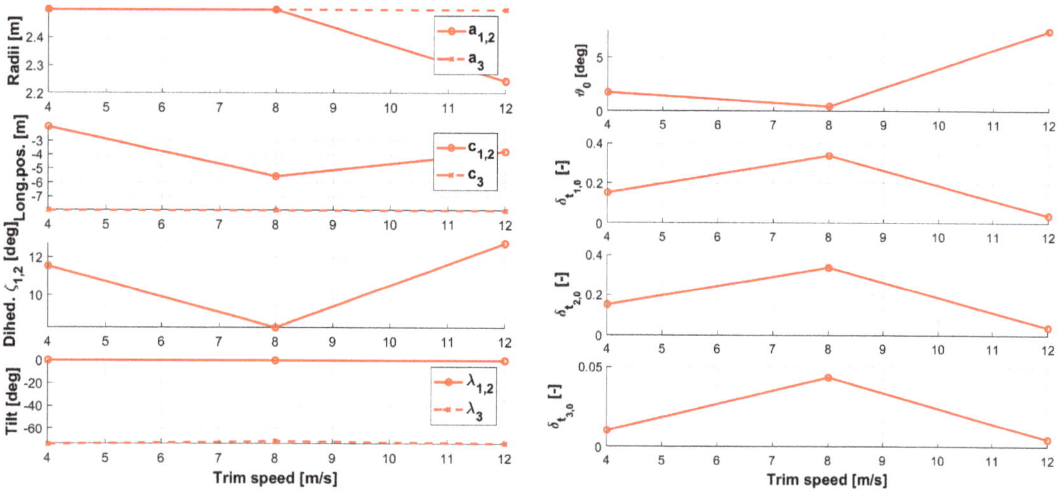

Figure 10. Optimal solutions (**left**) and trim results for optimizations at different airspeeds on layout (A).

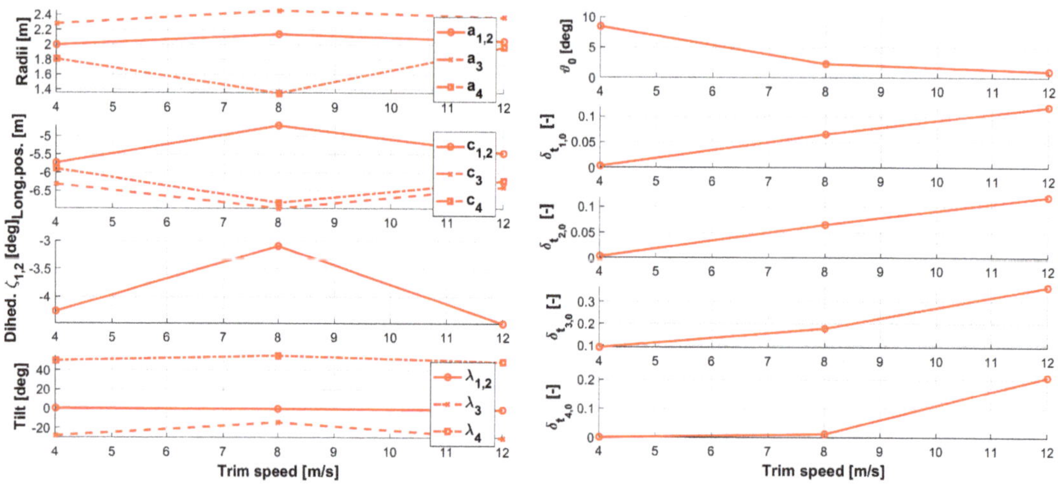

Figure 11. Optimal solutions (**left**) and trim results for optimizations at different airspeeds on layout (B).

Finally, concerning layout (C) (Figure 12), the results show an interesting homogeneity of the bottom (#1, #2) and top thrusters (#3, #4). In particular, the radii ($a_{1,2}$, $a_{3,4}$) are very similar, and slightly increasing towards the upper bound for higher airspeeds. The longitudinal position is moderately changing with respect to the airspeed, and the tilt angles are such to provide a top-bottom almost-symmetry. Compared to previous layouts, this one appears more promising in terms of roll control authority, being potentially capable of making use of direct control. The dihedral angles $\zeta_{1,2}$ and $\zeta_{3,4}$ are remarkably different for the top and bottom couples, especially for increasing values of the airspeed, and for the top couple $\zeta_{3,4}$ the corresponding optimal values are changing significantly over the considered airspeed domain.

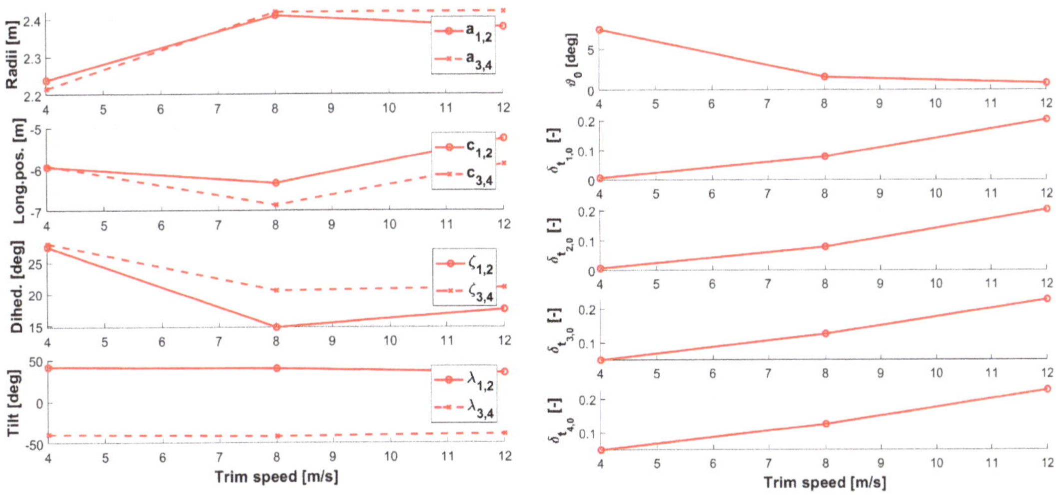

Figure 12. Optimal solutions (**left**) and trim results for optimizations at different airspeeds on layout (C).

From a numerical standpoint, checks on the robustness of the solution of the proposed optimization algorithm (gradient-based) have been performed before consolidating the results just shown for each configuration (Figures 10–12), in particular perturbing the starting condition of the optimization process, so as to reduce the chance to converge to possible local minima. No such issue has emerged however, in accordance with the indications obtained in the preliminary analysis of the cost function (Figure 9), which, albeit partial, shows (if any) only one minimum between the boundaries of the cost function domain.

Following the optimal analyses on each trim speed shown above, for each layout an optimization on the full airspeed domain has been carried out, where for the evaluation of the cost function in the optimal loop 18 simulations for as many initial conditions are performed for all 3 reference airspeeds, yielding a total of $N_s = 54$ simulations per function evaluation. The result of the corresponding analysis on each configuration is displayed in terms of the optimal value of the cost function in Figure 13, where these values are compared to the results of the optimizations carried out on each single trim condition for each layout (corresponding to the optimal results shown in Figures 10–12).

Two remarks can be proposed on the results in Figure 13. Firstly, considering each layout singularly, it can be noticed that the value of the cost function is not strongly dependent on the trim airspeed, and generally similar also for the corresponding global (i.e., over all airspeeds) optimization as well. This indicates that a single setting of the geometry may be close to optimal (i.e., slightly sub-optimal) irrespective of the airspeed regime encountered by the airship, at least considering forward flight (not hover). Secondly, the optimal values of the cost function achieved with layout (A), (B) and (C) are decreasing precisely in this order. In view of the structure of the cost function, thoroughly described in Section 4, this might indicate that an optimal (C)-type layout is more performing than an optimal (B)-type, and both are better than an optimal (A)-type layout. Of course, from a design perspective, this potentially relevant result should be weighed with respect to the adoption of a cost function specifically targeting dynamic performance, and which does not take into account—for instance—the potentially higher cost of having more thrusters (as in layouts (B) and (C), compared to (A)).

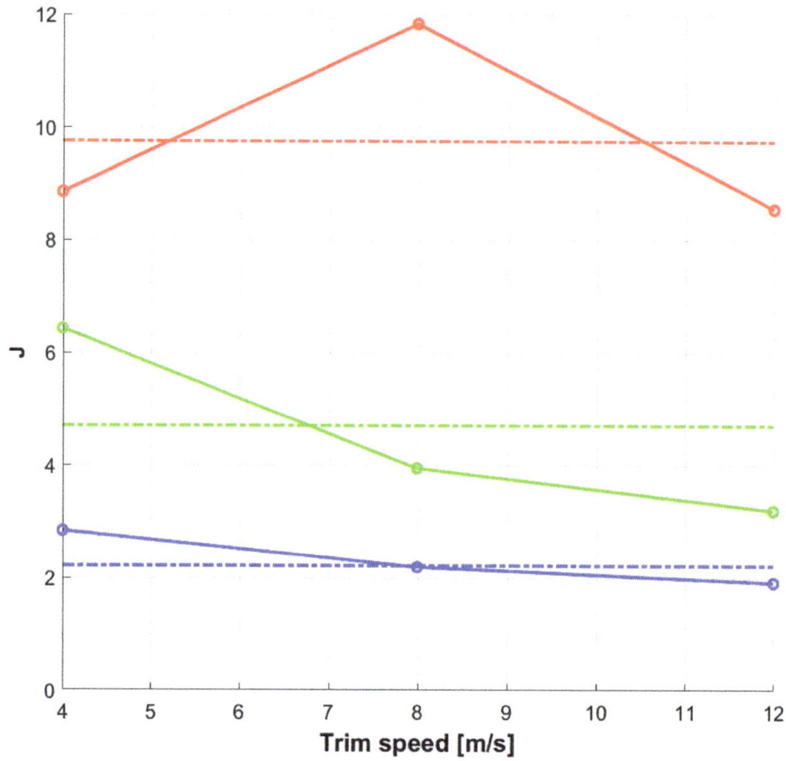

Figure 13. Optimal cost function for optimizations at different airspeeds (continuous line), and for global optimizations on all airspeeds (dash-dotted horizontal line). Red: layout (A). Green: layout (B). Blue: layout (C).

To better appreciate the features of the optimal configurations, Figure 14 provides a view of the sizing corresponding to the global optimal solutions for the three layouts, numerically reported in Table 4.

Table 4. Optimal results from global optimizations, for the three considered layouts.

Layout	(A)		(B)		(C)	
ID	Param.	Value	Param.	Value	Param.	Value
	$a_{1,2}$	2.24 m	$a_{1,2}$	2.10 m	$a_{1,2}$	2.48 m
	a_3	2.49 m	a_3	2.35 m	$a_{3,4}$	2.48 m
	ζ	13 deg	a_4	1.95 m	$\zeta_{1,2}$	14 deg
	$c_{1,2}$	3.77 m	ζ	−1 deg	$\zeta_{3,4}$	20 deg
	c_3	8.0 m	$c_{1,2}$	5.58 m	$c_{1,2}$	6.07 m
	$\lambda_{1,2}$	0 deg	c_3	5.98 m	$c_{3,4}$	7.79 m
	λ_3	−73 deg	c_4	5.78 m	$\lambda_{1,2}$	38 deg
			$\lambda_{1,2}$	0 deg	$\lambda_{3,4}$	−40 deg
			λ_3	−19 deg		
			λ_4	59 deg		

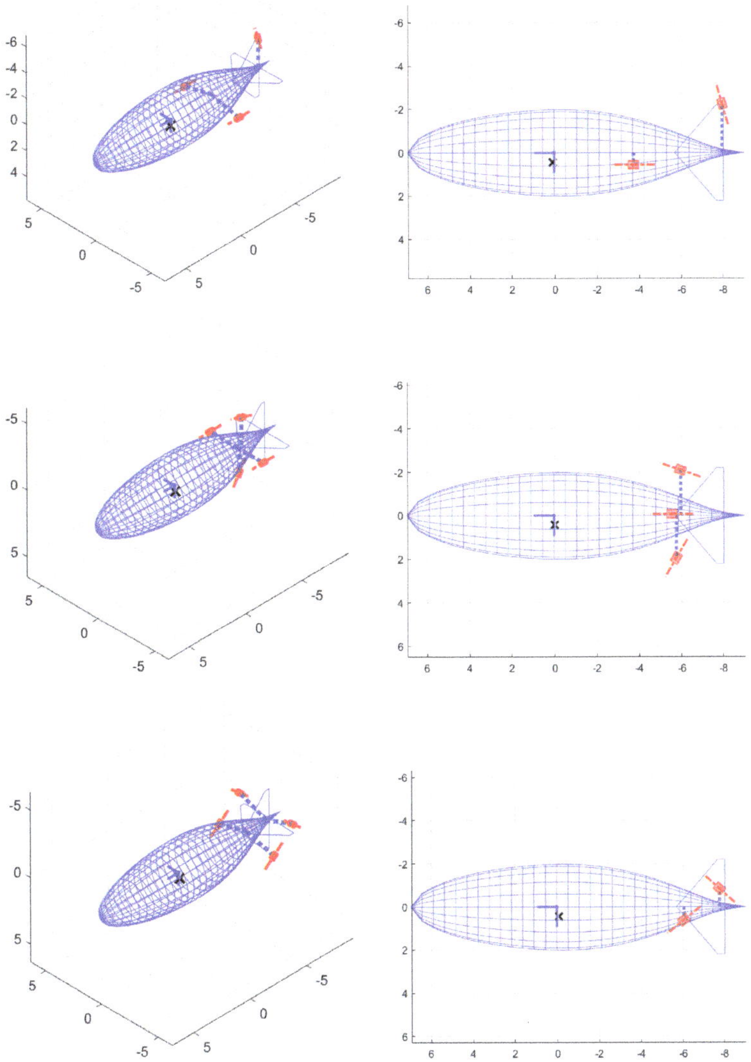

Figure 14. Geometric configuration resulting from global optimization (i.e., over the airspeed spectrum) on each configuration. **Top row**: layout (A). **Mid row**: layout (B). **Bottom row**: layout (C). **Left**: three-quarter view. **Right**: view from the airship left. Length unit: meters.

6. Conclusions

In this paper, we have studied and compared different configurations for thrust-based control on an airship. The major conceptual driver behind this interest, as an alternative to other forms of control (including thrust vectoring, currently deployed on some flying airships) is in the chance to increase design simplicity and reliability, by reducing the number of moving parts and actuators on-board.

The selected geometric arrangement of the airship is standard, and quantitative computations are based on an existing airship (i.e., the Lotte experimental airship [22,23]) as a baseline. However, with respect to the real airship, a virtual model lacking the control degrees of freedom on the horizontal and vertical tail, as well as the propulsion unit, has been synthesized and implemented. Control and propulsion have been delegated to thrusters, placed according to three possible layouts—of which one is a three-thrusters configuration, and the other two feature four thrusters. The latter have been shortlisted based on simplicity and cost minimization criteria, on the one hand, and control authority considerations on the other.

In view of prolonged, unmanned and autonomous flight missions, stability augmentation systems (SASs) have been designed for the proposed configurations, making use of rotational rate components as measures. Control systems have been designed for longitudinal and lateral-directional dynamics, deploying a proportional-integral control law, producing stabilizing values of the thrust settings for the thrusters.

In trying to separate the effect of layouts on dynamic performance from the effect of control on the same performance indices, care has been taken to reduce the set of gains to be tuned and to express the gain matrices for the SAS of all layouts as a function of a minimal, cross-layout set.

The very value of the gains has been selected based on an articulated trial and error procedure, where the eigenanalysis of the free and artificially stabilized systems, computed on linearized representations of the decoupled longitudinal and lateral-directional dynamics, have been compared, and also time-marching simulations have been carried out on the complete non-linear model of the airship, to further check the intensity (and therefore the realism) of the control action and state oscillations. The gain synthesis has been based on acceptable stabilization results obtained from the same minimal set of gains applied to all three layouts. The tuning process has turned out to be little time-consuming, bearing good results (i.e., stabilization and dampening of oscillations, at the price of a moderate control action), which tends to support the goodness of the proposed SAS structure.

Having assigned the gain matrices, the focus has been put on the definition of the geometry of the proposed layouts. In order to evaluate the performance on dynamics, an emended form of the classical state vs. control energy functional has been prepared. This cost function is intended to be computed from a set of dynamic simulations carried out on the non-linear system starting from a trimmed forward-flight condition, and corresponding to a variety of perturbed initial conditions. However, the cost function should not be touched by the specific settings of the simulation scenario, including obviously the number of simulations, as well as the number of thrusters in the layout—a crucial feature of the cost function, if a cross-layout comparison is of interest.

For each layout, a set of geometric variables capable of completely assigning the geometry has been defined. After showing that the proposed cost function is regular with respect to the geometric quantities selected as parameters, an optimal analysis has been carried out.

Firstly, each configuration has been optimized for three different airspeeds, spanning between the typical operational limits of the original testbed. Results from this trials show quite invariably for the three-thrusters layout (A) the need to markedly tilt the thruster in the plane of symmetry, while leaving the other two basically aligned with the longitudinal axis. The former thruster is, therefore, employed similarly to the elevator on the tail, whereas the other two mostly take the role of the only thruster on the original testbed, providing propulsion, plus achieving lateral-directional control. A similar result is obtained for the four-thrusters layout (B), where the bottom thruster is again markedly tilted with respect to the longitudinal axis of the airship, whereas the top thruster is more mildly tilted, and contributes more intensely to propulsion in trim. The last four-thrusters layout (C) bears a symmetrical tilting of the thrusters, yet they are not positioned at the same longitudinal station when moving significantly faster than hover—a likely effect of an optimal compensation of gravity moment by means of aerodynamic and thrust forces.

Finally, a global optimization of each layout, based on simulations carried out at different trim speeds, has produced corresponding sizing solutions. A comparison of the outcome of the performance of the layouts, according to the selected cost function, has been presented, bearing as a result an apparent performance hierarchy, where layout (A), (B), and (C) perform better with respect to one another in the cited order. This tends to suggest that—from a standpoint of dynamic performance, in the sense associated to the definition of the considered cost function—layout (C) is the most recommendable. It also features the slightest sensibility with respect to the trim speed, when considering airspeed-bound optimizations.

Besides producing interesting quantitative results, the proposed approach allows driving the design of the configuration following dynamic performance considerations, deploying a highly automatable algorithm, which may be of interest for speeding up the preliminary design phase. Clearly, the same procedure could be emended to carry out computations in more scenarios of special interests, e.g., including specific simulations of disturbances to initial conditions or induced by the pilot, or at different altitudes.

Of course, when finalizing the sizing of in a preliminary design, specific gains may be chosen to further increase dynamic performance, resulting from a finer tuning process. However, considering the standpoint adopted in this paper, where a quantitative comparison *among configurations* is in the focus, it was felt that the proposed approach could produce results of stronger value and easier to understand.

Author Contributions: Both Authors contributed equally to the production of the research and the preparation of the present article. All authors have read and agreed to the published version of the manuscript.

Funding: This research received no external funding.

Data Availability Statement: Data produced in this research are partially available upon request to the corresponding Author.

Conflicts of Interest: The Authors declare no conflicts of interest.

Appendix A. Matrices of the Linearized System

As pointed out in Section 5.1, linearized systems have been deployed to carry out the eigenanalysis of the airship dynamics for the decoupled longitudinal and lateral-directional dynamic equilibria. Based on the proposed formulation for the non-linear dynamic equations, decoupled linearized systems can be obtained around a symmetric, steady, horizontal trimmed flight condition. Considering the state arrays $\boldsymbol{x}_L = \{U, W, q, \vartheta\}$ for longitudinal dynamics and $\boldsymbol{x}_D = \{V, p, r, \varphi, \psi\}^T$ for lateral-directional dynamics, the corresponding linearized dynamic equations can be written in the form $\boldsymbol{M}_L\Delta\dot{\boldsymbol{x}}_L + \boldsymbol{G}_L\Delta\boldsymbol{x}_L = \boldsymbol{U}_L\Delta\boldsymbol{u}_L$ and $\boldsymbol{M}_D\Delta\dot{\boldsymbol{x}}_D + \boldsymbol{G}_D\Delta\boldsymbol{x}_D = \boldsymbol{U}_D\Delta\boldsymbol{u}_D$. Here, $\boldsymbol{u}_L$ and $\boldsymbol{u}_D$ are control arrays, which are composed of either three or four scalars δ_{T_i}, as described in the body of the text (Section 2.1), depending on the considered layout. The linearized equations of motion are based on the following matrices for longitudinal dynamics:

$$\boldsymbol{M}_L = \begin{bmatrix} m+\rho\mathcal{V}k_1 & 0 & mr_{CG_3} & 0 \\ 0 & m+\rho\mathcal{V}k_3 & -mr_{CG_1} & 0 \\ mr_{CG_3} & -mr_{CG_1} & J_{CB_y}(1+\rho k') & 0 \\ 0 & 0 & 0 & 1 \end{bmatrix} \tag{A1}$$

$$\boldsymbol{G}_L = -\begin{bmatrix} X_U & X_W & X_q - mW_0 & (\rho\mathcal{V}-m)g\cos\vartheta_0 \\ Z_U & Z_W & Z_q + mU_0 & (\rho\mathcal{V}-m)g\sin\vartheta_0 \\ M_{CB_U} & M_{CB_W} & M_{CB_q} - mU_0r_{CG_1} - mW_0r_{CG_3} & mg(r_{CG_1}\sin\vartheta_0 - r_{CG_3}\cos\vartheta_0) \\ 0 & 0 & 1 & 0 \end{bmatrix} \tag{A2}$$

$$\boldsymbol{U}_L = \begin{bmatrix} X_{\delta_{T_1}} & X_{\delta_{T_2}} & X_{\delta_{T_3}} & X_{\delta_{T_4}} \\ Z_{\delta_{T_1}} & Z_{\delta_{T_2}} & Z_{\delta_{T_3}} & Z_{\delta_{T_4}} \\ M_{CB_{\delta_{T_1}}} & M_{CB_{\delta_{T_2}}} & M_{CB_{\delta_{T_3}}} & M_{CB_{\delta_{T_4}}} \\ 0 & 0 & 0 & 0 \end{bmatrix}, \tag{A3}$$

where in Equation (A3), four scalar inputs have been considered. For a three-thruster layout, the last column would be simply removed.

For lateral-directional dynamics, the following matrices hold.

$$\boldsymbol{M}_D = \begin{bmatrix} m + \rho \mathcal{V} k_2 & -m r_{CG_3} & m r_{CG_1} & 0 & 0 \\ -m r_{CG_3} & J_{CB_x} & J_{CB_{xz}} & 0 & 0 \\ m r_{CG_1} & J_{CB_{xz}} & J_{CB_z}(1+\rho k') & 0 & 0 \\ 0 & 0 & 0 & 1 & 0 \\ 0 & 0 & 0 & 0 & 1 \end{bmatrix} \tag{A4}$$

$$\boldsymbol{G}_D = -\begin{bmatrix} Y_V & Y_p + mW_0 & Y_r - mU_0 & -(\rho\mathcal{V} - m) g \cos(\vartheta_0) & 0 \\ 0 & L_{CB_p} - mW_0 r_{CG_3} & mU_0 r_{CG_3} & -mg r_{CG_3} \cos\vartheta_0 & 0 \\ N_{CB_V} & mW_0 r_{CG_1} & N_{CB_r} - mU_0 r_{CG_1} & mg r_{CG_1} \cos\vartheta_0 & 0 \\ 0 & 1 & \tan\vartheta_0 & \tan\vartheta_0 q_0 & 0 \\ 0 & 0 & 1/\cos\vartheta_0 & q_0/\cos\vartheta_0 & 0 \end{bmatrix} \tag{A5}$$

$$\boldsymbol{U}_D = \begin{bmatrix} Y_{\delta_{T_1}} & Y_{\delta_{T_2}} & Y_{\delta_{T_3}} & Y_{\delta_{T_4}} \\ L_{CB_{\delta_{T_1}}} & L_{CB_{\delta_{T_2}}} & L_{CB_{\delta_{T_3}}} & L_{CB_{\delta_{T_4}}} \\ N_{CB_{\delta_{T_1}}} & N_{CB_{\delta_{T_2}}} & N_{CB_{\delta_{T_3}}} & N_{CB_{\delta_{T_4}}} \\ 0 & 0 & 0 & 0 \\ 0 & 0 & 0 & 0 \end{bmatrix}. \tag{A6}$$

Similarly to Equation (A3), Equation (A6) has been written accounting for four scalar inputs.

By defining the state matrices $\boldsymbol{A}_L = -\boldsymbol{M}_L^{-1}\boldsymbol{G}_L$, $\boldsymbol{A}_D = -\boldsymbol{M}_D^{-1}\boldsymbol{G}_D$, it is possible to readily perform the eigenanalysis of the free system, for which $\Delta\dot{\boldsymbol{x}}_L = \boldsymbol{A}_L \Delta \boldsymbol{x}_L$ and $\Delta\dot{\boldsymbol{x}}_D = \boldsymbol{A}_D \Delta \boldsymbol{x}_D$.

In order to carry out an eigenanalysis on the artificially stabilized system, due to the nature of the proposed SAS controllers, which are based on proportional and integral signals, it is necessary to augment the state array with the corresponding integral signals, yielding $\tilde{\boldsymbol{x}}_L = \{\boldsymbol{x}_L^T, \int q\}^T$ and $\tilde{\boldsymbol{x}}_D = \{\boldsymbol{x}_D^T, \int p, \int r\}^T$. Correspondingly, the state matrices can be augmented as follows.

$$\tilde{\boldsymbol{A}}_L = \begin{bmatrix} & \boldsymbol{A}_L & & & \begin{matrix} 0 \\ 0 \\ 0 \\ 0 \end{matrix} \\ 0 & 0 & 1 & 0 & 0 \end{bmatrix}, \quad \tilde{\boldsymbol{A}}_D = \begin{bmatrix} & & \boldsymbol{A}_D & & & \begin{matrix} 0 & 0 \\ 0 & 0 \\ 0 & 0 \\ 0 & 0 \\ 0 & 0 \end{matrix} \\ 0 & 1 & 0 & 0 & 0 & 0 \quad 0 \\ 0 & 0 & 1 & 0 & 0 & 0 \quad 0 \end{bmatrix}. \tag{A7}$$

Further defining $\boldsymbol{B}_L = -\boldsymbol{M}_L^{-1}\boldsymbol{U}_L$ and $\boldsymbol{B}_D = -\boldsymbol{M}_D^{-1}\boldsymbol{U}_D$ for the original systems (i.e., without accounting for integral states), for the augmented state, it is possible to introduce the following forms:

$$\tilde{\boldsymbol{B}}_L = \begin{bmatrix} \boldsymbol{B}_L \\ 0 \quad 0 \quad 0 \quad 0 \end{bmatrix}, \quad \tilde{\boldsymbol{B}}_D = \begin{bmatrix} \boldsymbol{B}_D \\ 0 \quad 0 \quad 0 \quad 0 \\ 0 \quad 0 \quad 0 \quad 0 \end{bmatrix}, \tag{A8}$$

where in Equation (A8) the last rows of zeroes would be of three elements for a three-thrusters layout. By applying the control laws in Equation (10) in this framework, the free dynamics of the artificially stabilized system are obtained, as $\Delta\dot{\tilde{\boldsymbol{x}}}_L = (\tilde{\boldsymbol{A}}_L + \tilde{\boldsymbol{B}}_L \boldsymbol{K}_L)\Delta\tilde{\boldsymbol{x}}_L$ and $\Delta\dot{\tilde{\boldsymbol{x}}}_D = (\tilde{\boldsymbol{A}}_D + \tilde{\boldsymbol{B}}_D \boldsymbol{K}_D)\Delta\tilde{\boldsymbol{x}}_D$. The latter expressions can be employed for the eigenanalysis of the corresponding systems.

References

1. Chu, A.; Blackmore, M.; Oholendt, R.G.; Welch, J.V.; Baird, G.; Cadogan, D.P.; Scarborough, S.E. A Novel Concept for Stratospheric Communications and Surveillance: Star Light. In Proceedings of the AIAA Balloon Systems Conference, Williamsburg, VA, USA, 21–24 May 2007. [CrossRef]
2. Miller, S.; Fesen, R.; Hillenbrand, L.A.; Rhodes, J.; Baird, G.; Blake, G.; Booth, J.; Carlile, D.E.; Duren, R.; Edworthy, F.G.; et al. *Airships: A New Horizon for Science*; Technical Report; Jet Propulsion Laboratory, Keck Institute for Space Studies: Pasadena, CA, USA, 2014.
3. Romeo, G.; Frulla, G.; Cestino, E. Design of a high-altitude long-endurance solar-powered unmanned air vehicle for multi-payload and operations. *J. Aerosp. Eng.* **2007**, *221*, 199–216. [CrossRef]
4. Colozza, A. *Initial Feasibility Assessment of a High Altitude Long Endurance Airship*; Technical Report NASA/CR–2003-212724; Analex Corp.: Brook Park, OH, USA, 2003.
5. Gonzalo, J.; López, D.; Domínguez, D.; García, A.; Escapa, A. On the capabilities and limitations of high altitude pseudo-satellites. *Prog. Aerosp. Sci.* **2018**, *98*, 37–56. [CrossRef]
6. D'Oliveira, F.A.; de Melo, F.C.L.; Devezas, T.C. High-Altitude Platforms—Present Situation and Technology Trends. *J. Aerosp. Technol. Manag.* **2016**, *8*, 249–262. [CrossRef]
7. Baraniello, V.R.; Persechino, G. Conceptual Design of a Stratospheric Hybrid Platform for Earth Observation and Telecommunications. In Proceedings of the Aerospace Europe CEAS 2017 Conference, Bucharest, Romania, 16–20 October 2017.
8. Riboldi, C.E.D.; Rolando, A.; Regazzoni, G. On the feasibility of a launcher-deployable high-altitude airship: Effects of design constraints in an optimal sizing framework. *Aerospace* **2022**, *9*, 210. [CrossRef]
9. Elfes, A.; Bueno, S.; Bergerman, M.; Ramos, J. A semi-autonomous robotic airship for environmental monitoring missions. In Proceedings of the 1998 IEEE International Conference on Robotics and Automation (Cat. No. 98CH36146), Leuven, Belgium, 20 May 1998; Volume 4, pp. 3449–3455.
10. Jon, J.; Koska, B.; Pospíšil, J. Autonomous airship equipped with multi-sensor mapping platform. *ISPRS-Int. Arch. Photogramm. Remote Sens. Spat. Inf. Sci.* **2013**, *XL-5/W1*, 119–124. [CrossRef]
11. Fedorenko, R.; Krukhmalev, V. Indoor autonomous airship control and navigation system. *MATEC Web Conf.* **2016**, *42*, 01006. [CrossRef]
12. Shah, H.N.M.; Rashid, M.Z.A.; Kamis, Z.; Mohd Aras, M.S.; Mohd Ali, N.; Wasbari, F.; Bakar, M.N.F.B.A. Design and develop an autonomous UAV airship for indoor surveillance and monitoring applications. *Int. J. Inform. Vis.* **2018**, *2*, 1–7.
13. Gili, P.; Civera, M.; Roy, R.; Surace, C. An unmanned lighter-than-air platform for large scale land monitoring. *Remote Sens.* **2021**, *13*, 2523. [CrossRef]
14. Riboldi, C.E.D.; Gualdoni, F. An integrated approach to the preliminary weight sizing of small electric aircraft. *Aerosp. Sci. Technol.* **2016**, *58*, 134–149. [CrossRef]
15. Riboldi, C.E.D. An optimal approach to the preliminary design of small hybrid-electric aircraft. *Aerosp. Sci. Technol.* **2018**, *81*, 14–31. [CrossRef]
16. Riboldi, C.E.D.; Trainelli, L.; Rolando, A.; Mariani, L. Predicting the effect of electric and hybrid electric aviation on acoustic pollution. *Noise Mapp.* **2020**, *7*, 35–36. [CrossRef]
17. Trainelli, L.; Riboldi, C.E.D.; Rolando, A.; Salucci, F. Methodologies for the initial design studies of an innovative community-friendly miniliner. *Iop Conf. Ser. Mater. Sci. Eng.* **2021**, *1024*, 012109. [CrossRef]
18. Riboldi, C.E.D.; Trainelli, L.; Biondani, F. Structural batteries in aviation: A preliminary sizing methodology. *J. Aerosp. Eng.* **2020**, *3*, 040200311–040200315. [CrossRef]
19. Carichner, G.E.; Nicolai, L.M. *Fundamentals of Aircraft and Airship Design*; AIAA Education Series; American Institute of Aeronautics and Astronautics, Inc.: Reston, VA, USA, 2013.
20. Pamadi, B.N. *Performance, Stability, Dynamics, and Control of Airplanes*; AIAA Education Series; American Institute of Aeronautics and Astronautics, Inc.: Reston, VA, USA, 2004.
21. Jones, S.P.; DeLaurier, J.D. Aerodynamic estimation techniques for aerostats and airships. *J. Aircr.* **1982**, *20*, 120–126. [CrossRef]
22. Kämpf, B. Flugmechanik und Flugregelung von Luftschiffen. Ph.D. Thesis, University of Stuttgart, Stuttgart, Germany, 2004. [CrossRef]
23. Kornienko, A. System Identification Approach for Determining Flight Dynamical Characteristics of an Airship from Flight Data. Ph.D. Thesis, University of Stuttgart, Stuttgart, Germany, 2006. [CrossRef]
24. Munk, M.M. *The Aerodynamic Forces on Airship Hulls*; Technical Report, No.184; National Advisory Committee for Aeronautics: Washington, DC, USA, 1924.
25. Mosca, E. *Optimal, Predictive and Adaptive Control*; Prentice Hall: Englewood Cliffs, NJ, USA, 1995.
26. Lamb, H. *Hydrodynamics*; Dover Publications: Mineola, NY, USA, 1945.

aerospace

MDPI

Article

Filtering and Estimation of State and Wind Disturbances Aiming Airship Control and Guidance

Apolo Silva Marton [1], José Raul Azinheira [2], André Ricardo Fioravanti [1], Ely Carneiro De Paiva [1,*], José Reginaldo H Carvalho [3] and Ramiro Romankevicius Costa [4]

[1] School of Mechanical Engineering, University of Campinas, R. Mendeleyev, 200, Campinas 13083-180, SP, Brazil
[2] Department of Mechanical Engineering, Instituto Superior Técnico, Av. Rovisco Pais 1, 1049-001 Lisboa, Portugal
[3] Institute of Computing of the Federal University of Amazonas, Av. General Rodrigo Octavio Jordão Ramos, 1200-Coroado I, Manaus 69067-005, AM, Brazil
[4] Federal Institute of Education, Science and Technology of São Paulo, Campus Campinas, R. Heitor Lacerda Guedes, 1000, Campinas 13059-581, SP, Brazil
* Correspondence: elypaiva@unicamp.br

Abstract: Good state and wind estimation is a requirement for the development of guidance and control techniques for airships. However, usually this information is not directly available from the airship sensors. The typical solution applies filtering, estimation and sensor fusion methods. This paper presents a comparative study, evaluating three solutions for the state estimation of NOAMAY airship. We also present alternative versions for the crucial estimation of the wind velocity, combining Kalman filters with a data-driven Neural Network. Finally, we present special solutions to usual problems encountered in filtering implementation as the mitigation of delays caused by second-order filters. The sensors set considered is composed of a global positioning system, an inertial measurement unit and a one-dimensional Pitot probe. Comparative simulation results are presented with the use of a realistic nonlinear model of the airship.

Keywords: wind estimation; guidance; kalman filter; neural network

1. Introduction

Airships with autonomous operation capability are in the focus of worldwide investigation. Airships are particularly suited to scenarios that demand long endurance, high payload and low operational risk, such as surveillance and environmental monitoring [1].

Airship control is a big challenge for the control systems designers as they are under-actuated vehicles with strong nonlinear undamped dynamics [2]. Further, many nonlinear control approaches like sliding mode, backstepping and incremental dynamic inversion require the knowledge of vehicle velocities and accelerations, that may not be available from sensors [3]. Thus, an accurate estimation of the state and its derivatives is fundamental to the design of efficient control and guidance techniques.

Another crucial point is the wind estimation, which is particularly important for lighter-than-air vehicles [4,5]. The work of [6] showed that a good wind knowledge can result in more accurate geolocation of a ground target. Additionally, the crab-angle between ground track and heading, obtained from estimating the wind conditions, can improve the control performance of an airship in trajectory following. Finally, when the objective is ground-hover at constant altitude, the blimp must align itself against the wind and use its forward propulsion to balance the aerodynamic drag [7]. Therefore, by knowing the wind speed and direction control performance can be improved.

For both state and wind estimation, the solution relies on filtering, estimation and fusion of the sensory data. Common approaches are the Extended Kalman Filter (EKF) and the Unscented Kalman Filter (UKF). As an example, in [8], it is proposed an approach for estimating the angle of attack and sideslip angle by the kinematic equations of motion of an aerobatic UAV. Meanwhile, with the same kinematic equations, in [6] an extended Kalman flter (EKF) is proposed for estimating the wind heading and velocity using an aircraft with a single GPS and Pitot tube. In [9], a wind velocity observer also based on the kinematics is proposed for small UAVs with experimental results. Similarly, in [10] is also proposed an EKF for wind velocity estimation, however applied to a stationary stratospheric airship in simulation environment. Then in [11] are presented four model-based solutions considering an aircraft with four diferent possible configurations of sensors, as is shown in Figure 1.

Figure 1. NOAMAY Airship.

This work is placed in the context of project InSAC https://www2.eesc.usp.br/insac accessed on 25 May 2022. The project aims to develop an autonomous airship for performing environmental monitoring tasks in remote Amazon rain forest areas. Such tasks include aerodynamic flights and ground-hover (i.e., keep a stationary position with respect to a ground target).

Firstly, the paper presents a comparative study evaluating different different solutions for the state/derivatives estimation of NOAMAY airship, which is instrumented with a pitot tube and an Inertial Measurement Unit (IMU) including: accelerometer, gyroscopes, GPS, barometer and thermometer.

In a second moment, for the wind estimation, this work presents an alternative version of a model based on wind velocity estimator using the EKF combined with a data-driven approach of estimation using a Neural Network (NN). The main tool to validate the proposed estimation approaches is a dynamical realistic nonlinear model of an airship in Simulink/MATLAB. This tool is a result of the research group efforts since the seminal project AURORA, which was improved during the projects DIVA and DRONI [12].

Finally, we present possible solutions to mitigate common problems arriving in filtering implementation like the control input redundancy and the delays caused by second-order filters [3]. For the input redundancy, related to an overactuated control system, we propose filtering the less important commanded input, an approach that is known as Washout Filter (WOF) [13]. Another critical point is the drawback that appears in the second-order filters used to estimate state derivatives. The natural dynamics delay produced by these filters may harm the control system feedback. One possible solution to mitigate this delay is the so called Input Scaling Gain (ISG), that was first proposed by our research group in 2015 as a scalar factor gain [14]. In this paper, we generalize the input scale gain to the MIMO case. To illustrate these two design solutions we present briefly the concept of incremental controllers, that are used to show the benefits of these filtering techiques, although not limited to these kind of controllers.

2. Airship Model

The main tool to validate the proposed control approach is a dynamical realistic nonlinear model of a robotic airship, which can be expressed in a state-space form as following:

$$\dot{\xi} = g(\xi, \mathbf{x}, \mathbf{d}), \tag{1a}$$

$$\dot{\mathbf{x}} = f(\xi, \mathbf{x}, \mathbf{d}, \mathbf{u}), \tag{1b}$$

where:

- the kinematic states $\xi = [\mathbf{P}^{\mathbf{T}} \ \ \mathbf{\Phi}^{\mathbf{T}}]^{\mathbf{T}}$ include the cartesian position $\mathbf{P} = [P_N \ P_E \ P_D]^T$ and angular position $\mathbf{\Phi} = [\phi \ \theta \ \psi]^T$ in the North-East-Down (NED) frame (SI units);
- the dynamic states $\mathbf{x} = [\mathbf{V}^{\mathbf{T}} \ \ \mathbf{\Omega}^{\mathbf{T}}]^{\mathbf{T}}$ include the linear speed $\boldsymbol{V} = [u \ \ v \ \ w]^T$ and angular speed $\mathbf{\Omega} = [p \ \ q \ \ r]^T$ in the body frame (SI units);
- the input vector $\boldsymbol{u} = [\delta_e \ \delta_a \ \delta_r \ \delta_0 \ \delta_q \ \mu_0]^T$ includes: δ_e, δ_a and δ_r which are elevator, aileron and rudder deflections; δ_0 as the normalized thrusters voltage; δ_q as the differential voltage between the front-back thrusters; μ_0 as the common vectoring angle of the thrusters;
- and, finally, the disturbance vector $\mathbf{d}$ that includes wind velocities and gust variables (SI units), as is shown in Figure 2.

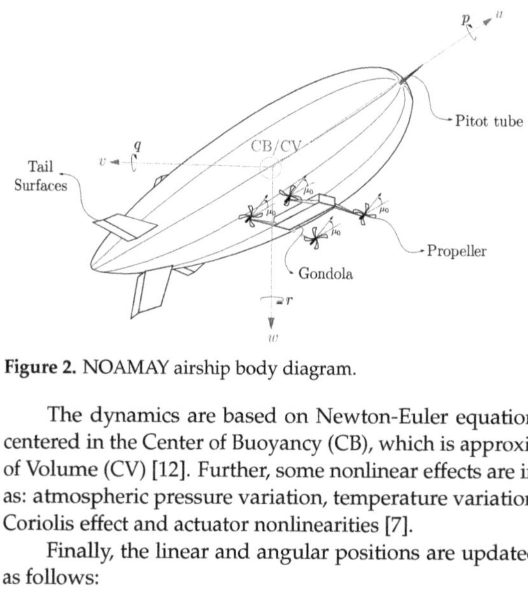

Figure 2. NOAMAY airship body diagram.

The dynamics are based on Newton-Euler equations considering the body frame centered in the Center of Buoyancy (CB), which is approximately equivalent to the Center of Volume (CV) [12]. Further, some nonlinear effects are included in the simulation, such as: atmospheric pressure variation, temperature variation, lifting gas pressure variation, Coriolis effect and actuator nonlinearities [7].

Finally, the linear and angular positions are updated through kinematic equations as follows:

$$g(\xi, \mathbf{x}) = \begin{bmatrix} \mathbf{S}_{\mathbf{\Phi}}^{\mathbf{T}} & 0_{3\times3} \\ 0_{3\times3} & \mathbf{R}_{\mathbf{\Phi}}^{\mathbf{T}} \end{bmatrix} \begin{bmatrix} \mathbf{V}_{\mathrm{g}} \\ \mathbf{\Omega} \end{bmatrix}, \tag{2}$$

where $0_{3\times3}$ is a null matrix of 3th order, $\mathbf{S}_{\mathbf{\Phi}} \in \mathbf{R}^{3\times3}$ is the rotational matrix from airship body to NED frame, and $\mathbf{R}_{\mathbf{\Phi}} \in \mathbf{R}^{3\times3}$ is the angular transformation matrix, both given by [7]:

$$\mathbf{S}_{\mathbf{\Phi}} = \begin{bmatrix} c_\psi c_\theta & s_\psi c_\theta & -s_\theta \\ c_\psi s_\theta s_\phi - s_\psi c_\phi & s_\psi s_\theta s_\phi + c_\psi c_\phi & c_\theta s_\phi \\ c_\psi s_\theta c_\phi + s_\psi s_\phi & s_\psi s_\theta c_\phi - c_\psi s_\phi & c_\theta c_\phi \end{bmatrix},$$

$$\mathbf{R}_{\mathbf{\Phi}} = \begin{bmatrix} 1 & s_\phi t_\theta & c_\phi t_\theta \\ 0 & c_\phi & -s_\phi \\ 0 & s_\phi / c_\theta & c_\phi / c_\theta \end{bmatrix},$$

where $c_x = \cos(x)$, $s_x = \sin(x)$ and $t_x = \tan(x)$.

3. Sensors Modeling

In this section, a summary of the sensors modeling is presented, including manufacturer specifications, such as bias and sample frequency. The NOAMAY robotic airship is instrumented with a set of sensors including a Pitot Tube, responsible for measuring the airspeed (V_{pitot}) and a multi-sensor Xsens Mti-G 700 which provides the following information [15]:

- absolute position $\mathbf{P}_{\mathbf{NED}}$ [m] measured by GPS;
- absolute orientation in the world $\mathbf{\Phi}$ [rad], as result of a sensor fusion between Gyroscope and Magnetometer;
- inertial linear velocity ($\mathbf{V}_\mathbf{g}$) in three-axis [m/s], estimated by the GPS;
- inertial angular velocity ($\mathbf{\Omega}$) in three-axis [rad/s], estimated by the IMU;
- inertial acceleration in three-axis ($\boldsymbol{a} = [a_x \ a_y \ a_z]^T$) [m/s^2], measured by the Accelerometer;
- atmospheric pressure P_h [hPa], by the barometer;
- and temperature T_h [K], measured by the thermometer.

The sample frequency of the sensors are shown in the Table 1. Further, a first order model for the Gauss Markov error of each sensor is given as [16]:

$$y = \hat{y} + b + w_y \tag{3a}$$

$$\dot{b} = -\frac{1}{\tau} b + w_b \tag{3b}$$

where b is the sensor bias with Gaussian noise w_b, w_y is a Gaussian noise for the measure as shown in Table 2, $\hat{y}$ is the true simulated value and y is the corresponding sensor output. For the sensors presenting bias, the parameter σ_{RMS} (The noise power is given by $(\sigma_{RMS})^2/(\text{sample frequency})$) is specified in Table 3. In the case of Accelerometer, Barometer and Thermometer there are some special parameters shown below.

Table 1. Sample frequency specification.

Sensor	Frequency [Hz]	Sampled data
IMU	100	$\mathbf{\Phi}, \mathbf{\Omega}, \boldsymbol{a}$
GPS	4.00	$\mathbf{V}_\mathbf{g}, \mathbf{P}_{\mathbf{NED}}$
Barometer	50.0	P_h
Thermometer	1.00	T_h

Table 2. Sensor noise specification (w_y).

Data	σ_{RMS}	Data	σ_{RMS}
ϕ, θ [rad]	0.00520	$\boldsymbol{a}$ [m/s^2]	0.0040
ψ [rad]	0.10000	P_N, P_E [m]	2.5000
P_D [m]	5.00000	V_{pitot} [m/s]	0.0020
$\mathbf{\Omega}$ [rad/s]	0.00035	$\mathbf{V}_\mathbf{g}$ [m/s]	0.4000
P_h [hPa]	0.01000	T_h [K]	1.0000

Table 3. Sensor bias specification (w_b).

Sensor	σ_{RMS}
Gyroscope [°/s]	$1.3{\cdot}10^{-4}$
Accelerometer [m/s^2]	$2.3{\cdot}10^{-5}$
Barometer [hPa]	$1.4{\cdot}10^{-5}$

For the case of pose measurement a coordinate transformation is applied to the data (with translational and rotational components). The GPS output is modeled as following:

$$\mathbf{P_{gps_k}} = \mathbf{S_{xsens}}(\mathbf{P_{NED_k}} + \mathbf{P_{xsens}}) + \boldsymbol{\varphi}_{\mathbf{pos_k}}, \tag{4}$$

where: $\mathbf{S_{xsens}}$ is the rotational matrix from the CB angular pose to the Xsens angular pose; and $\mathbf{P_{xsens}}$ is the Xsens position in the body frame; $\boldsymbol{\varphi}_{pos_k}$ is a position Gaussian noise vector at instant k; and $\mathbf{P_{gps_k}}$ is the sampled position given by the GPS at instant k.

The orientation output is modeled as following:

$$\boldsymbol{\Phi}_{imu_k} = \boldsymbol{\Phi}_{xsens} + \boldsymbol{\Phi}_k + \boldsymbol{\varphi}_{ori_k}, \tag{5}$$

where: $\boldsymbol{\Phi}_k$ is the orientation given in simulation at instant k; $\boldsymbol{\Phi}_{xsens}$ is the Xsens angular pose in the body frame; $\boldsymbol{\varphi}_{ori_k}$ is an orientation Gaussian noise vector at instant k; and $\boldsymbol{\Phi}_{imu_k}$ is the sampled angular position given by IMU at instant k.

For the velocity measurement, we apply the coordinate transformation to the data (with rotational components).

The resulting linear velocity output is as follows:

$$\mathbf{V_{gps_k}} = \mathbf{S_{xsens}}(\mathbf{V_{g_k}} + \boldsymbol{\Omega}_\mathbf{k} \times \tilde{\mathbf{O}}_\mathbf{xsens}) + \boldsymbol{\varphi}_{\mathbf{vel_k}}, \tag{6}$$

where: $\boldsymbol{\Omega}_k$ is the angular velocity given in simulation at instant k; $\tilde{\mathbf{O}}_\mathbf{xsens}$ is the position vector from the CB to the Xsens position; $\mathbf{S_{xsens}}$ is the rotational matrix from the CB angular pose to the Xsens angular pose; and $\mathbf{P_{xsens}}$ is the Xsens position in the body frame; $\boldsymbol{\varphi}_{vel_k}$ is a position Gaussian noise vector at instant k; and $\mathbf{P_{gps_k}}$ is the sampled position given by the GPS at instant k.

The angular velocity output is modeled as following:

$$\boldsymbol{\Omega}_{imu_k} = \mathbf{S_{xsens}}\boldsymbol{\Omega}_\mathbf{k} + \boldsymbol{\varphi}_{\mathbf{rate_k}}, \tag{7}$$

where: $\boldsymbol{\varphi}_{rate_k}$ is an angular velocity Gaussian noise vector at instant k; and $\boldsymbol{\Omega}_{imu_k}$ is the sampled angular velocity given by the IMU at instant k.

The accelerometer signal includes the additional components of centripetal acceleration (a_c) and gravitational acceleration (a_g), as:

$$\boldsymbol{a} = \dot{\mathbf{V}}_g + \dot{\boldsymbol{\Omega}} \times \tilde{\mathbf{O}}_\mathbf{xsens} - \mathbf{S}_{\boldsymbol{\Phi}}\mathbf{a_g} + \boldsymbol{a_c}, \tag{8}$$

where $\tilde{\mathbf{O}}_\mathbf{xsens}$ is the vector from the CB to the sensor location, S_{Φ} is the rotation matrix from the body frame to the North-East-Down frame and $\boldsymbol{a}_c = \boldsymbol{\Omega} \times (\boldsymbol{\Omega} \times \tilde{\mathbf{O}}_\mathbf{xsens})$.

In the case of the static pressure at a given altitude, we use the International Standard Atmosphere (ISA):

$$P_h = P_0\left(\frac{T_h}{T_h - 0.0065P_D}\right)^{5.257}, \tag{9}$$

where P_h is the static pressure, T_h is temperature in Kelvin, P_D is the negative of the altitude (in meters) in the inertial frame and P_0 is the pressure at zero altitude (at the sea level $P_0 = 101.325$ hPa).

The temperature is given by the classical model, as linearly varying with the altitude:

$$T_h = T_0 + 0.0065P_D, \tag{10}$$

where T_0 is the temperature at zero altitude, recalling that P_D is negative in our model.

4. Filtering and Estimation

We present here three possible solutions for the estimation of the state variables of an airship, with particular analysis for the NOAMAY airship. The first one is the classical second-order filter, and the other two are Kalman Filters (EKF and UKF). The performances of the three approaches are then analyzed and compared. The algorithms developed in C/C++ using the Robotic Operating System (ROS) [17], are available for consulting.

4.1. Low-Pass Filter

The first estimator is the classical second-order Low-pass filter (LPF), defined as:

$$\frac{\chi(s)}{Y(s)} = \frac{\omega_f^2}{s^2 + 2\omega_f \zeta s + \omega_f^2}, \tag{11}$$

where $\chi(s)$ is the estimated state, $Y(s)$ is the sensor measure, ω_f is the cut-off frequency and ζ is the damping coefficient. For each estimated state and corresponding sensor, the filter parameters used here are given in the Table 4. These values were chosen during simulation by manual tuning.

Table 4. Configuration for Low-pass filter.

State	Sensor	ω_f [rad/s]	ζ
$\mathbf{P}_{\mathbf{NED}}, \mathbf{V}_{\mathbf{g}}$	GPS	3.33	0.96
$\mathbf{\Phi}, \mathbf{\Omega}$	IMU	10	0.96

4.2. Extended Kalman Filter

The EKF algorithm is probably the most popular estimator used in the literature. In this work we use the EKF ROS package (Available https://wiki.ros.org/robot_localization accessed on 25 May 2022) known as `robot_localization`) [18]. This package is related to the configuration of a generic 6DOF mobile robot with 6-DOF, yielding the filtered estimation of positions, velocities and accelerations, both linear and angular.

We want to estimate here the complete airship state vector ($\mathbf{P}_{\mathbf{NED}}, \mathbf{\Phi}, \mathbf{V}_{\mathbf{g}}, \mathbf{\Omega}, \boldsymbol{a}$). Thus, the measurement update stage uses the sensors output, as described in Table 5, where 1 indicates the presence of the sensor in the corresponding measured variable while 0 is the opposite.

Table 5. Configuration for measurement update stage.

Data\Sensor	GPS	IMU	Barometer
P_{NED}	1	0	0
Φ	0	1	0
u, v	1	0	0
w	1	0	1
Ω	0	1	0
a	0	1	0

One important feature of `robot_localization` package is the sensor-based frame reference in the East-North-Upper (ENU) frame, while we use the Norht-East-Down (NED) frame, requiring a conversion. Further, regarding the acceleration signal, we implement a cascaded approach in order to remove the centripetal acceleration component, as:

$$\hat{\mathbf{V}}_{\mathbf{g}} = \boldsymbol{a} - \hat{\mathbf{\Omega}}_0 \times (\hat{\mathbf{\Omega}}_0 \times \tilde{\mathbf{O}}_{\mathbf{xsens}}), \tag{12}$$

The weights of the diagonal covariance matrix (15 $\times$ 15) of the process are shown in Table 6. The gains were chosen empirically, considering that the vehicle has slow dynamics and that the airship can be assumed to be a rigid body. The hypothesis of enough stiffness of the hull is true if we consider that the envelope is under full operational volume and pressure [19–21]. The rigidity assumption of a conventional airship is broken for extremely high speeds or if the envelope is made of thin films to reduce the weight [22,23], which is not the case of the NOAMAY airship.

Table 6. Process covariance matrix.

State	Weight	State	Weight
Φ [rad]	0.010	a [m/s^2]	0.100
Ω [rad/s]	0.100	u, v[m/s]	0.050
P_N [m]	0.001	P_E [m]	0.001
P_D [m]	0.010	w [m/s]	0.001

4.3. State Estimation Results

We show here the simulation results comparing the three approaches, for the state estimation in a trajectory tracking control case, as developed in [3]. The sampling rate of the estimators is 32 Hz and the wind is blowing from the North. A visualization of the estimation results, made in Simulink/Matlab, are available in youtu.be/VL5dvCyOZwY (EKF) and youtu.be/B26xaKtAyWo (UKF) accessed on 25 May 2022.

Figure 3(above) shows the estimated trajectories for the different estimators, from which we can see that EKF appears to be smoother and with less errors. The signals of the longitudinal (u), vertical (w) and angular velocity in z-axis (r) appear in Figure 3(down). We can see that the classical LPF provides good filtering, but with a significant delay, while UKF is faster though with a bigger error. The EKF filter has a time response similar to the UKF case, while displaying lower errors.

Thus, for the airship model considered here, the EKF showed a better performance in comparison to UKF and LPF. For further details, including the RMS values of the position errors, please refer to [3]. These results show the natural compromise between noise attenuation versus estimation delay, found in an estimator design.

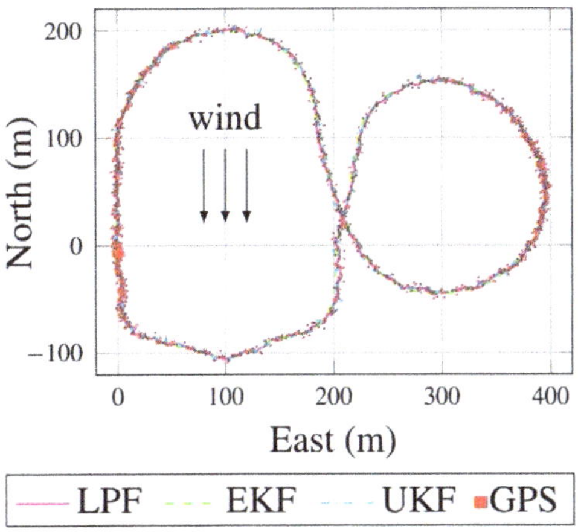

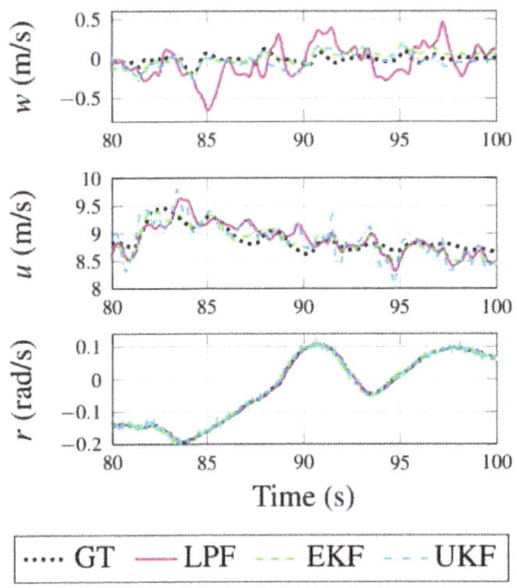

Figure 3. Estimated trajectory (**above**); Estimated vertical velocity (w), longitudinal velocity (u) and angular velocity (r) (**down**).

5. Wind Velocity Estimators

This section addresses the problem of wind velocity estimation. Firstly, we detail the equations related to the wind ($\mathbf{V_w}$) and inertial airship velocities ($\mathbf{V_g}$), as well as the airship orientation ($\mathbf{\Phi}$), which are related to the Pitot probe sensor.

Let us define the airship's relative velocity ($\mathbf{V_a}$), or airspeed, as:

$$\mathbf{V_a} = \mathbf{V_g} - \mathbf{V_w}, \tag{13}$$

Whose modulus is called *true airspeed* (V_t) and is given by:

$$V_t = ||\mathbf{V}_a||_2 = \sqrt{u_a^2 + v_a^2 + w_a^2}. \tag{14}$$

Let us recall also the aerodynamic angles, the sideslip β and the angle of attack α, as shown in Figure 4, defined by:

$$\beta = \sin^{-1} \frac{v_a}{V_t}, \tag{15}$$

$$\alpha = \tan^{-1}\frac{w_a}{u_a}. \quad (16)$$

Or, equivalently:

$$w_a = u_a\frac{\sin\alpha}{\cos\alpha} \quad \text{and} \quad v_a = V_t\sin\beta. \quad (17)$$

Finally, we obtain:

$$V_t = \frac{u_a}{\cos\alpha\cos\beta}. \quad (18)$$

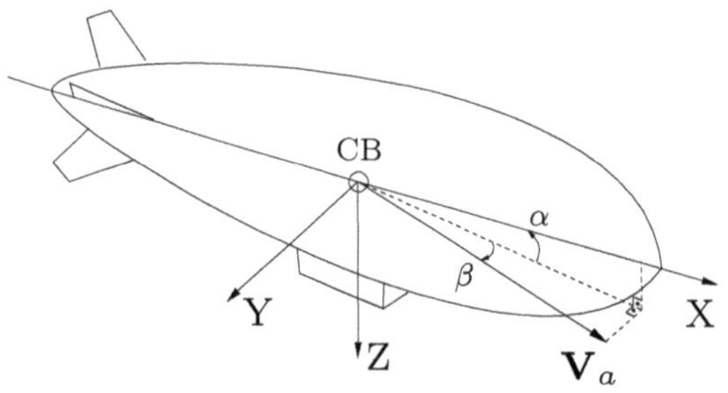

Figure 4. Sideslip angle (β) and angle of attack (α).

The Pitot tube is located at the airship nose (see Figure 2). Also, the airship has a pressure sensor connected to the Pitot Tube. Thus, as the airship moves forward, the longitudinal dynamic pressure ΔP will be measured by the pressure sensor. The relation between the measured pressure and the airship longitudinal airspeed can be derived from Bernoulli's equation as shown by [6], which can be rewritten as below:

$$\Delta P = \eta(u_a)^2, \quad (19)$$

where η is the calibrating factor related to the air density and pitot efficiency. Now, if we use the variable transformation below:

$$V_{pitot} = \sqrt{\Delta P}, \quad (20)$$

And using the statements (18) and (20), we can rewrite (19) as below:

$$V_t = \frac{V_{pitot}}{\sqrt{\eta}\cos\alpha\cos\beta}. \quad (21)$$

The unknown parameters in η together with the inaccurate angle values α and β are then estimated together in a single variable, or scale factor c_f, given by:

$$c_f = \sqrt{\eta}\cos\alpha\cos\beta, \quad (22)$$

And thus (21) is now given by:

$$V_t = \frac{1}{c_f}V_{pitot}. \quad (23)$$

Now, recalling that $\mathbf{V_g}$ is written in the body, as well as $\mathbf{V_a}$, if we take these vectors written in the inertial frame (NED), we can write the relative velocity equation (13) using the rotational matrix $\mathbf{S_\Phi}$ from airship body to NED frame, we have:

$$\mathbf{V_{NEDa}} = \mathbf{V_{NED}} - \mathbf{V_{NEDw}} = \mathbf{S_\Phi^T}\mathbf{V_a}. \quad (24)$$

As the rotational operation does not change the vector modulus, and assuming a wind speed in the North-East plane, we have:

$$V_{pitot}^2 = c_f^2((V_N - V_{N_w})^2 + (V_E - V_{E_w})^2 + (V_D)^2). \tag{25}$$

Assuming initial conditions of $\alpha \approx \beta \approx 0$, we have $u_a = V_t$ and $v_a = w_a = 0$, and in the global frame we have:

$$V_{N_a} = u_a \cos\psi \cos\theta, \tag{26}$$

where ψ and θ are the yaw and pitch angles, respectively, which together with (13) give:

$$V_{E_a} = u_a \sin\psi \cos\theta, \tag{27}$$

$$V_N = \frac{V_{pitot}}{c_f} \cos\psi \cos\theta + V_{N_w}, \tag{28}$$

$$V_E = \frac{V_{pitot}}{c_f} \sin\psi \cos\theta + V_{E_w}. \tag{29}$$

The measures of V_{pitot}, V_N and V_E come from the Pitot tube plus the GPS. In the same way, the Euler angles (ϕ, θ and ψ) are obtained from the IMU. Therefore (25), (28) and (29) can be used as observation equations, while c_f, V_{N_w} and V_{E_w} are the estimated states in the EKF Kalman Filter. Although these equations are linear dependent, we are introducing redundant information of different sensors by adding IMU combined with GPS measurements, which may lead to faster convergence and better filtering of noise.

5.1. Extended Kalman Filter

In this section we propose an EKF filter to estimate the wind speed in the airship body. Supposing that the wind is blowing in the North-East plane, the objective here is to estimate the wind velocity in the horizontal plane as well as the Pitot scale factor. By introducing both (28) and (29) in the measurement update stage, we expect to improve the results, in contrast when using just (25), as in the aproach of [6].

We consider here that the wind is constant with an additional Gaussian variance component. In addition, as the uncertain and time varying factor c_f is estimated at the same time, we have the following model:

$$\chi_{k+1} = \mathbf{F}\chi_\mathbf{k} + \nu_\mathbf{k}, \tag{30a}$$

$$\mathbf{z_k} = \mathbf{h}(\chi_\mathbf{k}) + \boldsymbol{v}_\mathbf{k} \tag{30b}$$

where: $\chi_k = [V_{N_{wk}}\ V_{E_{wk}}\ c_{fk}]^T$ is the state vector in the instant $t = kt_s$; $\mathbf{z_k} = [V_{pitot_k}^2\ V_{N_k}\ V_{E_k}]^T$ is the system output in the instant $t = kt_s$; t_s is the sample time in seconds; $h(\chi_k)$ is the output function, computed through (25), (28) and (29), and from which the Jacobian matrix is derived. Further, $F = I_3$, is the identity matrix, $\nu_k \sim \mathbf{N}(\mathbf{0}, \mathbf{Q})$ is the process noise with Gaussian distribution and covariance $\mathbf{Q}$; and, finally, $v_k \sim \mathbf{N}(\mathbf{0}, \mathbf{R})$ is the measurement noise also with Gaussian distribution and covariance $\mathbf{R}$. Given the model described in (30a), we can use the traditional EKF algorithm, as presented in [3].

5.2. Neural Network

This section presents the design of a Neural Network (NN) algorithm for the estimation of the wind speed and angle, supposing an horizontal wind profile. One of the advantages of the use of a Neural Network is that it is trained by measured data, and it is thus able to detect abrupt variations in the wind velocity. However, in contrast, it is usually more sensitive to measurement errors. We will show that combining NN with the classical estimators, using Kalman filters, we can benefit of the advantages of both.

As presented above in equations (30a)–(30b), the wind model output is composed by (25), (28) and (29), which are nonlinear equations in the model states, vehicle velocity, orientation and Pitot pressure. To obtain an easier and more tractable model, the measured data vector is transformed into 8 inputs given by:

$$\mathbf{z_{nn}} = \begin{bmatrix} V_{pitot}^2 \\ V_D^2 \\ V_N \\ V_E \\ V_E^2 \\ V_N^2 \\ V_{pitot} \cos\psi \cos\theta \\ V_{pitot} \sin\psi \cos\theta \end{bmatrix}.$$

Meanwhile, the output vector χ_{nn} of the Neural Network contains the estimated wind speed components and the scale factor c_f as:

$$\chi_{nn} = \begin{bmatrix} V_{N_w} \\ V_{E_w} \\ c_f \end{bmatrix}.$$

The proposed Neural Network consists of a three-layer fitting NN, which has three nonlinear hidden layers containing 24 neurons each and three linear outputs. The activation function of the nonlinear neurons is sigmoidal. It was designed with the help of the MATLAB Neural Network Toolbox^{TM}. For more details, the reader is referred to [3].

5.3. Hybrid Estimator

Following the developments in [3], we propose now a hybrid estimator that is a combination of both estimators, the EKF designed in Section 5.1 and the Neural Network designed in Section 5.2. The idea is to change the measure update stage of the EKF approach, when the output of the NN χ_{nn} is added to the measurement vector of the EKF as a redundant measure. Thus, resulting in the new measurement vector $\mathbf{z_{h_k}}$, updating function $\mathbf{h_{h_k}}(\chi_\mathbf{k})$ and its respective Jacobian $\mathbf{H_{h_k}}$ given below:

$$\mathbf{z_{h_k}} = \begin{bmatrix} \mathbf{z_k} \\ \chi_{nn} \end{bmatrix}, \mathbf{h_{h_k}}(\chi_\mathbf{k}) = \begin{bmatrix} \mathbf{h}(\chi_k) \\ \chi_k \end{bmatrix} \text{ and } \mathbf{H_{h_k}} = \begin{bmatrix} \mathbf{H} \\ \mathbf{I_3} \end{bmatrix},$$

where $\mathbf{I_3}$ is the identity matrix. Then the EKF standard algorithm is used by updating the matrices $\mathbf{C_k}$, $\mathbf{K_k}$ and $\mathbf{R}$, as shown in [3].

5.4. Wind Estimation Results

In this section, we run a simulation to evaluate the three approaches presented, establishing a comparison with the traditional model-based approach proposed by [6]. The sensors models in the simulation environment use sample frequency as follows: IMU (100 Hz), GPS (4 Hz) and Pitot tube (18 Hz). An online repository (https://github.com/leve-fem/airship_estimatoraccessed on 25 May 2022) is available containing all the approaches.

As shown in Figure 5, the simulation considers wind with absolute value $|\vec{V}_w| = 2m/s$ and heading $\psi_w = \frac{\pi}{2}$ rad (blowing from East to West). Then in the instant $t = 160s$ the wind is intensified to $|\vec{V}_w| = 3m/s$ and its heading is changed to $\psi_w = \pi$ (blowing from South to North). In the same Figure we highlighted five special moments with gray background in order to establish further comparisons with the results in Figure 6. Moreover, we introduced results using the estimator proposed by [6] as "Cho2011" for comparison.

From Figure 6, we can see that under an abrupt wind variation, as in instant (III), the classical estimators (EKF and UKF) do not converge immediately as they are model-based estimators depending directly on the Pitot information. Meanwhile the NN clearly, which is data driven, shows an instantaneous wind correction in this situation, although exhibiting a biased value. The hybrid approach was also able to yield a correct estimation of the wind speed before the instants (IV) and (V), when the Model-based estimators finally converges for the acceptable values.

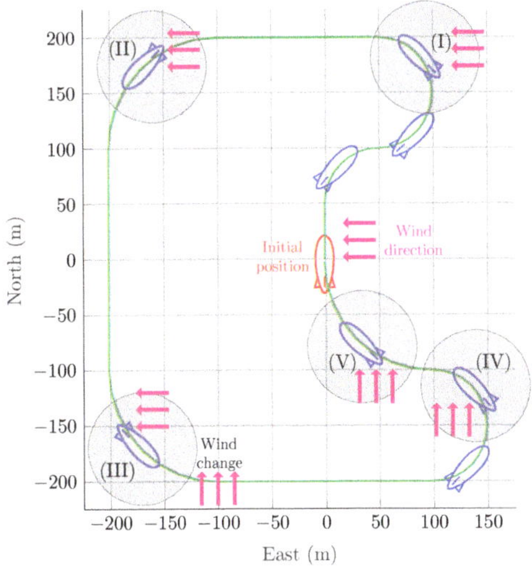

Figure 5. Simulation trajectory for wind estimation.

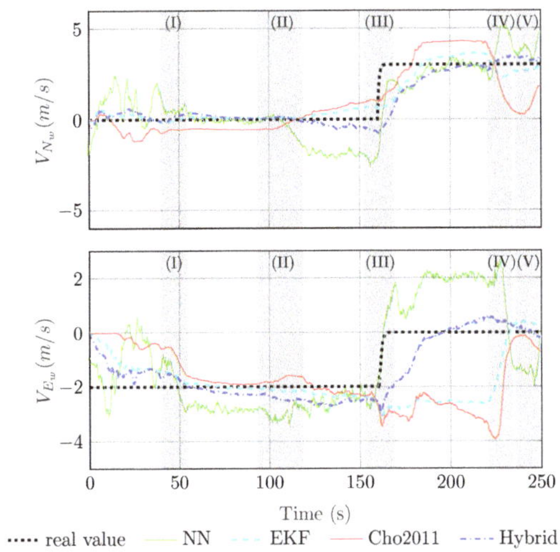

Figure 6. Simulation using the airship nonlinear model with realistic sensor noise: estimated wind velocity components in NE frame.

In conclusion, we can observe that the NN had a better estimation of V_{E_w} in comparison to the Model-based approaches. However, for the component V_{N_w}, the Model-based approaches presented a better performance. Meanwhile the "Hybrid" approach had the best performance in the estimation of V_{E_w} and an acceptable estimation for V_{N_w}.

6. Mitigating Filtering Design Issues

In this section, we present possible solutions to mitigate two common problems arriving in filtering implementation that are the control input redundancy and the delay appearing in second-order filters. Input redundancy is related to an overactuated control system. A common issue is that the redundant inputs can cancel each other when the system achieves a stationary state condition which results in more energy consumption. One solution for this problem is to perform a filtering in the commanded redundant input which is less important for maintaining the equilibrium point, which is known as Washout Filter (WOF) [13].

Another critical point is the drawback that appears in the second-order filters used to estimate state derivatives. The natural dynamics delay produced by these filters may impose serious problems to the control system feedback. One possible solution to mitigate this delay is the so called Input Scaling Gain (ISG), that was first proposed by our research group as a scalar factor gain [14]. In this section, we generalize the input scale gain to the MIMO case. To illustrate these two design solutions we present briefly the concept of Incremental Controllers, that are used to show the benefits of these filtering techiques, although not limited to these kind of controllers.

6.1. Incremental Controllers

In order to illustrate the application of these two design techniques (WOF and ISG), we present first the concept of incremental dynamics (ID) and incremental controllers.

In the last two decades, researchers have studied the classical Nonlinear Dynamics Inversion (NDI), also known as feedback linearization, as a promising approach to unify the control scheme of an Unmanned Aerial Vehicles (UAV) during the different phases of a standard flight plan [24–26].

To cancel model nonlinearites, NDI controllers require a complete and precise model of the system. However, in real-world systems, realistic dynamic models with accurate parameters are almost impossible to be obtained. Firstly presented in [27], the incremental controlleres (ID) was used for designing a simplified version of Nonlinear Dynamic Inversion. Later this strategy was named as Incremental Nonlinear Dynamic Inversion (INDI) [28]. Since then, several works use Incremental Dynamics (ID) for designing nonlinear control laws. As some examples, in [29], the authors use ID for designing Incremental Backstepping (IBKS). In [30], ID is used for design the Incremental Sliding-Mode (ISM) control law.

As a first step, let us present the concept of incremental dynamics [28]. Consider a control affine nonlinear system in state space representation:

$$\dot{\mathbf{x}} = f(\mathbf{x}) + g(\mathbf{x})\mathbf{u}, \tag{31a}$$

$$\mathbf{y} = h(\mathbf{x}). \tag{31b}$$

where $\mathbf{x} \in \mathbb{R}^n$ is the vector of state variables, $\mathbf{u} \in \mathbb{R}^m$ is the vector of control inputs, $\mathbf{y} \in \mathbb{R}^l$ is the output vector, and $f,\ g,\ h$ are real analytic Lipschitz continuous functions.

The system dynamics (31a) can be approximated by its Taylor series expansion around $\mathbf{x} = \mathbf{x}_0$ and $\mathbf{u} = \mathbf{u}_0$:

$$\dot{\mathbf{x}} = \dot{\mathbf{x}}_0 + \mathbf{A}_0(\mathbf{x} - \mathbf{x}_0) + \mathbf{B}_0(\mathbf{u} - \mathbf{u}_0) + \mathcal{O}((t - t_0)^2), \tag{32}$$

where $(\mathbf{x}, \dot{\mathbf{x}}, \mathbf{u})$ and $(\mathbf{x}_0, \dot{\mathbf{x}}_0, \mathbf{u}_0)$ are respectively the state, the state derivative, and input at current time t and some previous time $t_0 < t$, $\mathcal{O}((t - t_0)^2)$ includes the higher order terms of the Taylor expansion, and

$$\mathbf{A}_0 = \frac{\partial}{\partial \mathbf{x}}[f(\mathbf{x}) + g(\mathbf{x})\mathbf{u}_0]\bigg|_{\mathbf{x}_0}, \quad \mathbf{B}_0 = g(\mathbf{x}_0), \tag{33}$$

are state-dependent matrices that capture the linear system dynamics relationship with the state and input variables, respectively.

One of the key points in incremental controllers is that we assume that the time interval $t_s = t - t_0$ elapsed between $\mathbf{x}_0$ and $\mathbf{x}$ is sufficiently small, such that we can suppose $\mathbf{x}_0 \approx \mathbf{x}$, and thus the system dynamics (32) can be approximated by the so-called incremental dynamics formulation:

$$\dot{\mathbf{x}} \approx \dot{\mathbf{x}}_0 + \mathbf{B}_0(\mathbf{u} - \mathbf{u}_0), \tag{34}$$

yielding the current state derivative from the knowledge of its value at the previous time step and the input increment $\Delta\mathbf{u} = \mathbf{u} - \mathbf{u}_0$.

Incremental controllers such as IBKS [29], ISM [30] and INDI [14] are sensor-based controllers, taking advantage of the simplified dynamics (34), where the use of state dependent dynamics is replaced by the measurement of the previous time derivative $\dot{\mathbf{x}}_0$.

INDI is the equivalent of the well known NDI control applied to the incremental dynamics (34).

Let us impose a desired dynamics $\nu \in \mathbb{R}^n$. Then, the incremental dynamic inversion results in the following control law:

$$\mathbf{u} = \mathbf{u}_0 + \mathbf{B}_0^+(\nu - \dot{\mathbf{x}}_0), \tag{35}$$

where $\mathbf{B}_0\mathbf{B}_0^+ = \mathbf{I}_n$ is the identity matrix of order n.

Note that if the matrix inversion is perfect then, replacing the control law in the incremental dynamic equation gives

$$\dot{\mathbf{x}} \approx \dot{\mathbf{x}}_0 + \mathbf{B}_0\Delta\mathbf{u} = \dot{\mathbf{x}}_0 + \mathbf{B}_0\mathbf{B}_0^+(\nu - \dot{\mathbf{x}}_0) = \nu \tag{36}$$

which shows that:

- the system modes are decoupled, allowing the design of independent linear controllers for each of them.
- the state derivative tracks the dynamics imposed by ν;
- the previous state derivatives, and consequently their nonlinearities, are canceled (DI is also called feedback linearization);

Taking advantage of the above, we can define ν as a pseudo-control signal, which is usually taken as a linear state feedback. Imposing a linear dynamics, the feedback gain places the closed loop poles, designing the desired response. As it is common sense in cascade control, a Time Scale Separation Principle (TSSP) must be respected, and the INDI loop must converge faster than the linear control loop. In addition, the implementation of incremental controller considers the following assumptions:

A6.1 The system is output controllable (34), and any internal dynamics are intrinsically stable in closed-loop.

A6.2 States are sampled at a sufficiently high frequency when compared with system dynamics.

A6.3 Fast control action in comparison to the system modes.

A6.4 The control signals and state references are measurable, continuous and bounded. Additionally, accurate information on the state derivatives and actuator variables is available.

A6.5 The input matrix B_0 has known coefficient signals, and it is non-singular around the region of interest.

Since all assumptions of Incremental controllers are satisfied we can define the INDI control law (35) for the system in (31a). Figure 7 shows a typical block diagram of a sensor based INDI control, where z^{-1} represents a delay of one sample time.

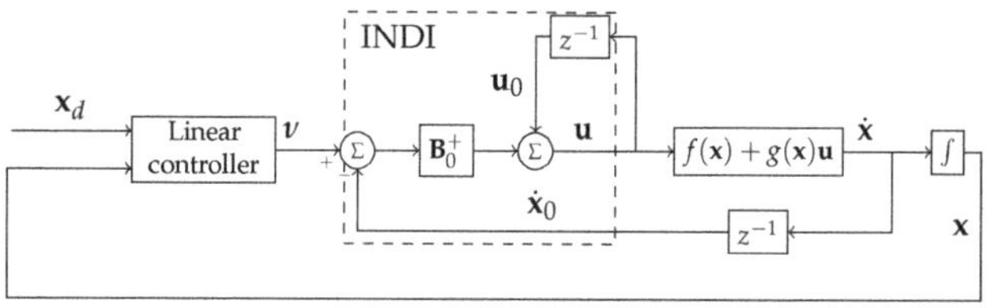

Figure 7. INDI closed loop block diagram.

One of the most important advantages of using incremental controllers, which makes it a successfull candidate for real applications, is the great robustness to model parameters. The most common control approaches are said to be model based, for their design being strongly dependent on the system dynamics model. This aspect requires the definition of an accurate mathematical representation, as well as a careful system identification process. Nevertheless, the resulting model will still be subject to noise, disturbances and remaining model uncertainties.

The approach based on the incremental dynamics on its hand requires only the identification of the input matrix, neglecting the parameters that depend exclusively on the internal states and it is expected to be robust to model uncertainties. Furthermore, the incremental formulation is very simple and intuitive, supported on well established mathematical fundamentals, with only a few parameters to set-up.

On the other hand, restrictions on the use of incremental controllers arise out of assumptions *A6.1* to *A6.5*. Initially, as a consequence of the model simplifications, all the needed information about the system states is obtained from measurements. Thus, to ensure quality in the measurements, the feedback states and state derivatives are to be updated at a sufficiently high sampling rate, with good quality sensors. Furthermore, *A6.3* implies that input control signal must have dominance on the system dynamics, demanding fast actuation when compared with system modes. These assumptions are satisfied in the most of UAVs since the control actuation has the greater influence over the dynamics.

In the following subsections we present and analyze the two techniques (WOF and ISG) proposed to mitigate the filtering design problems of input redundancy and second-order filtering delay. The results are illustrated through numerical examples considering an application for the INDI control.

6.2. Input Redundancy Treatment

If we consider that the input matrix $\mathbf{B}_0$ from (35) may not be square, we can expect an input redundancy to occur. Therefore, there are various possible solutions of inputs $\mathbf{u}_e$ that achieve a given equilibrium point $\mathbf{x}_e$. A common issue is that the redundant inputs can cancel each other when the system achieves a stationary state condition, resulting in more energy consumption.

As an example, consider a simplified vehicle dynamics with a single state variable x given by the longitudinal velocity and two inputs u_1 and u_2. This system dynamics can be represented by the following mathematical modeling:

$$\dot{x} = f(x) + g_1(x)u_1 + g_2(x)u_2. \tag{37}$$

Both control inputs u_1 and u_2 have influence over the vehicle velocity dynamics. Now consider an equilibrium point $x_e > 0$, thus $\dot{x}_e = 0$ and the following can be stated:

$$\begin{bmatrix} b_1 & -b_2 \end{bmatrix} \begin{bmatrix} u_1 \\ u_2 \end{bmatrix} = -f(x_e), \tag{38}$$

where $b_1 = g_1(x_e)$ and $b_2 = -g_2(x_e)$ are positive constants.

In this case, there are many choices for u_1 and u_2 resulting in the same constant longitudinal velocity x_e. However, this choice will impact in the energy consumption. The ideal solution is to use the minimal control effort in order to save energy.

One solution for this problem is to perform a filtering in the commanded redundant input which is less important for maintaining the equilibrium point. As a result we obtain a system dynamics described in the following form:

$$\dot{x} = f(x) + g_1(x)u_1 + g_2(x)\hat{u}_2, \tag{39a}$$

$$\dot{\hat{u}}_2 = \frac{-1}{\tau}\hat{u}_2 + u_2. \tag{39b}$$

where $\hat{u}_2$ is the filtered commanded input for the brake. By imposing this dynamics, u_2 will naturally converge to zero and u_1 will also reduce, once u_2 is no longer canceling it. Note that u_2 is still useful for the transient state condition when the vehicle needs to slowdown fast. This strategy is commonly referenced in the literature as Washout Filter (WOF) [13].

The solution can be extended for systems with Multiple Inputs and Multiple Outputs (MIMO). Consider the following generalized MIMO system dynamics:

$$\dot{\mathbf{x}} = f(\mathbf{x}) + g(\mathbf{x})\mathbf{u}, \tag{40}$$

where $\mathbf{x} \in \mathbb{R}^n$, $\mathbf{u} \in \mathbb{R}^m$ and $m \geq n$.

Thus, applying the WOF to the redundant inputs we obtain the following extended dynamics:

$$\begin{bmatrix} \dot{\mathbf{x}} \\ \dot{\hat{\mathbf{u}}}_s \end{bmatrix} = \begin{bmatrix} \mathbf{I}_n & g_s(\mathbf{x}) \\ \mathbf{0}_{l\times n} & \mathbf{T}_s \end{bmatrix} \begin{bmatrix} f(\mathbf{x}) \\ \hat{\mathbf{u}}_s \end{bmatrix} + \begin{bmatrix} g_m(\mathbf{x}) & \mathbf{0}_{n\times l} \\ \mathbf{0}_{l\times n} & \mathbf{I}_l \end{bmatrix} \begin{bmatrix} \mathbf{u}_m \\ \mathbf{u}_s \end{bmatrix} \tag{41}$$

where: $l = m - n$, $\mathbf{x} \in \mathbb{R}^n$ is the vector of states; $\mathbf{u}_m \in \mathbb{R}^n$ is the vector of main actuators; $\mathbf{u}_s \in \mathbb{R}^l$ is the vector of secondary (or redundant) actuators; $\hat{\mathbf{u}}_s \in \mathbb{R}^l$ is the vector of filtered input signals, $f(\mathbf{x})$ is the function of state dynamics; $g_m(\mathbf{x})$ is a function which describes the influence of $\mathbf{u}_m$ in the state dynamics; $g_s(\mathbf{x})$ is a function which describes the influence of $\mathbf{u}_s$ in the state dynamics; $\mathbf{T}_r = \text{diag}([-1/\tau_1; -1/\tau_2; -1/\tau_3; \cdots; -1/\tau_l])$ is a diagonal matrix with positive constants $\tau_1, \cdots, \tau_l$ chosen by the designer; $\mathbf{I}_n$ is the identity matrix of order n; and $\mathbf{0}_{n\times l}$ is a matrix full of zeros with n lines and l columns.

The vector of redundant actuators can be chosen by analyzing the input function $g(\mathbf{x})$ through a systematic procedure. First, the designer must linearize $g(\mathbf{x})$ in a chosen point $\mathbf{x}_0$, obtaining the following:

$$\mathbf{B}_0 = \frac{\partial}{\partial \mathbf{x}} g(\mathbf{x})\Big|_{x_0}. \tag{42}$$

By analyzing the matrix $\mathbf{B}_0 \in \mathbb{R}^{n\times m}$, the designer must identify the linearly dependent columns, which indicates the redundant inputs. After identifying the redundant actuators, the designer must choose and separate between main ($\mathbf{u}_m$) and secondary ($\mathbf{u}_s$) actuators, by also defining the functions $g_m(\mathbf{x})$ and $g_s(\mathbf{x})$. Then the extended dynamics (41) can be applied. This solution can be represented by a block diagram in a cascaded form as shown in Figure 8, where $\hat{\mathbf{u}} = [\mathbf{u}_m \ \hat{\mathbf{u}}_s]^T$ and $g(\mathbf{x}) = [g_m(\mathbf{x}) \ g_s(\mathbf{x})]$.

Numerical Example

As an example consider the following linear system with two redundant inputs:

$$\dot{x} = x + 0.2u_1 - 0.8u_2 \tag{43}$$

where x is the state, u_1 is the first input and u_2 is the second input.

By applying INDI, we obtain the following control law:

$$\mathbf{u} = \mathbf{u}_0 + \mathbf{B}_0^{+}(\nu - \dot{x}_0) \tag{44}$$

where $\mathbf{u} = [u_1 \ \ u_2]^T$, $\mathbf{B}_0 = [0.2 \ \ -0.8]$, $\mathbf{u}_0$ is the previous input and $\nu = k(x_d - x)$ is a desired dynamics of first order. The block diagram in Figure 8 illustrates the closed loop system, where z^{-1} represents a delay of one sample time $t_s = 0.01$ seconds, $\hat{\mathbf{u}} = [u_1 \ \ \hat{u}_2]^T$ is the control signal filtered by the WOF, $k = 20s^{-1}$ is a linear gain and $\tau = 5$ seconds is the time constant of the WOF. We suppose the presence of actuator saturation ($-0.5 \leq u_1 \leq 0.5$ and $-0.5 \leq u_2 \leq 0.5$,) which is treated with an anti-windup strategy [3].

Figures 9 and 10 show the results for a step in x_d using WOF denominated case (a) (or "INDI+WOF") and without WOF denominated case (b) (or "INDI"). For sake of comparison consider the following quadratic cost function:

$$J_u = \int_0^{t_f} u_1^2(t) + u_2^2(t)dt \tag{45}$$

where $t_f = 15$ seconds is the final time of simulation.

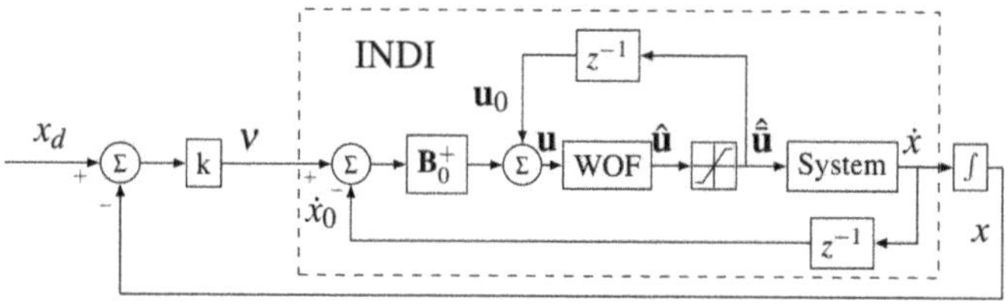

Figure 8. Numerical Example: Block diagram of the closed loop system of INDI+WOF.

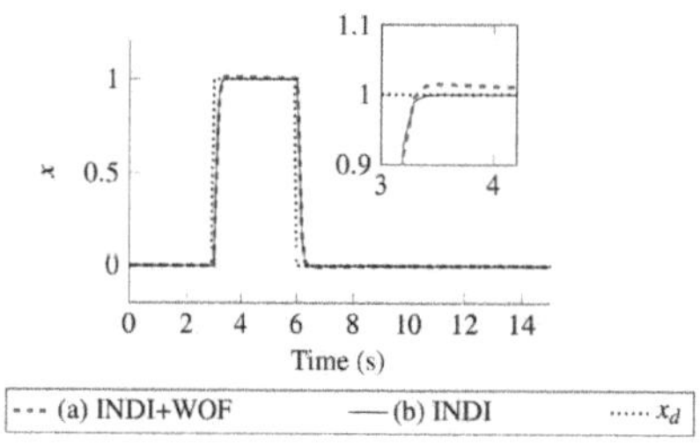

Figure 9. Numerical example: comparison of tracking performance between INDI+WOF and INDI

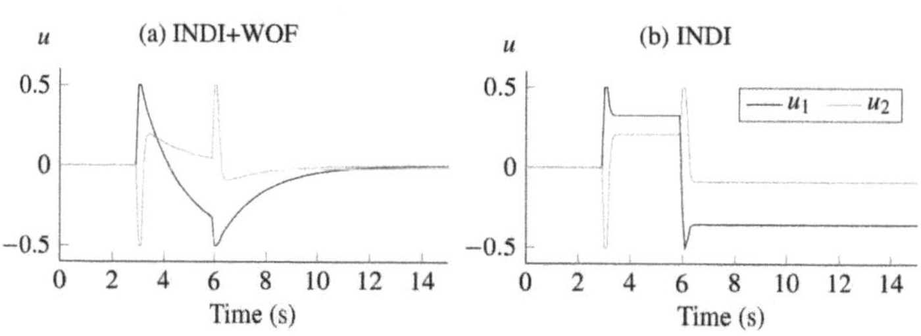

Figure 10. Numerical example: comparison of input signals between INDI+WOF and INDI

From the figures, we see that the tracking performance is similar for both cases, which does not happen for the control effort. In Figure 10, the saturation is achieved in both control inputs for both cases (a) and (b). In case (b) the commanded signals u_1 and u_2 converge to values different from zero after the transient state ($t > 6$ seconds). Consequently the system has higher energy consumption in case (b) with $J_u = 17.85$. In case (a) the energy consumption is reduced to $J_u = 5.74$. Therefore, the WOF appears as a simple and advantageous solution for systems with redundant actuation with actuator saturation, such as aerial vehicles i.e. multirotor drones, aircrafts and airships.

6.3. The Second-Order Differentiator and the Estimation of Derivatives

One important problem in control design, not limited to incremental controllers, is the need for the derivative of a given state. They may be difficult to obtain or even impossible, as is the case of the derivative of angular rates. Further, direct numerical differentiation of the output signal may lead to noise amplification and abrupt variations.

A common solution to this problem is the use a Second-Order Differentiator (SOD) to obtain the derivative of the state. The SOD works as a low-pass filter for the state yielding a filtered output of the measure $\mathbf{y}$. While providing an estimation of the state derivative, the SOD simultaneously attenuates the high-frequency noise in the measure signal.

If we call the state derivative vector as $\mathbf{w} = \dot{\mathbf{x}}$, the estimation of the derivative $\hat{\mathbf{w}}$, using the second-order-derivative (SOD), is given by:

$$\hat{\mathbf{w}}(s) = \left[\frac{\omega_n^2 s}{s^2 + 2\zeta\omega_n s + \omega_n^2}\right]\mathbf{y}(s)\,. \tag{46}$$

In the design procedure, the filter parameters like the natural frequency ω_n and the damping ratio ζ are used to adjust the passband mid-frequency, as well as the passband size.

6.4. Input Scaling Gain

A drawback that comes with the SOD filter is the natural delay produced by the second order dynamics, which may harm the control system feedback. Thus, it is important to mitigate this delayed estimation in order to robustify the controller.

One possible solution to this issue is the so called Input Scaling Gain (ISG), that was first proposed by our research group in 2015 [14], as a scalar factor. In this paper, we propose the generalization of the scale gain to MIMO systems, using a diagonal matrix $\mathbf{\Lambda} \in \mathbb{R}^{n\times n}$. In the case of the incremental controller design, the ISG is used to scale down the difference $(\nu - \dot{\mathbf{x}}_0)$ in (35), reducing the bandwidth of the closed-loop system. The matrix diagonal elements λ_{ii} are real numbers in the interval $]0,\, 1]$, such that the modified INDI control law is now given as:

$$\mathbf{u} = \mathbf{u}_0 + \mathbf{B}_0^{+}\mathbf{\Lambda}(\nu - \dot{\mathbf{x}}_0)\,, \tag{47}$$

Note that with $\Lambda = \mathbf{I}_n$ we have the traditional INDI control law (35).

Substituting (47) in the closed loop Incremental Dynamics (34) results in:

$$\dot{\mathbf{x}} = \dot{\mathbf{x}}_0 + \mathbf{\Lambda}(\nu - \dot{\mathbf{x}}_0)\,. \tag{48}$$

Considering the discrete implementation, where the time interval between t_0 and t is $\Delta t = T_s$, we can rewrite (48) as:

$$\frac{\Delta\dot{\mathbf{x}}}{\Delta t} = \frac{\Delta\dot{\mathbf{x}}}{T_s} = \frac{1}{T_s}\mathbf{\Lambda}(\nu - \dot{\mathbf{x}}_0)\,, \tag{49}$$

Assuming a fast sampling rate (small Δt), and denoting $\dot{\mathbf{x}}$ as $\mathbf{w}$, (49) approximates to:

$$\dot{\mathbf{w}} \simeq \frac{1}{T_s}\mathbf{\Lambda}(\nu - \mathbf{w}_0)\,. \tag{50}$$

Considering $t_0 = t - T_s$, (50) can be rewritten as:

$$\dot{\mathbf{w}}(t) = \frac{1}{T_s}\mathbf{\Lambda}[\nu(t) - \mathbf{w}(t - T_s)]\,. \tag{51}$$

Applying Laplace transform to (51), provides the transfer function between the state derivative and the pseudo-control as:

$$\mathcal{L}\left\{\dot{\mathbf{w}}(t)+\frac{1}{T_s}\mathbf{\Lambda}\mathbf{w}(t-T_s)\right\}=\frac{1}{T_s}\mathbf{\Lambda}\mathcal{L}\{\boldsymbol{\nu}(t)\}\Leftrightarrow$$
$$\Leftrightarrow T_s\mathbf{\Lambda}^{-1}s\mathbf{w}(s)+e^{-sT_s}\mathbf{w}(s)=\boldsymbol{\nu}(s)\Leftrightarrow$$
$$\Leftrightarrow \mathbf{w}(s)=\left(T_s\mathbf{\Lambda}^{-1}s+e^{-sT_s}\mathbf{I}\right)^{-1}\boldsymbol{\nu}(s)=\mathrm{diag}\left\{\frac{1}{\frac{T_s}{\lambda_{ii}}s+e^{-sT_s}}\right\}\boldsymbol{\nu}(s)\,. \tag{52}$$

Some important points to be remarked are:

- each component of ν_i of pseudo-control $\boldsymbol{\nu}$ has an independent time constant $\frac{T_s}{\lambda_{ii}}$, as the scale matrix $\mathbf{\Lambda}$ is a diagonal matrix.
- Under fast sampling, we can assume $e^{-sT_s}\simeq 1$ implying that (52) simplifies to a simple first order low-pass filter, attenuating high-frequency perturbations in the pseudo-control ν;
- as a possible drawback, the ISG reduces the bandwidth of the closed-loop dynamics, eventually decreasing the overall performance;

6.5. Combining ISG and SOD: Closed-Loop Analysis

In order to investigate the stability properties as well as the performance improvements against delays in the closed-loop system, we analyze here the proposed SOD+ISG solution for a first-order SISO system, as shown in Figure 11.

Although illustrated here for a SISO system, the approach can be extended to multiple-input-multiple-output (MIMO) systems as the diagonal structure of the ISG yields independent components z_i of the INDI-loop error $\mathbf{z}=\boldsymbol{\nu}-\mathbf{w}_0$, and the SOD is also independent for each state x_i.

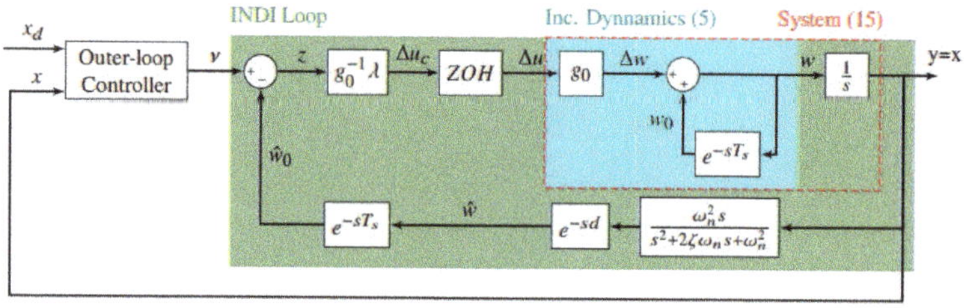

Figure 11. Block diagram of INDI controller applied to a first-order plant.

Firstly, to consider continuous systems, the discrete sample delays z^{-1} of INDI (see Figure 7) are substituted by time delay components e^{-sT_s}. Further, to simulate the discrete feature of INDI we add a Zero-Order Hold (ZOH) in the control loop, as well as an extra transport delay d to investigate the effects of additional unmodeled delays in the loop.

6.5.1. Analytical Formulation

The system to be controlled is defined as:

$$\begin{cases}\dot{x}=\dot{x}_0+g(x_0)\Delta u\\ y=x\end{cases}, \tag{53}$$

where its dynamics is described using ID formulation (34), and its output is the state itself.

In this section we only analyze the local behavior of the system, where the control effectiveness function $g(x_0)$ is approximated by a constant g_0.

The transfer function $H_1(s)$ relating the state derivative w with the input increment Δu is given (34) by:

$$H_1(s) = \frac{w(s)}{\Delta u(s)} = g_0\left(\frac{1}{1-e^{-sT_s}}\right), \tag{54}$$

and, thus, the transfer function $H_2(s)$ is given (53) by:

$$H_2(s) = \frac{y(s)}{\Delta u(s)} = \frac{1}{s}H_1(s). \tag{55}$$

One important feature of the INDI controller, as shown in (36), is that the state derivative w should follow a given pseudo-control ν. Therefore, it is important to analyze the transfer function from ν to w, denoted here as $H_3(s)$, shown in the block diagram of Figure 11, and given by:

$$\begin{aligned} H_3(s) = \frac{w(s)}{\nu(s)} &= \frac{g_0^{-1}\lambda Z(s)H_1(s)}{1+g_0^{-1}\lambda Z(s)H_2(s)\frac{\omega_n^2 s}{s^2+2\zeta\omega_n s+\omega_n^2}e^{-s(T_s+d)}} \\ &= \frac{g_0^{-1}\lambda Z(s)H_1(s)(s^2+2\zeta\omega_n s+\omega_n^2)}{s^2+2\zeta\omega_n s+\omega_n^2+g_0^{-1}\lambda\omega_n^2 e^{-s(T_s+d)}Z(s)H_2(s)s}, \end{aligned} \tag{56}$$

where $Z(s) = \frac{1-e^{-sT_s}}{sT_s}$ is the zero-order-holder.

Substituting (54), (55), and $Z(s)$ into (56), we finally come to the transfer function from ν to w as:

$$H_3(s) = \frac{\frac{\lambda}{T_s}(s^2+2\zeta\omega_n s+\omega_n^2)}{s^3+2\zeta\omega_n s^2+\omega_n^2 s+\frac{\lambda}{T_s}\omega_n^2 e^{-s(T_s+d)}}. \tag{57}$$

Note that (57) is a delayed system where the delay appears in the constant term of the characteristic equation, such that the simple first-order Padé approximation

$$e^{-sT} \approx \frac{1-sT/2}{1+sT/2}. \tag{58}$$

can be used to investigate the stability analysis of the closed-loop system.

The resulting transfer function $H_3(s)$ is finally approximated to:

$$H_3(s) = \frac{\frac{\lambda}{T_s}\left(1+s\frac{(T_s+d)}{2}\right)(s^2+2\zeta\omega_n s+\omega_n^2)}{\frac{(T_s+d)}{2}s^4+(\zeta\omega_n(T_s+d)+1)s^3+\left(2\zeta\omega_n+\frac{(T_s+d)}{2}\omega_n^2\right)s^2+\left(\frac{(2-\lambda)T_s-\lambda d}{2T_s}\omega_n^2\right)s+\frac{\lambda}{T_s}\omega_n^2}. \tag{59}$$

From which we can conclude that:

- As desired in the INDI approach, w will follow a pseudo control ν, due to the fact that the transfer function has unitary dc-gain;
- Considering the Routh–Hurwitz criterion, $H_3(s)$ will be stable for positive coefficients in the characteristic polynomial, which implies that $(2-\lambda)T_s - \lambda d > 0$;
- Further, the ISG is also effective to mitigate the effects of additional delays in the closed-loop. For example, without the use of ISG ($\lambda = 1$), the extra delay should never be greater than the sample time T_s.

6.5.2. Numerical Example

A numerical example is presented here to investigate the effect of the ISG gain and the SOD differentiator approach in the control loop. To investigate the role of the delay in the INDI control, let us analyze the scenario in which there is an extra delay $d > T_s$ in the INDI feedback loop. Figure 12 shows the results for the step inputs applied to $H_3(s)$ and the corresponding Bode plots, assuming a sampling frequency of 50 Hertz.

The simulations were done for three diferent conditions: a) Varying ISG scale gains for a given measured state derivative (note that lower λ implies lower natural frequency and bigger damping ratio); b) Varying ISG scale gains for a differentiator with a given natural frequency of $\frac{15}{T_s}$; and c) Varying the natural frequencies of the differentiators for a fixed ISG gain equal to 0.5.

The step response plots, without using the ISG (equivalent to $\lambda = 1$), shows that the system is unstable for this plant that has an extra delay d larger than the sample time T_s. However, the system becomes stable if a lower value of λ is considered. For the case where the bandwidth of the differentiator satisfies the relation $\omega_n T_s = 15$, the extra delay d added by the filter destabilizes the system even if we use an ISG gain of 0.8, while an ISG gain of 0.5 is sufficient to make it stable for different SOD frequencies.

From these results, we conclude that the combined approach SOD + ISG yields a useful and practical derivative estimator, since the ISG can mitigate the effects of the delay generated by the second-order differentiator. Additionally, we remark that it can even mitigate other possible delays appearing in the feedback loop.

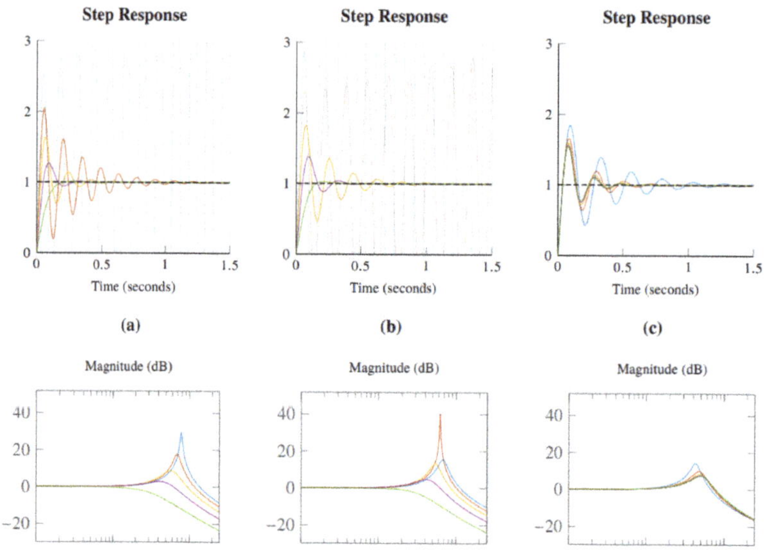

Figure 12. Step responses and bode plots of the INDI loop with exta loop delay $d = 1.1T_s$ ($22ms$), comparing: (**a**) ISG values with ideal state feedback; (**b**) ISG values with $\omega_n T_s = 15$ in the SOD; (**c**) SOD frequencies with ISG of $\lambda = 0.5$.

7. Conclusions

In this paper we presented a comparative study evaluating different solutions for the state estimation of a robotic airship as well as its derivatives. For the state estimation we used a second order filter, an Extended Kalman Filter and an Unscented Kalman Filter. We performed all simulations using a high-fidelity airship dynamic model for Simulink platform. For the state estimation, the EKF showed a better performance when compared to the other estimation techniques.

In a second moment, we presented a Model-based (EKF) and a Data-driven (Neural Network) approaches for the estimation of the wind velocity. Also, we proposed a novel Hybrid approach, by performing a fusion between the designed Model-based and Data-driven approaches in a cascaded structure. While the NN approach alone tends to estimate the wind speed with an offset (bias), the EKF estimation has long delays for sudden wind changes. However, the Hybrid approach tends to yield a faster response with small bias for the estimated wind components.

Finally, we presented possible solutions to mitigate two common problems arriving in control implementation that are the control input redundancy and the delay introduced by second-order filters. The first one is solved with the use of the Wash Out Filter (WOF). The second one is mitigated with the use of the Input Scaling Gain (ISG) which is generalized here for the MIMO case. To illustrate the design solutions we applied the techniques for the design of incremental controllers, although they can be used also for other nonlinear control approaches in the same situation.

Author Contributions: A.M.: formal analysis and investigation; J.A.: investigation and supervision; A.F.: supervision; E.P.: supervision, validation and writing—review & editing; J.C.: methodology and resource; R.C.: methodology and writing—review & editing. All authors have read and agreed to the published version of the manuscript.

Funding: This work was supported by Project INCT-SAC (CNPq 465755/2014-3;FAPESP 2014/508510).

Data Availability Statement: The estimation results are available for visualization as video at youtu.be/VL5dvCyOZwY (EKF) and youtu.be/B26xaKtAyWo (UKF) accessed on 25 May 2022.

Acknowledgments: The authors gratefully acknowledge the agencies FAPESP and CNPq for the financial support. From Fapesp INCT-SAC Project (2014/50851-0), co-funded by CNPq (465755/2014-3).

Conflicts of Interest: The authors declare no conflict of interest.

Abbreviations

The following abbreviations are used in this manuscript:

AURORA	Autonomous Unmanned Remote Monitoring Robotic Airship
CB	Center of Buoyancy
CV	Center of Volume
DIVA	Instrumented Airship for Air Surveillance
DRONI	Robotic Airship of Innovative Design
EKF	Extended Kalman Filter
ENU	East-North-Upper
ID	Incremental Dynamics
INDI	Incremental Nonlinear Dynamic Inversion
ISG	Input Scale Gain
GPS	Global Positioning System
IMU	Inertial measurement unit
InSAC	National Institute of Science and Technology for Cooperative Autonomous Systems
LPF	Low-pass filter
NDI	Nonlinear Dynamic Inversion
NED	North-East-Down
NN	Neural Network
ROS	Robotic Operating System
SOD	Second-Order Differentiator
UAV	Unmanned Aerial Vehicle
UKF	Unscented Kalman Filter
WOF	Wash Out Filter

References

1. Liu, Y.; Pan, Z.; Stirling, D.; Naghdy, F. Control of autonomous airship. In Proceedings of the 2009 IEEE International Conference on Robotics and Biomimetics (ROBIO), Guilin, China, 13–19 December 2009 ; Institute of Electrical & Electronics Engineers (IEEE): Piscataway, NJ, USA, 2009.
2. Liu, S.Q.; Gong, S.J.; Li, Y.X.; Lu, Z.R. Vectorial backstepping method–based trajectory tracking control for an under-actuated stratospheric airship. *Aeronaut. J.* **2017**, *121*, 916–939. [CrossRef]
3. Marton, A.S. Control Architecture for the Navigation System of Robotic Airship using Incremental Controllers. Ph.D. Thesis, Universidade de Campinas, Campinas, Brazil, 2021.

4. Recoskie, S.; Lanteigne, E.; Gueaieb, W. A High-Fidelity Energy Efficient Path Planner for Unmanned Airships. *Robotics* **2017**, *6*, 28. [CrossRef]
5. Yang, Y.; Wu, J.; Zheng, W. Positioning Control for an Autonomous Airship. *J. Aircr.* **2016**, *53*, 1638–1646. [CrossRef]
6. Cho, A.; Kim, J.; Lee, S.; Kee, C. Wind Estimation and Airspeed Calibration using a UAV with a Single-Antenna GPS Receiver and Pitot Tube. *IEEE Trans. Aerosp. Electron. Syst.* **2011**, *47*, 109–117. [CrossRef]
7. Moutinho, A.; Azinheira, J.R.; de Paiva, E.C.; Bueno, S.S. Airship robust path-tracking: A tutorial on airship modelling and gain-scheduling control design. *Control. Eng. Pract.* **2016**, *50*, 22–36. [CrossRef]
8. Perry, J.; Mohamed, A.; Johnson, B.; Lind, R. Estimating Angle of Attack and Sideslip Under High Dynamics on Small UAVs. In Proceedings of the 21st International Technical Meeting of the Satellite Division of The Institute of Navigation (ION GNSS 2008), Savannah, Georgia, 16–19 September 2008; Institute of Electrical and Electronics Engineers (IEEE): Piscataway, NJ, USA, 2008; pp. 1165–1173.
9. Johansen, T.A.; Cristofaro, A.; Sorensen, K.; Hansen, J.M.; Fossen, T.I. On estimation of wind velocity, angle-of-attack and sideslip angle of small UAVs using standard sensors. In Proceedings of the 2015 International Conference on Unmanned Aircraft Systems (ICUAS), Denver, CO, USA, 9–12 June 2015; Institute of Electrical & Electronics Engineers (IEEE): Piscataway, NJ, USA, 2015.
10. Shen, S.; Liu, L.; Huang, B.; Lin, X.; Lan, W.; Jin, H. Wind Speed Estimation and Station-Keeping Control for Stratospheric Airships with Extended Kalman Filter. In *Proceedings of the 2015 Chinese Intelligent Automation Conference*; Springer Science Business Media: Berlin/Heidelberg, Germany, 2015; pp. 145–157.
11. Rhudy, M.B.; Gu, Y.; Gross, J.N.; Chao, H. Onboard Wind Velocity Estimation Comparison for Unmanned Aircraft Systems. *IEEE Trans. Aerosp. Electron. Syst.* **2017**, *53*, 55–66. [CrossRef]
12. Vieira, H.S.; de Paiva, E.C.; Moriguchi, S.K.; Carvalho, J.R.H. Unified Backstepping Sliding Mode Framework for Airship Control Design. *IEEE Trans. Aerosp. Electron. Syst.* **2020**, *56*, 3246–3258. [CrossRef]
13. Hassouneh, M.; Lee, H.C.; Abed, E. Washout filters in feedback control: Benefits, limitations and extensions. In Proceedings of the 2004 American Control Conference, Boston, MA, USA, 30 June–2 July 2004; IEEE: Piscataway, NJ, USA, 2004.
14. Azinheira, J.; Moutinho, A.; Carvalho, J. Lateral Control of Airship with Uncertain Dynamics using Incremental Nonlinear Dynamics Inversion. *IFAC-PapersOnLine* **2015**, *48*, 69–74. [CrossRef]
15. Rueda, M.; Mirisola, L.; Nogueira, L.; Fonseca, G.; Ramos, J.; Koyama, M.; Azinheira, J.; Carvalho, R.; Bueno, S.; de Paiva, E. Uma Infraestrutura, de Hardware, Software e Comunicação para a Robotização de Plataformas Rádio-Controladas: Aplicação a um Dirigível Robótico. In Proceedings of the 2017 SBAI-XIII Simpósio Brasileiro de Automação Inteligente, Porto Alegre, Brazil, 1–4 October 2017.
16. Meyer, J.; Sendobry, A.; Kohlbrecher, S.; Klingauf, U.; von Stryk, O. Comprehensive Simulation of Quadrotor UAVs using ROS and Gazebo. In Proceedings of the 3rd International Conference on Simulation, Modeling and Programming for Autonomous Robots (SIMPAR), Tsukuba, Japan, 5–8 November 2012.
17. Quigley, M.; Conley, K.; Gerkey, B.P.; Faust, J.; Foote, T.; Leibs, J.; Wheeler, R.; Ng, A.Y. ROS: An open-source Robot Operating System. In Proceedings of the ICRA Workshop on Open Source Software, Kobe, Japan, 12–17 May 2009.
18. Moore, T.; Stouch, D. A Generalized Extended Kalman Filter Implementation for the Robot Operating System. In *Proceedings of the 13th International Conference on Intelligent Autonomous Systems (IAS-13)*; Springer: Berlin/Heidelberg, Germany, 2014.
19. Bestaoui, Y. *Lighter than Air Robots*; Springer: Berlin/Heidelberg, Germany, 2011.
20. Gomes, S. An Investigation of the Flight Dynamics of Airships with Application to the YEZ-2A. Ph.D. Thesis, Cranfield Institute of Technology, Cranfield, UK, 1990.
21. Li, Y.; Nahon, M.; Sharf, I. Airship dynamics modeling: A literature review. *Prog. Aerosp. Sci.* **2011**, *47*, 217–239. [CrossRef]
22. Li, Y. Dynamics Modeling and Simulation of Flexible Airships. Ph.D. Thesis, McGill University, Montreal, QC, Canada, 2008.
23. Azouz, N.; Bestaoui, Y.; Lemaitre, O. Dynamic analysis of airships with small deformations. In Proceedings of the Third International Workshop on Robot Motion and Control (RoMoCo '02), Bukowy Dworek, Poland, 11 November 2002; pp. 209–215. [CrossRef]
24. Lane, S.H.; Stengel, R.F. Flight control design using non-linear inverse dynamics. *Automatica* **1988**, *24*, 471–483. [CrossRef]
25. Benallegue, A.; Mokhtari, A.; Fridman, L. Feedback Linearization and High Order Sliding Mode Observer For A Quadrotor UAV. In Proceedings of the International Workshop on Variable Structure Systems, 2006. (VSS06), Alghero, Italy, 5–7 June 2006; Institute of Electrical and Electronics Engineers (IEEE): Piscataway, NJ, USA, 2006.
26. Voos, H. Nonlinear control of a quadrotor micro-UAV using feedback-linearization. In Proceedings of the 2009 IEEE International Conference on Mechatronics, Malaga, Spain, 14–17 April 2009; Institute of Electrical & Electronics Engineers (IEEE): Piscataway, NJ, USA, 2009.
27. Smith, P. A simplified approach to nonlinear dynamic inversion based flight control. In Proceedings of the 23rd Atmospheric Flight Mechanics Conference, Boston, MA, USA, 10–12 August 1998; American Institute of Aeronautics and Astronautics (AIAA): Reston, VA, USA, 1998.
28. Sieberling, S.; Chu, Q.P.; Mulder, J.A. Robust Flight Control Using Incremental Nonlinear Dynamic Inversion and Angular Acceleration Prediction. *J. Guid. Control. Dyn.* **2010**, *33*, 1732–1742. [CrossRef]

9. Acquatella, P.; van Kampen, E.; Chu, Q.P. Incremental backstepping for robust nonlinear flight control. In Proceedings of the EuroGNC 2013, 2nd CEAS Specialist Conference on Guidance, Navigation & Control, Delft University of Technology, Delft, The Netherlands, 10–12 April 2013.
0. Wang, X.; Kampen, E.J.V.; Chu, Q.; Lu, P. Incremental Sliding-Mode Fault-Tolerant Flight Control. *J. Guid. Control. Dyn.* **2019**, *42*, 244–259. [CrossRef]

aerospace MDPI

Article

Research on Trajectory Prediction of a High-Altitude Zero-Pressure Balloon System to Assist Rapid Recovery

Jiwei Tang [1], Shumin Pu [2,*], Peixi Yu [2], Weicheng Xie [3], Yunfei Li [4] and Binxing Hu [3]

1 School of Aeronautics & Astronautics, Shanghai Jiao Tong University, Shanghai 200240, China
2 Shanghai Jiao Tong University Chongqing Near Space Innovation R&D Center, Chongqing 401135, China
3 Aerospace System Engineering Shanghai, Shanghai 200240, China
4 North Institute of Nuclear Technology, Xi'an 710000, China
* Correspondence: psm446734413@mail.nwpu.edu.cn; Tel.: +86-18281029537

Abstract: A comprehensive simulation model is established to predict the trajectory of a high-altitude zero-pressure balloon flight system with no parachute that is required to carry the load floating at the designated altitude for several hours or less. A series of mathematical models, including thermal dynamic, atmospheric, earth, wind, geometry, and exhaust models, are developed to predict the trajectory of the balloon flight system. Based on these models, the uncertainties of the launch parameters and the corresponding flight performance are simulated. Combined with the control strategy, the entire flight trajectory is simulated and discussed in detail, including the ascending, floating, and descending phases. The results show that the vertical velocity takes on a W shape during the ascent process. Furthermore, the balloon begins to gradually descend with weakening solar radiation after noon. Moreover, the landing vertical speed of the balloon flight system can approach zero with the control strategy applied, whereas the lateral drift range is more limited relative to the uncontrolled flight mode. The results and conclusions presented herein contribute to the design and operation of a zero-pressure balloon flight system within limited airspace to improve the rapid recovery ability of the flight system.

Keywords: zero-pressure balloon; trajectory prediction; thermal dynamic model; control strategy; rapid recovery

1. Introduction

With the potential for civil and military applications, including communication, observation, and scientific research, there is growing interest in studying platforms in near space [1–3]. The high-altitude zero-pressure balloon has become an indispensable test platform in near space owing to its accessibility and affordability, attracting the attention of researchers worldwide [4–8].

In order to accomplish the mission of a balloon flight system, it is necessary to predict the system's trajectory in advance to prevent the balloon from floating out of the designated airspace with the wind. Throughout the flight mission, uncertainties with respect to the launch and environmental parameters play an important role in predicting flight performance, which influences the safety of the balloon flight. The failure of helium mass prediction can result in the balloon not reaching the designated altitude. The failure of helium temperature prediction can affect trajectory planning. The effectiveness of trajectory prediction depends on the accuracy of the relevant parameters that are necessary to optimize the balloon design and launch process.

In recent years, many investigations have been conducted on high-altitude balloon trajectory predictions. Rodger E. Farley designed a software program without considering the uncertainties during the flight process [9]. Qiumin Dai established a thermal dynamic model of a super-pressure balloon, emphasizing the importance of the film radiation property and clouds with respect to the thermal behaviors of the balloon [10]. András Sóbester developed a trajectory prediction model incorporating a new stochastic drag model based on empirical data derived from thousands of radiosonde flights to improve prediction accuracy [11]. Yi Zhang took the influences of the initial launch conditions into consideration to simulate the ascending process of a balloon [12]. Ö. Kayhan designed a gas-compress-release system with PID control for a zero-pressure balloon system, which helps the balloon system reach its target altitude and land at a safe vertical speed [13]. Yongseon Lee studied a trajectory prediction model that considers the uncertainties during the flight process [14]. Sherif Saleh developed an ascending performance prediction model that optimizes the necessary inflating quantity of helium to improve the ascending performance [15]. R. Waghela developed a balloon trajectory control system capable of passively guiding a high-altitude balloon that uses a given sail as an actuator to control the balloon system [16]. Xiaolong Deng detailed the Google Loon super-pressure balloon system, which uses a sub-ballonet for altitude control and an electric propulsion propeller system for enhanced operational performance [17]. Previous studies have mainly focused on the uncertainties with respect to thermal environmental parameters and launch parameters, which influence the accuracy of trajectory prediction. Furthermore, the previously proposed control strategies have mainly focused on altitude control.

Based on previous works, in this paper, we focus on trajectory prediction for a high-altitude zero-pressure balloon within limited airspace to optimize the rapid recovery of the balloon flight system. First, a comprehensive simulation model is established to predict flight performance and trajectory during the ascending, floating, and descending phases by considering various uncertainties. Second, taking control strategies into consideration improves the ability to rapidly recover the balloon flight system, preventing the balloon from flying out of the designated airspace in order to avoid possible accidents. Because the balloon is unpowered and there is no way to control its horizontal motion, it is necessary to predict its trajectory after launching from the ground.

The balloon flight system considered in this study descends without the separation of the balloon from the load, which enables the reuse of the balloon capsule, representing cost savings, compared with a system in which the balloon bursts, relying on a parachute to recover the load. Figure 1 shows the entire process of the flight mission of a balloon flight system, including the dynamic launching phase from the ground, the ascending phase, the floating phase, and the descending phase.

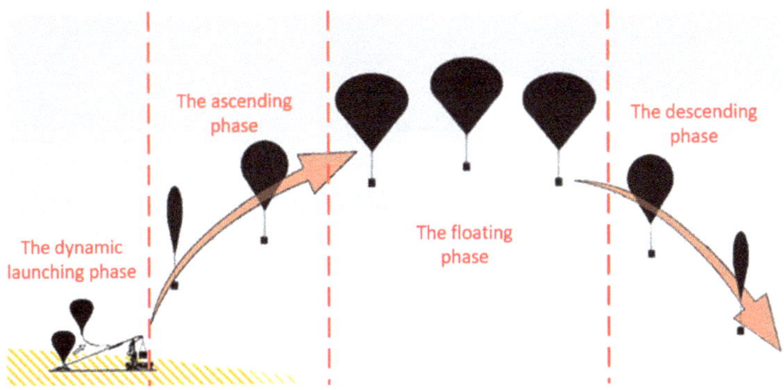

Figure 1. Flight mission of a balloon flight system.

2. Environment Model

During the flight process of a zero-pressure balloon, the differential temperature magnitude between the internal helium and the ambient atmosphere directly affects the net buoyancy of the balloon. Changes in environmental parameters, such as the surrounding atmospheric temperature, air pressure, air density, thermal radiation, and acceleration of gravity during the flight process, also directly or indirectly affect the internal helium temperature and helium mass of the zero-pressure balloon. All of these variations should be taken into account to improve the accuracy of balloon system trajectory prediction.

2.1. Atmospheric Model

Atmospheric temperature, pressure, and density vary with altitude. In this study, the standard atmospheric model is used, which was released in 1976 [18]. The atmospheric temperature is defined as

$$T_{air} = \begin{cases} 288.15 - 0.0065 * z & 0\,m < z \le 11,000\,m \\ 216.65 & 11,000\,m < z \le 20,000\,m \\ 216.65 + 0.0010 * (z - 20,000) & 20,000\,m < z \le 32,000\,m \end{cases}. \quad (1)$$

The atmospheric pressure is defined as

$$P_{air} = \begin{cases} 101,325 * ((288.15 - 0.0065 * z)/288.15)^{5.25577} & 0\,m < z \le 11,000\,m \\ 22,632^{(-(z-11,000)/6341.62)} & 11,000\,m < z \le 20,000\,m \\ 5474.87 * ((216.65 + 0.0010 * (z - 20,000))/216.65)^{-34.163} & 20,000\,m < z \le 32,000\,m \end{cases}. \quad (2)$$

According to the ideal gas law, the density is defined as

$$\rho_{air} = P_{air} / \ (T_{air} * R_{air}), \quad (3)$$

where z is the altitude, and R_{air} refers to the gas constant of air.

2.2. Thermal Model

Throughout the entire flight process of a zero-pressure balloon, the balloon system is in a thermal environment. Research [19] has shown that the main heat sources are direct solar radiation, ground-reflected radiation, sky-scattered radiation, Earth's infrared radiation, atmospheric infrared radiation, surface thermal radiation, and convective heat transfer. Figure 2 shows the heat sources for a high-altitude balloon system. Direct solar radiation, ground-reflected radiation, and sky-scattered radiation are examples of solar short-wave radiation. Earth's infrared radiation, atmospheric infrared radiation, and surface thermal radiation are examples of long-wave radiation.

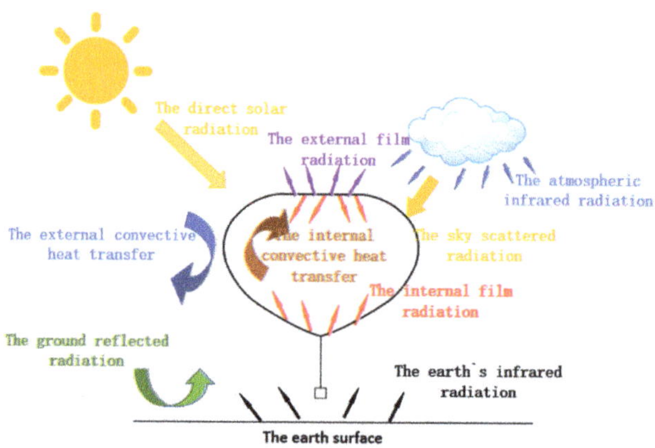

Figure 2. Heat sources for a high-altitude balloon system.

Through the heat transfer mechanism of the thermal model, heat is transferred among the balloon film, the internal helium, and the atmosphere [20]. Figure 3 shows the heat transfer mechanism. The following content illustrates the heat sources' models and the heat transfer relationships among the atmosphere, the balloon film, and the internal helium in detail.

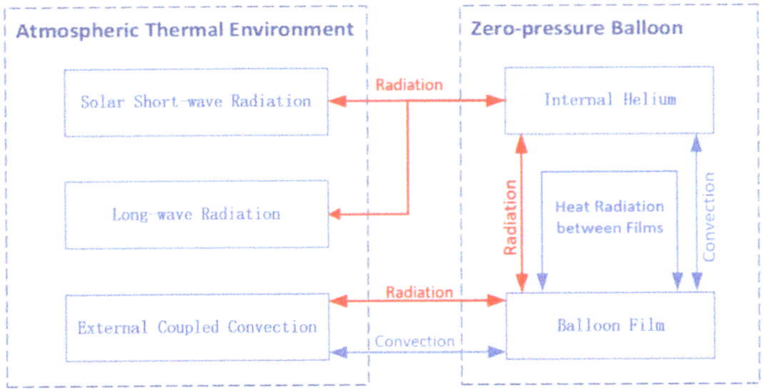

Figure 3. Heat transfer mechanism of the thermal model.

2.2.1. Direct Solar Radiation

The absorption of direct solar radiation by the balloon consists of absorption by the outer surface and absorption by the inner surface through the balloon film, and it is defined as follows [9]:

$$Q_{Direct} = \alpha * A_{projected} * I_{sun} * \tau_{atm} * \left[1 + \tau * \left(1 + r_{effective}\right)\right], \tag{4}$$

where α indicates the film absorption factor of solar radiation, $A_{projected}$ is the illuminated projected area, I_{sun} is the solar radiation constant number, τ_{atm} is the atmospheric transmissivity factor of solar radiation, τ is the film transmissivity factor of solar radiation, and $r_{effective}$ is the solar radiation effective reflectivity factor of the balloon film.

τ_{atm} is specified as [9]

$$\tau_{atm} = 0.5 * \left[e^{-0.65 * Air_{Mass}} + e^{-0.95 * Air_{Mass}}\right], \tag{5}$$

where Air_{Mass} is the absolute atmospheric optical quality. As the altitude increases, the radius of Earth's curvature changes, meaning that absolute atmospheric optical quality needs to be represented in two forms [21]:

- solar elevation angle < 5°:

$$Air_{Mass} = \left(\sin h + 0.457 * (90 - h)^{0.07}\right) * (6.484 + h)^{-1.697} * \left(\frac{P_{air}}{P_0}\right)^{(0.488 + 0.15 * h)} \tag{6}$$

- solar elevation angle > 5°:

$$Air_{Mass} = \left(\sin h + 0.457 * (90 - h)^{0.07}\right) * \left((6.484 + h)^{-1.697}\right)^{-1} * \frac{P_{air}}{P_0}, \tag{7}$$

where h stands for the solar elevation angle, and P_0 is for the standard atmospheric pressure. The solar elevation angle varies with the region, season, and time of day, but it ranges from 0° to 90°. For a specified location, the solar elevation angle equals 0° from sunset to sunrise. During the daytime, it reaches 90° at noon, which is the greatest intensity of solar radiation for the day. The solar elevation angle is formulated as

$$\sin h = \sin \varphi * \sin \zeta + \cos \varphi * \cos \zeta * \cos \omega, \tag{8}$$

where φ refers to the local latitude, ζ refers to the solar declination angle, and ω refers to the solar hour angle. The variation of the solar declination angle is the main cause of the variation of solar radiation intensity on Earth's surface, which is formulated as [21]

$$\zeta = 23.45 * \sin\left(2 * \pi * \frac{Nday + 284}{365}\right), \tag{9}$$

where $Nday$ stands for the day's serial number, and it is computed from the first of January each year. The solar hour angle is defined as

$$\omega = (TH - 12) * 15, \tag{10}$$

where TH represents the true solar time. There are two reasons causing the time difference between true solar time and mean solar time. First, the distance and the relative position between the Sun and Earth vary with time. Secondly, Earth's equator does not coincide with the plane of its orbit around the Sun. The true solar time is specified as [21]

$$TH = TH_s \pm \frac{La - La_s}{15} + \frac{\Delta TH}{60}, \tag{11}$$

where TH_s stands for the regional standard time, which depends on which time zone the balloon flight system is in. La stands for the local longitude, and La_s refers to the regional standard time position's longitude, which is equivalent to the longitude of the centerline of the time zone. ΔTH stands for the time difference. The time difference is formulated as [21]

$$\Delta TH = -0.00167 + 7.36629 * \cos(W + 1.4981) - 9.92745 * \cos(2 * W - 1.26155) - 0.32123 * \cos(3 * W - 1.15710), \tag{12}$$

where W represents Earth's orbit time angle and is indicated by the following formula [21]:

$$W = \frac{2 * \pi * Nday}{365} + 2 * e * \sin\left(\frac{2 * \pi * Nday}{365}\right) + 1.25 * e^2 * \sin\left(2 * \frac{2 * \pi * Nday}{365}\right), \tag{13}$$

where e stands for Earth's orbit eccentricity. For use in the equations below, $r_{effective}$ is formulated as [10]

$$r_{effective} = r + r^2 + r^3 + r^4 + r^5 + \ldots, \tag{14}$$

where r stands for the solar radiation reflectivity factor of the balloon film.

2.2.2. Ground-Reflected Radiation

Ground-reflected radiation is specified as [9]

$$Q_{Albedo} = \alpha * \rho_{earth} * A_{surf} * I_{sun} * \sin h * \tau_{ViewFactor} * \left[1 + \tau * \left(1 + r_{effective}\right)\right], \tag{15}$$

where ρ_{earth} indicates Earth's surface reflectance factor, $\tau_{ViewFactor}$ is the angular coefficient from the balloon's surface to Earth's surface, and A_{surf} stands for the surface area of the zero-pressure balloon.

2.2.3. Sky-Scattered Radiation

Sky-scattered radiation is formulated as [22]

$$Q_{Scatter} = 0.5 * \alpha * A_{surf} * I_{sun} * \sin h * \tau_{ViewFactor} * \frac{Air_{Mass} * (1 - \tau_{atm})}{Air_{Mass}^{-1.4*\ln}\tau_{atm}} * \left[1 + \tau * \left(1 + r_{effective}\right)\right]. \tag{16}$$

2.2.4. Earth's Infrared Radiation

Earth's infrared radiation is indicated by the following formula [22]:

$$Q_{IR,\ earth} = \alpha_{IR} * \varepsilon_{earth} * \sigma * T_{earth}^4 * A_{surf} * \tau_{ViewFactor} * \tau_{atm,IR} * \left[1 + \tau_{IR} * \left(1 + r_{effective,\ IR}\right)\right], \quad (17)$$

where α_{IR} represents the balloon film absorption factor of infrared radiation, ε_{earth} refers to the ground's average infrared emissivity factor, σ is the Stefan–Boltzmann constant, τ_{IR} is the balloon film transmissivity factor of infrared radiation, $r_{effective,IR}$ is the infrared radiation effective reflectivity factor of the balloon film, T_{earth} represents Earth's surface temperature, and $\tau_{atm,IR}$ is the atmospheric transmissivity factor of infrared radiation.

2.2.5. Atmospheric Infrared Radiation

Atmospheric infrared radiation is formulated as [22]

$$Q_{IR,\ Sky} = \alpha_{IR} * \sigma * \varepsilon_{sky} * T_{sky}^4 * A_{surf} * (1 - \tau_{ViewFactor}) * \left[1 + \tau_{IR} * \left(1 + r_{effective,\ IR}\right)\right], \quad (18)$$

where ε_{sky} stands for the atmospheric average infrared emissivity factor. T_{sky} refers to the sky equivalent temperature, which is defined as [23]

$$T_{sky} = 0.052 * T_{air}^{1.5} \quad (19)$$

2.2.6. Surface Thermal Radiation

The infrared radiation of the balloon film includes the film's external surface and the film's internal surface's infrared radiation. The total surface thermal radiation is calculated by the following formula [24]:

$$Q_{IR,\ film} = Q_{IR,out} - Q_{IR,\ absorb} = \varepsilon * \sigma * 2 * A_{surf} * T_{film}^4 - \sigma * \varepsilon * \alpha_{IR} * A_{surf} * T_{film}^4 * \left(1 + r_{effectivce,\ IR}\right), \quad (20)$$

where the ε stands for the average infrared emissivity factor of the film material, and T_{film} represents the temperature of the film.

2.2.7. Convective Heat Transfer

The convective heat transfer of a high-altitude balloon includes the convective heat transfer between the balloon film and the external atmosphere and the convective heat transfer between the film and the internal helium.

For the convective heat transfer between the film and the external atmosphere, there are two types of convective modes: natural convection and forced convection. These are defined as [9]

$$Q_{ConvExt} = H_{external} * A_{surf} * \left(T_{air} - T_{film}\right), \quad (21)$$

where $H_{external}$ represents the external convection heat transfer coefficient. For the convection heat transfer between the film and the external atmosphere, it is usually a coupled heat transfer process, so $H_{external}$ is specified as [10]

$$H_{external} = \left(H_{force}^3 + H_{free}^3\right)^{1/3}, \quad (22)$$

where H_{force} represents the forced convection heat transfer coefficient, and H_{free} is the natural convection heat transfer coefficient.

For the convection heat transfer between the film and the internal helium, its convective mode is natural convection, which is specified as [9]

$$Q_{ConvInt} = H_{internal} * A_{surf} * \left(T_{film} - T_{helium}\right), \quad (23)$$

where T_{helium} is the temperature of the helium.

2.2.8. Heat Transfer on Balloon Film

In this study, the temperature of the film is assumed to be uniform, and it is formulated as [9]

$$\frac{dT_{film}}{dt} = \frac{Q_{Direct} + Q_{Albedo} + Q_{Scatter} + Q_{IR,\ earth} + Q_{IR,\ sky} + Q_{ConvExt} - Q_{ConvInt} - Q_{IR,\ film}}{C_f * m_{film}}, \quad (24)$$

where C_f is the specific heat capacity of the film material, Q_{Direct} stands for direct solar radiation, Q_{Albedo} is ground-reflected radiation, $Q_{Scatter}$ is sky-scattered radiation, $Q_{IR,earth}$ is Earth's infrared radiation, $Q_{IR,sky}$ stands for atmospheric infrared radiation, $Q_{ConvExt}$ and $Q_{ConvInt}$ represent convective heat transfer, and $Q_{IR,film}$ refers to surface thermal radiation.

2.2.9. Heat Transfer on Internal Helium

This study also assumes that the temperature and pressure of the internal helium are uniform. According to the first law of thermodynamics, the temperature change rate of the internal helium is specified as [9]

$$\frac{dT_{helium}}{dt} = \frac{Q_{ConvInt}}{C_v * m_{helium}} + (\gamma - 1) * T_{helium} * \left(\frac{dm_{helium}}{dt} * \frac{1}{m_{helium}} - \frac{dVolume_{helium}}{dt} * \frac{1}{Volume_{helium}} \right), \quad (25)$$

where C_v represents the specific heat capacity at a constant volume of helium. Here, γ means the specific heat ratio, which equals C_p/C_v. C_p is the specific heat capacity at a constant pressure of helium.

2.3. Earth Model

As the balloon flight system moves, the local Earth radius and gravity change with its location. In this study, the WGS-84 Earth model is used to offer real-time data on Earth's radius and the acceleration of gravity at the current balloon location. The corresponding equations are defined as [25]

$$R_{meridian} = \frac{r_e * (1 - \epsilon^2)}{\left(1 - \epsilon^2 * (\sin \varphi)^2\right)^{3/2}} \quad (26)$$

$$R_{normal} = \frac{r_e}{\left(1 - \epsilon^2 * (\sin \varphi)^2\right)^{1/2}} \quad (27)$$

$$R_{equiv} = \sqrt{R_{meridian} * R_{normal}} \quad (28)$$

$$g = g_{WGS_0} \frac{1 + g_{WGS_1} * (\sin \varphi)^2}{\left(1 - \epsilon^2 * (\sin \varphi)^2\right)^{1/2}}, \quad (29)$$

where $R_{meridian}$ stands for meridian radius, R_{normal} means normal radius, R_{equiv} refers to equivalent radius, g is the acceleration of gravity, r_e stands for equatorial radius, ϵ represents first eccentricity, g_{WGS_0} is gravitational acceleration at the equator, and g_{WGS_1} is the gravity formula constant.

3. Trajectory Prediction Model

To simulate the motion of the balloon, the following hypotheses are recognized:

(1) The horizontal components of the balloon's speed are equal to the horizontal components of the wind's speed.
(2) The balloon is assumed to be a point mass when considering the external forces acting on the balloon.
(3) The high-altitude wind speed magnitude and direction are assumed to be constant for at least 24 h.

3.1. Wind Model

As the altitude increases, the wind speed's magnitude and direction change with it. Wind direction is usually divided into latitudinal wind and longitudinal wind. For latitudinal wind, the wind blowing from the west is positive, while for longitudinal wind, the wind blowing from the south is positive.

In this study, the wind data are measured on August 8, 2019 (UTC+16) in China, which include the wind speed magnitude and direction from the ground to 32 km in altitude. After data processing, these discrete wind data are fitted to a series of continuous curves by the sum of sines models, which is specified as [26]

$$y = \sum_{i=1}^{n} a_i * \sin(b_i * x + c_i), \tag{30}$$

where a represents amplitude, b refers to frequency, and c is the phase constant for each sine wave term. n is the number of terms in the series, and $1 \leq n \leq 8$. Figure 4 shows the discrete wind data and the corresponding fitted curve.

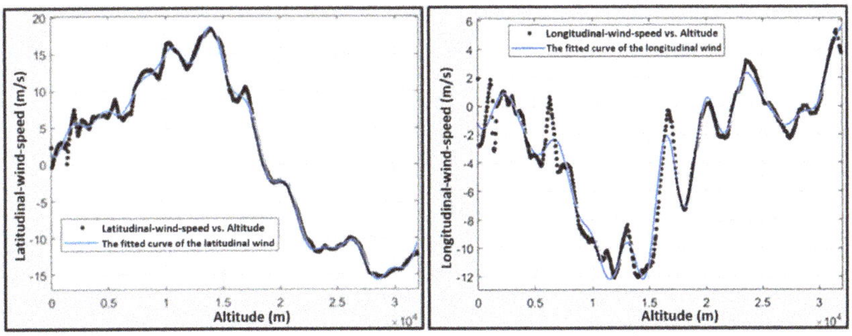

(a) **(b)**

Figure 4. Discrete wind data and the corresponding fitted curve. (**a**) shows the discrete latitudinal wind data and the fitted curve of the latitudinal wind. (**b**) shows the discrete longitudinal wind data and the fitted curve of the longitudinal wind.

Based on these fitting results, the corresponding fitted curves are defined as

(1) the fitted curve of the latitudinal wind speed (m/s) vs. altitude (m):

$$f(z)_{latitude} = \sum_{i=1}^{8} a_i * \sin(b_i * z + c_i), \tag{31}$$

where z indicates altitude, $f(z)_{latitude}$ is the latitudinal wind speed at the corresponding altitude, $a_i = [23.39, 30.97, 7.256, 35.02, 1.212, 1.159, 34.26, 0.9401]$, $b_i = [7.183e-05, 3.544e-05, 0.00031, 0.00099, 0.001989, 0.000675, 0.000992, 0.00165]$, and $c_i = [0.4929, 3.151, 3.723, 5.658, -1.421, -2.213, 2.584, -2.243]$; and

(2) the fitted curve of the longitudinal wind speed (m/s) vs. altitude (m):

$$f(z)_{longitude} = \sum_{i=1}^{8} a_i * \sin(b_i * z + c_i), \tag{32}$$

where z indicates altitude, $f(z)_{longitude}$ is the longitudinal wind speed at the corresponding altitude, $a_i = [80.21, 3.932, 80.58, 1.416, 1.007, 0.2995, 0.8878, 1.157]$, $b_i = [8.839e-05, 0.00036, 9.238e-05, 0.0018, 0.00078, 0.00123, 0.0029, 0.001476]$, and $c_i = [-0.7307, 0.3148, 2.454, -2.866, 1.572, -0.4936, -1.095, 2.805]$.

3.2. Geometry Model

A zero-pressure balloon with a shape factor of 0.2 is employed as the research object. The detailed geometric equations are as follows [9]:

(1) diameter from the top view:

$$Diameter = 1.383 * Volume^{1/3} \tag{33}$$

(2) projected area from the top view:

$$A_{top} = \frac{\pi}{4} * Diameter^2 \tag{34}$$

(3) During the balloon flight process, the illuminated projected area varies with the solar elevation angle, which is defined as

$$A_{projected} = A_{top} * [0.9125 + 0.0875 * \cos(\pi - 2 * h)] \tag{35}$$

(4) surface area of the zero-pressure balloon:

$$A_{surf} \approx 4.94 * Volume^{2/3} \tag{36}$$

(5) When the balloon is not fully inflated, the surface of the balloon is presented in crenellated form. The maximum crenellated surface area is defined as follows:

$$A_{surf1} = 4.94 * Volume_{design}^{2/3} * (1 - \cos(\pi * L_{goreB} / L_{goreDesign})), \tag{37}$$

where L_{goreB} indicates the actual length of gore exposed in the balloon, $L_{goreDesign}$ indicates the designed gore length of the balloon, and $Volume_{design}$ indicates the maximum volume of the balloon.

(6) To better approximate a realistic situation for convection and film mass calculations, $A_{effective}$ is used to describe the effective exposed surface area:

$$A_{effective} = 0.65 * A_{surf} + 0.35 * A_{surf1} \tag{38}$$

(7) A zero-pressure balloon has a distinctive feature: the internal helium pressure is almost equal to the surrounding atmospheric pressure. The average pressure difference between the internal helium pressure and the surrounding atmospheric pressure is defined as

$$\Delta P = 0.517 * g * (\rho_{air} - \rho_{helium}) * Volume^{1/3}, \tag{39}$$

where ρ_{helium} is the density of helium.

Furthermore, based on the ideal gas law, the balloon's volume can be formulated as

$$Volume = \frac{m_{helium} * R_{helium} * T_{helium}}{P_{helium}} \tag{40}$$

$$P_{helium} = P_{air} + \Delta P, \tag{41}$$

where $Volume = Volume_{balloon} = Volume_{helium}$, R_{helium} stands for the helium gas constant, T_{helium} represents helium temperature, P_{helium} refers to helium pressure, and P_{air} is the surrounding atmosphere pressure.

3.3. Dynamic Model

The motion of a zero-pressure balloon is divided into vertical motion and horizontal motion. For the vertical motion, the gross external force consists of buoyant force, gross weight, and aerodynamic drag force. The vertical dynamic equation of motion is defined as [27]

$$m_{virtual} * \frac{dv_z}{dt} = (\rho_{air} - \rho_{helium}) * Volum * g - m_{gross} * g - \frac{1}{2} * C_d * \rho_{air} * A_{top} * v_z^2, \tag{42}$$

where v_z is the vertical speed of the balloon flight system; m_{gross} is the gross mass of the payload, film, and ballast; $m_{virtual}$ is the gross mass of the payload mass, film mass, ballast mass, helium mass, and air virtual mass. Air virtual mass is specified as

$$m_{air,\ virtual} = C_{vir} * \rho_{air} * Volume_{helium}, \tag{43}$$

where C_{vir} stands for the virtual mass coefficient, and $C_{vir} \approx 0.5$. C_d is the drag coefficient, which is defined as [28]

$$\begin{cases} C_d = 1.5 * \left(\frac{24}{Re} + \frac{6}{1+\sqrt{Re}} + 0.4\right), \quad Re \le 2.7 * 10^5 \\ \log C_d = 1.5 * (25.821 - 4.825 * \log Re), \quad 2.7 * 10^5 < Re \le 3.7 * 10^5 \\ \log C_d = 1.5 * \left(-0.699 - 0.347 * e^{-38.533*(\log \frac{Re}{3.7*10^5})^{5.306}}\right), \quad 3.7 * 10^5 < Re \le 10^6 \\ C_d = 0.3, \quad Re > 10^6 \end{cases} \tag{44}$$

For the horizontal motion, the balloon is regarded as moving with the wind, assuming that the horizontal velocity is equal to the wind speed's magnitude. The horizontal dynamic equation is defined as

$$v_{east} = \frac{dx_{latitude}}{dt} = v_{wind} * \sin\theta \tag{45}$$

$$v_{north} = \frac{dx_{longitude}}{dt} = v_{wind} * \cos\theta, \tag{46}$$

where v_{east} and v_{north} are the longitudinal and latitudinal balloon speed, respectively, and $x_{latitude}$ and $x_{longitude}$ are the longitudinal and latitudinal distance, respectively. Here, θ indicates the angle between the direction of the horizontal wind and due north. In Earth's coordinate system, the balloon's horizontal displacements are defined as [29]

$$\dot{Lat} = \frac{v_{north}}{R_{meridian} + ALT} \tag{47}$$

$$\dot{Lon} = \frac{v_{east}}{(R_{normal} + ALT) * \cos Lat}, \tag{48}$$

where ALT indicates altitude. Lat and Lon are the latitude and longitude, respectively.

3.4. Exhaust Model

3.4.1. Automatic Exhaust Model

The zero-pressure balloon's ascending process includes two phases: one with the balloon partially expanded and the other with the balloon fully expanded.

The balloon fully expanded means that the balloon has expanded to its designed volume, then the overfilled helium is automatically vented from the exhaust pipes installed at the bottom of the balloon, preventing the balloon from exploding. This helium exhaust process is called the automatic exhaust process. The corresponding helium mass loss is defined as

$$\frac{dm_{helium}}{dt} = -\rho_{helium} * \frac{Volume_{altitude} - Volume_{design}}{dt}, \tag{49}$$

where ρ_{helium} stands for the density of the helium, and $Volume_{altitude}$ is the balloon volume with the corresponding altitude. The equation works on the premise that the balloon has expanded to its designed volume.

3.4.2. Active Exhaust Model

Considering realistic flight mission requirements and some emergencies, the balloon flight system needs to be forced to descend. To accomplish this, a helium exhaust valve installed at the apex of the balloon can be opened to exhaust helium. This helium exhaust process is called the active exhaust process. The corresponding helium mass loss is specified as [30]

$$\frac{dm_{helium}}{dt} = -\rho_{helium} * c * A * \sqrt{\frac{2 * \Delta P}{\rho_{helium}}}, \tag{50}$$

where c indicates a helium flow coefficient that depends on the number and shape of the helium exhaust valves, and A is the total opening area of the exhaust valves.

3.5. Ballast Model

After local noon, the intensity of solar radiation gradually weakens, causing the internal helium temperature to drop, and the volume of the balloon starts to shrink. Eventually, the zero-pressure balloon flight system begins to descend. To keep the designed altitude stable or to control the vertical landing velocity, a corresponding mass of ballast needs to be dropped.

In this study, granular iron sand is used as ballast, and the ballast discharge rate is determined experimentally by dropping ballast before the balloon flight mission. The ballast model is defined as

$$m_{ballast_{remained}} = m_{ballast_{actual}} - \int_0^t k_{ballast} dt, \tag{51}$$

where $m_{ballast_{actual}}$ stands for the original mass of the ballast, $m_{ballast_{remained}}$ refers to the mass of the ballast after dropping, and $k_{ballast}$ is the ballast flow rate.

4. Results and Discussion

The simulation program used Matlab/Simulink, and the corresponding simulation program structure diagram is shown in Figure 5. Figure 5 shows the above models, including the atmospheric, thermal, Earth, wind, geometry, exhaust, dynamic, and ballast models. The simulation process was realized through the data flow among the models, and the nonlinear differential equations were solved by the Runge–Kutta methods.

4.1. Model Validation

The best way to evaluate the simulation program was to compare the simulation results with real flight data. However, the flight mission data, balloon specification, atmospheric information, and film surface characteristics were almost inaccessible. In this paper, NASA's scientific balloon analysis model (SINBAD) and Palumbo et al.'s analysis model for high-altitude balloons (ACHAB) were used to validate the present model [31,32]. Both models have been proven to offer highly accurate trajectory predictions and were compared with real flight data [14,32]. The corresponding input data and output data can be found in [32]. Based on the previous works, the drag coefficient was taken into consideration for model validation. The varying drag coefficient was used for ACHAB and the present model, which was defined by the corresponding functions. A constant drag coefficient of 0.45 was used for ACHAB, SINBAD, and the present model. The simulation results are shown in Figure 6.

On the left side of Figure 6, the vertical velocity profile of the present model seems to be similar to that of ACHAB and SINBAD with the constant drag coefficient. The reason for the differences at the initial time and the time of reaching the designed altitude was that the ground surface temperature model and atmospheric model were different, which affected the ascent flight performance. For the varying coefficient, the time of reaching the designed altitude for both models was the same. All of the velocity profiles took on a W shape prior to float. On the right side of Figure 6, the float altitude of the present model is slightly lower than the previous programs. The reason for this was that atmosphere density and gravity acceleration differences mainly affected the net buoyancy at the designed altitude. In the present model, the variation of gravity due to the position was considered, while for the other models, it was not considered.

In summary, the present model appeared to be similar to the previous simulation programs and could offer enough accuracy for trajectory prediction. However, we still needed to optimize the simulation program with an interpolation model that considers the relationship between height and drag coefficient, since the balloon's geometry changes with the altitude during the ascent phase. Furthermore, to improve its accuracy, the simulation model needs to be further optimized by the flight test data in the following research.

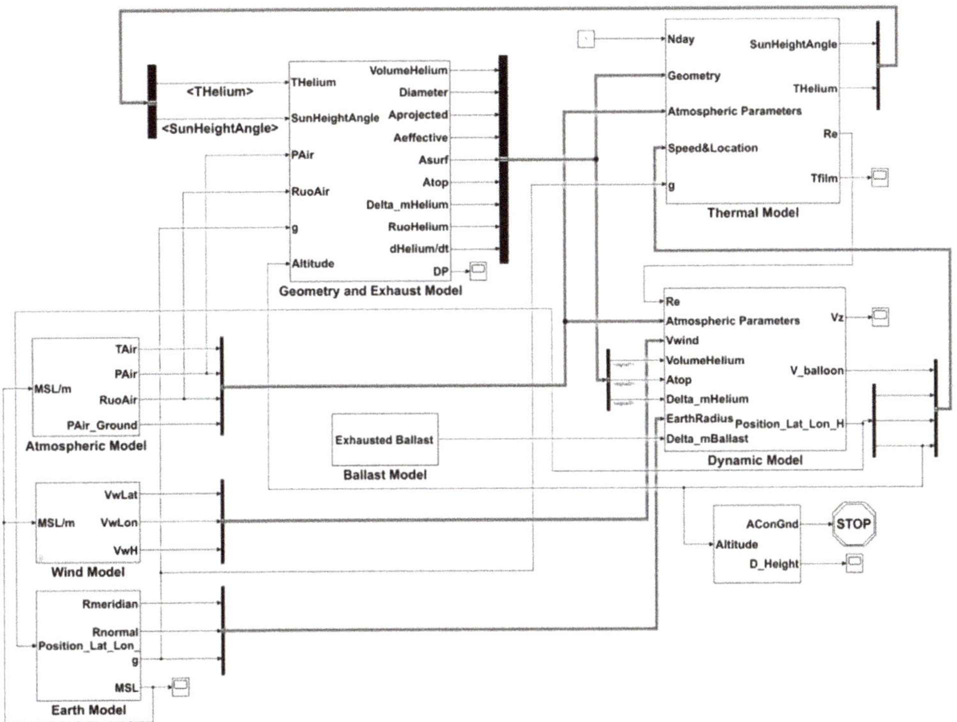

Figure 5. Matlab/Simulink simulation program structure diagram.

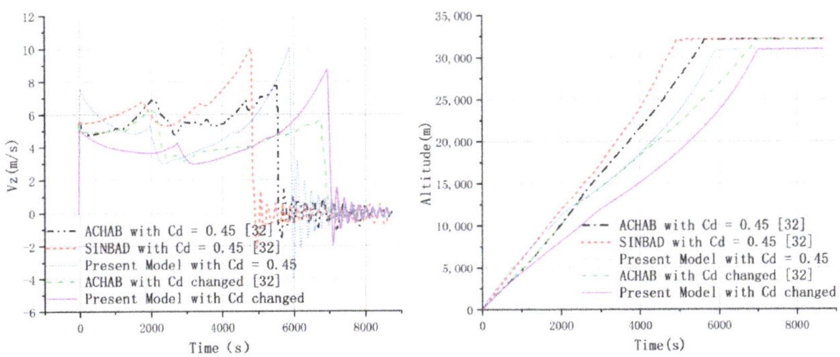

(a) (b)

Figure 6. Simulation results. (**a**) shows the comparison of vertical speed vs. time. (**b**) shows the comparison of altitude vs. time.

4.2. *Flight Simulation*

The zero-pressure balloon's specifications and the corresponding flight simulation information are presented in Table 1. The flight performance was simulated during the entire trajectory in detail.

Table 1. Balloon's specifications and flight simulation information.

Parameter	Value
Maxi volume	44,930.5 m^3
Lifting gas (helium) mass	626.5 kg
Balloon mass	487 kg
Suspending mass (payload)	2930.87 kg
Initial film temperature	286.95 K
Initial helium temperature	286.95 K
Number of the helium exhaust valves	2
Film absorption factor of solar radiation	0.33
Film transmissivity factor of solar radiation	0.65
Film absorption factor of infrared radiation	0.75
Film transmissivity factor of infrared radiation	0.20
Film average infrared emissivity factor	0.75
Earth surface reflectance factor	0.3
Ground average infrared emissivity factor	0.95
Launch date and time	8 August 2019 (UTC +16)
Launch base	Lat: 39°, Lon: 105°
Target altitude	20 km
Ground air pressure	101,325 Pa
Ground air temperature	288.15 K

UTC is the abbreviation for Universal Time Coordinated.

The variations of film temperature, internal helium temperature, atmospheric temperature, altitude, exhausted helium mass, vertical velocity, and trajectory throughout the flight mission are shown in Figures 7–10.

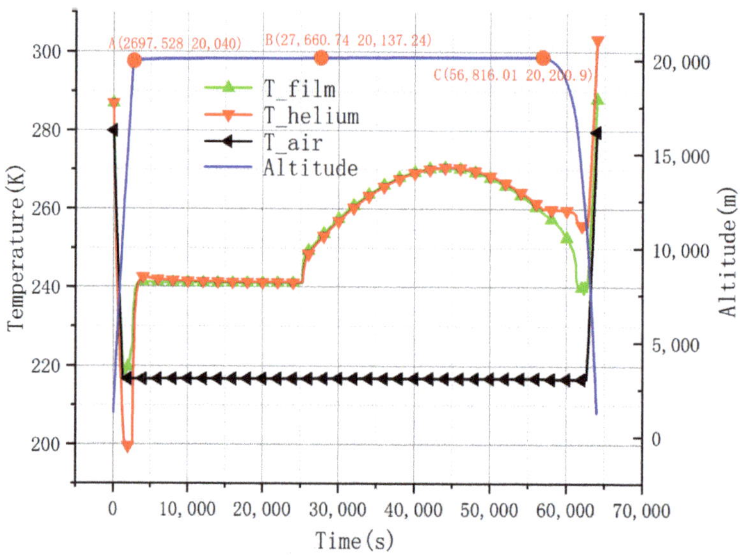

Figure 7. Temperature change of the film, helium, and air vs. time, and altitude change vs. time.

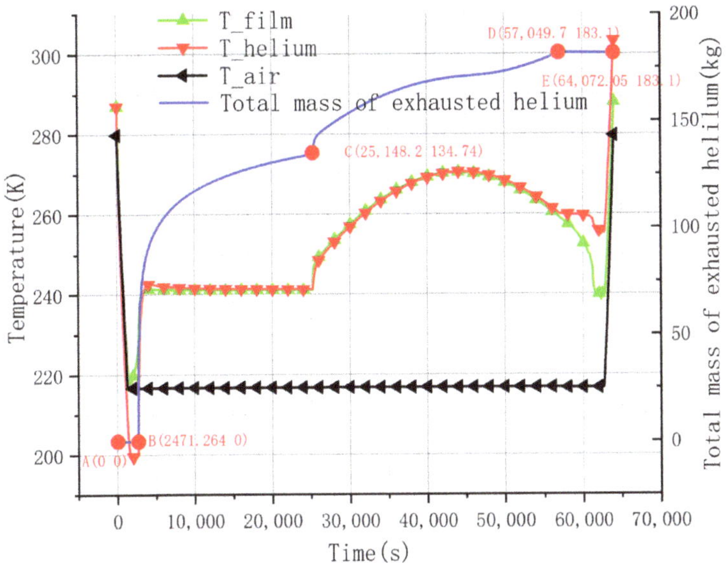

Figure 8. Temperature change of the film, helium, and air vs. time, and the change of the total mass of exhausted helium vs. time.

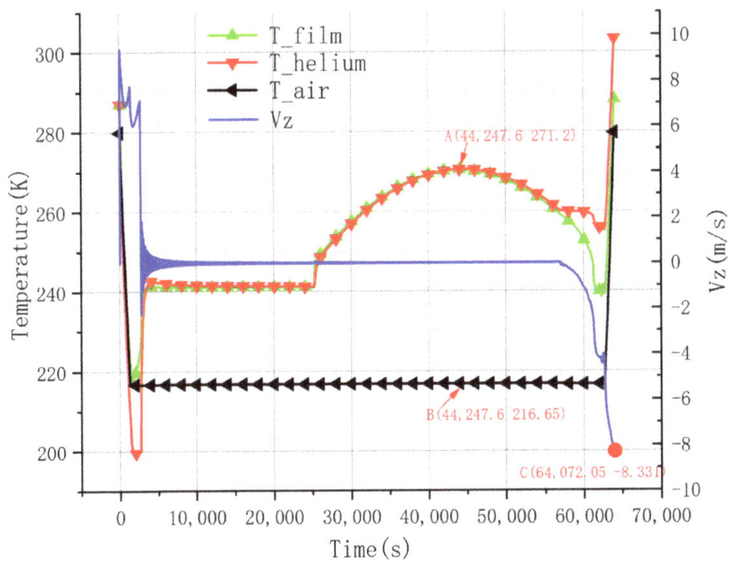

Figure 9. Temperature change of the film, helium, and air vs. time, and the change of the vertical speed of the balloon flight system vs. time.

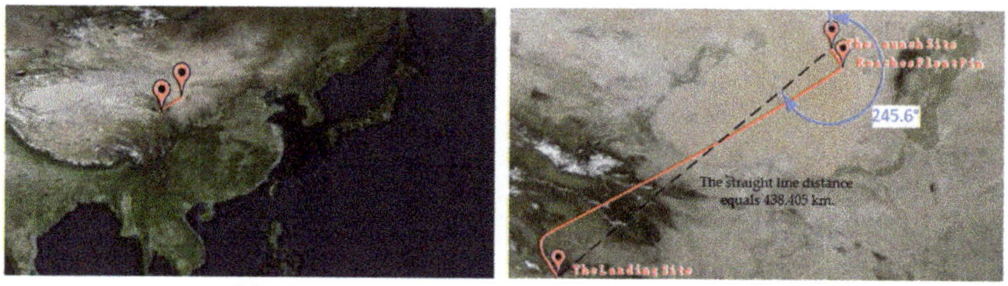

(a) **(b)**

Figure 10. Trajectory of the balloon flight system. (**a**) shows the relative position of the trajectory on the world map. (**b**) shows the trajectory in detail.

1. Change Pattern of Temperature

During the ascending phase, the temperatures of the film and helium decreased simultaneously with the atmospheric temperature. In addition, the temperatures of both the film and helium were lower than the atmospheric temperature, causing the phenomenon of supercooling. The maximum temperature difference reached −17.2°C between the helium and the surrounding atmosphere. The reason for this was that the volume of the balloon expanded rapidly, and the internal helium did positive work on the external atmosphere.

After the balloon flight system reached the designed float altitude, the temperatures of the film and helium gradually increased and came to equilibrium with the ambient thermal environment. According to Figure 7, the trend of temperature increase in the balloon film and internal helium became stable, and the increase patterns tended to be similar. After sunrise, the internal helium and film temperatures rose rapidly again due to solar radiation, with the temperature difference peaking at noon. The corresponding temperature difference between the helium and the surrounding atmosphere was +54.55°C.

In the afternoon, the temperature of the internal helium began to drop as the solar radiation decreased, resulting in a decrease in balloon volume and net buoyancy. The balloon began to move downward. During the descent, the external atmosphere did work on the helium expansion together with solar radiation, effectively compensating for the surface of the balloon so that helium temperature changed very little in the stratospheric range. After entering the tropospheric range, the temperature of the helium rose rapidly, causing the phenomenon of superheating. By the time the balloon system reached the ground, the helium temperature could reach 43°C.

2. Change Pattern of Altitude

The ascent and descent processes of the balloon flight system were manipulated by the vertical velocity. As shown in Figures 9 and 11, the ascent time of the system was almost 75 min faster than the descent process. In addition, during the floating phase, the balloon reached its designed volume. The increase in helium temperature caused the internal helium to expand, and a corresponding amount of helium was continuously discharged from the bottom exhaust pipe, causing the balloon flight system to move upward due to the buoyancy force being greater than gravity. The maximum altitude eventually reached 20,200.9 m. After reaching that altitude, the balloon flight system started to move downward.

3. Change Pattern of Vertical Speed

Throughout the ascent process, the vertical velocity changed in a W pattern. Initially, the balloon flight system rose rapidly, and its vertical velocity increased sharply to 9.285 m/s. Then, the vertical velocity decreased due to the weakening of the upward net buoyancy caused by the supercooling properties of the internal helium. After reaching a certain altitude, the atmospheric density was reduced to a minimum magnitude, resulting in the drag going down to a relatively small value and the vertical speed increasing.

When the balloon flight system climbed above 11 km, the vertical velocity decreased significantly and then increased again. Furthermore, when the system was about to reach the designed altitude, the vertical velocity dropped sharply and oscillated around 0 m/s until it finally dropped gradually to 0 m/s. In the afternoon, influenced by the surrounding thermal environment, the balloon flight system started to descend, and its downward vertical velocity continuously increased, reaching 8.331 m/s by the time it reached the ground.

4. Change Pattern of the Total Exhausted Mass of Helium

A high-altitude zero-pressure balloon has a unique and distinctive feature, which is exhaust pipes installed at the bottom of the balloon for exhausting helium. This is the reason why a zero-pressure balloon flight system cannot float for a long time. Therefore, it was imperative to figure out the helium-exhausting process in detail to plan the flight mission.

From the results described in Figures 7 and 9, the ascent process could be divided into two phases: the non-forming and forming geometry phases. The non-forming geometry phase occurred mainly at the initial time of the ascent process. As the external atmospheric pressure decreased, the volume of the balloon kept expanding while the helium mass remained constant. When the balloon flight system was about to reach the designed altitude, the forming geometry phase appeared. In this phase, the volume of the balloon reached its maximum and remained unchanged. Due to the momentum and vertical velocity, the balloon flight system continued to rise, causing excess helium to be discharged from the corresponding tubes. The helium mass began to decrease.

During the floating phase, the loss of helium mass continued, which was proportional to the solar radiation. After the balloon flight system began to descend, the balloon started to shrink, but the helium mass stayed constant until it reached the ground. During the entire flight process, the total exhausted mass of helium was 183.1 kg.

5. Trajectory Prediction

As shown in Figure 10, the straight-line distance between the launch site and the landing site was 438.405 km, and the azimuth between the launch site and the landing site was 245.6°. Moreover, the flight mission lasted for nearly 18 h. The high-altitude zero-pressure balloon flew in a large volume of airspace without a corresponding control strategy, and the trajectory took on a Z shape for this flight.

Based on these results, the balloon's performance during the entire flight mission could be discussed in detail, helping the balloon system's design and operation. Furthermore, it could be concluded that the vertical speed during landing has a serious impact on the safe recovery of a balloon flight system, but the flight time and flight distance are not conducive to its rapid recovery. Thus, it was imperative to better understand the impact of the uncertainty of the launch parameters on trajectory prediction and to develop the corresponding control strategies to assist with rapid recovery and achieve the efficient and safe application of a high-altitude zero-pressure balloon flight system in limited airspace.

4.2.1. Launch Time

Different launch times, including UTC + 16 on 8 August 2019 (Chinese standard time 00:00 on 9 August), UTC + 22 on 8 August 2019 (Chinese standard time 06:00 on 9 August), and UTC + 04 on 9 August 2019 (Chinese standard time 12:00 on 9 August), were used to simulate the flight mission. As shown in Figures 11 and 12, the launch time can play an important role in trajectory prediction. This is because direct solar radiation, ground-reflected radiation, sky-scattered radiation, and other thermal radiation vary periodically, causing a great difference in radiation for flights launched at different times of the day.

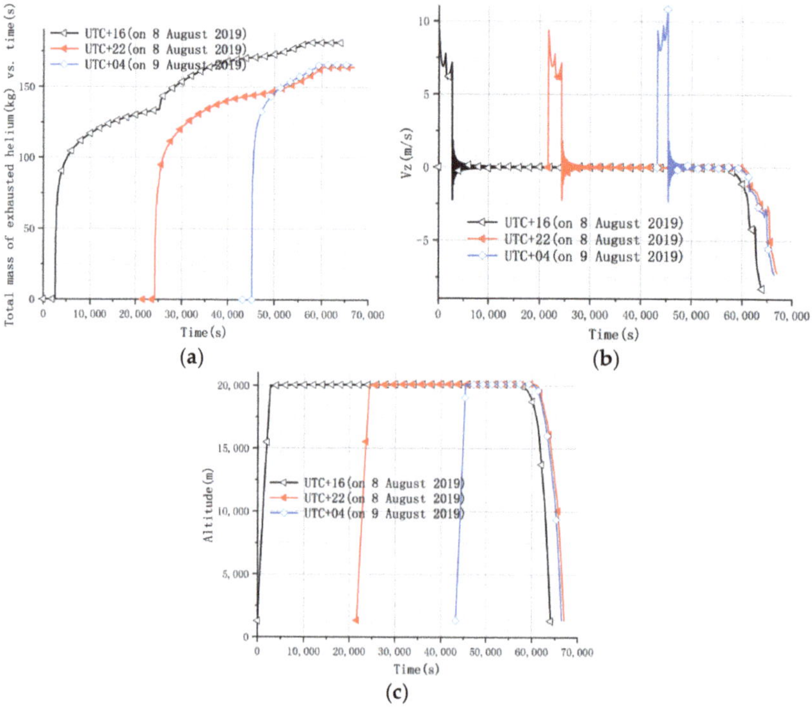

Figure 11. Diagrams corresponding to different launch times. (**a**) shows the total mass of exhausted helium vs. time, (**b**) shows the change in vertical velocity, and (**c**) shows the change in altitude.

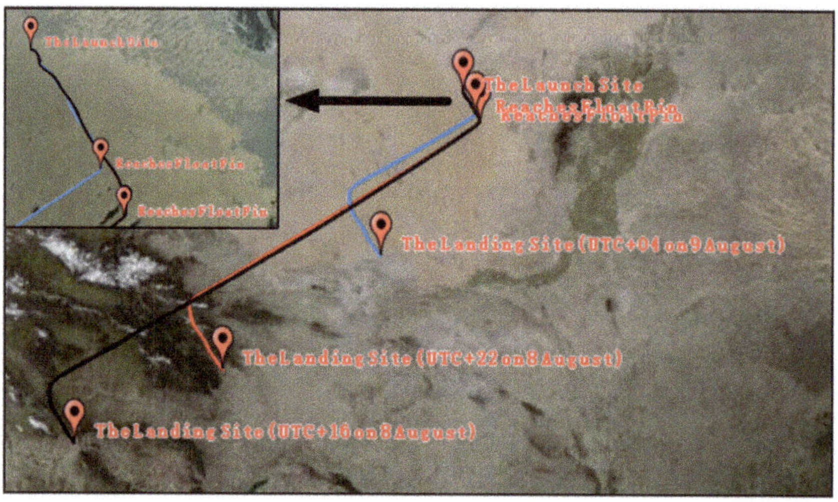

Figure 12. Trajectory predictions at different launch times of the day.

By comparing the flight performance of the zero-pressure balloon system at different times of the day, the following conclusions could be made:

1. The earlier the launch time, the more the total mass of helium is exhausted finally.

2. The earlier the launch time, the smaller the vertical velocity during the accent phase, resulting in relatively low requirements for balloon structure design. This is because the faster the vertical speed, the more severe the supercooling phenomenon.
3. The launch time has almost no effect on the floating altitude and landing vertical velocity.
4. The earlier the launch time, the longer the flight time and flight distance. However, the maximum flight time is less than 24 h.

Using these results, a suitable launch time can be chosen to start a flight mission and meet the mission requirements.

4.2.2. Initial Helium Mass

Different initial helium masses, such as 626 kg, 676 kg, and 726 kg, were used to simulate the flight mission on 8 August 2019 (UTC + 16). As shown in Figures 13 and 14, the corresponding results were as follows:

1. The more the initial mass of helium, the more the mass of helium was exhausted. However, the mass of helium inside the balloon was the same when landing.
2. The more the initial mass of helium, the faster the balloon system ascended.
3. The initial mass of helium had no effect on the floating altitude and trajectory prediction.

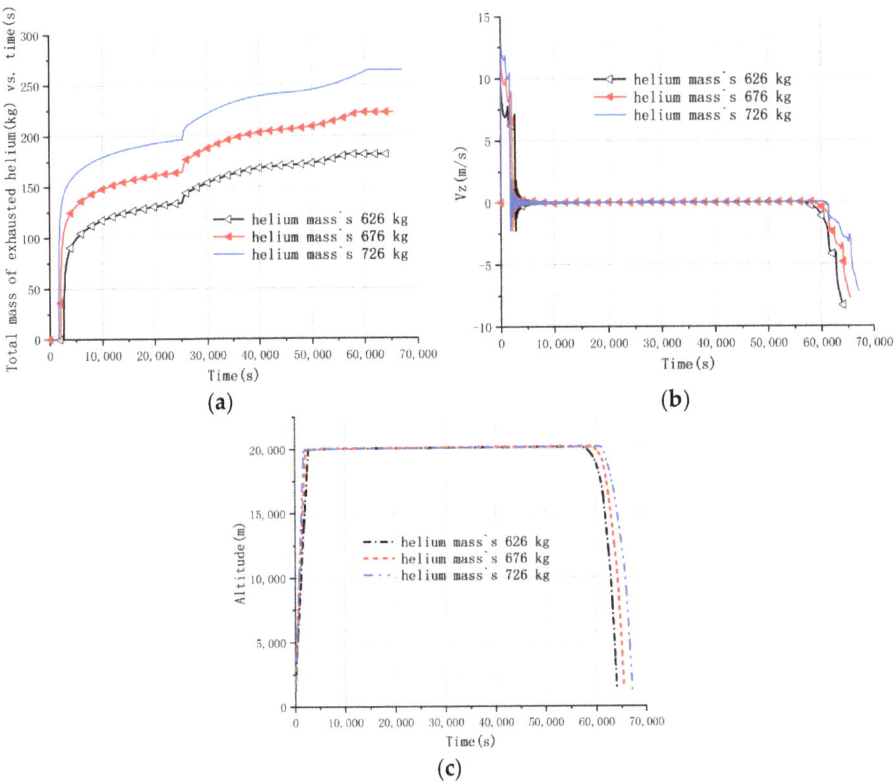

Figure 13. Diagrams corresponding to different initial masses of helium. (**a**) shows the total mass of exhausted helium vs. time, (**b**) shows the change in vertical velocity and (**c**) shows the change in altitude.

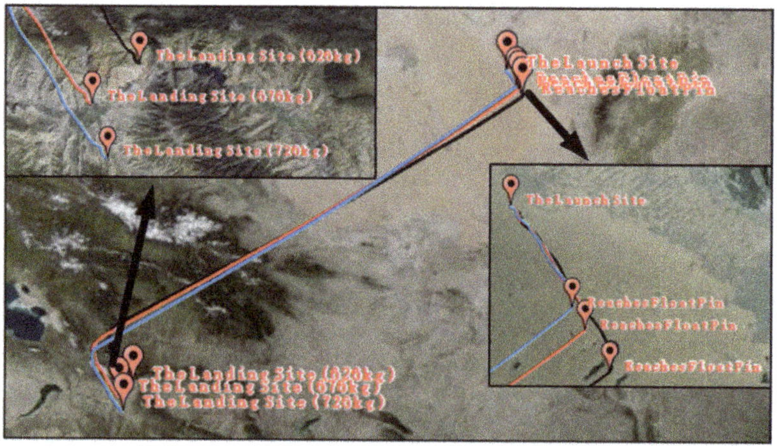

Figure 14. Trajectory predictions for different initial helium weights.

These results showed that the initial mass of helium could affect the flight performance during the ascent phase, increasing the ascent rate and decreasing the time needed to reach float, but it had no effect on the floating and descending phases. Therefore, the selection of the initial helium mass can be determined by the desired ascent velocity.

4.3. Trajectory Prediction with Control Strategy

These results showed that a zero-pressure balloon flight system flying in a huge amount of airspace makes it difficult to recover the balloon system, even considering the launch time. Furthermore, the landing speed of the balloon system is detrimental to safe recovery. Based on these considerations, the corresponding control strategy was employed to plan the trajectory.

In this study, the active exhaust model and ballast model were used together to control the trajectory. The control strategy also took the floating time and wind speed profile into consideration, which is helpful for real flight. The control strategies were as follows:

Most of the balloon flight missions only needed to float at the designed altitude for a few hours, sometimes less. It was clear that the zero-pressure balloon was in the automatic exhaust phase while in its float phase. Then, the balloon system began to use the active exhaust procedure to accelerate the transition to descent after finishing the floating mission. Once the balloon system started to descend, the vertical speed increased. Figure 15 shows the wind speed profile measured on 8 August 2019. This diagram illustrates that the horizontal wind speed was excessively large from an altitude of 10 km to an altitude of 15 km, causing excessive horizontal motion of the balloon system while passing through that altitude range. There was no doubt that traversing these height layers as quickly as practicable was reasonable. After the active exhaust procedure was finished, the landing speed needed to be optimized by the ballast model to ensure a safe landing.

In summary, the control strategies comprised two phases. First, enforcing the active exhaust procedure allowed the balloon system to descend until the designed ballast drop altitude, and then the active exhaust process was finished. Second, it was imperative to start dropping ballast immediately at the end of the active exhaust process, which helped the balloon system reach the ground at a vertical speed close to zero.

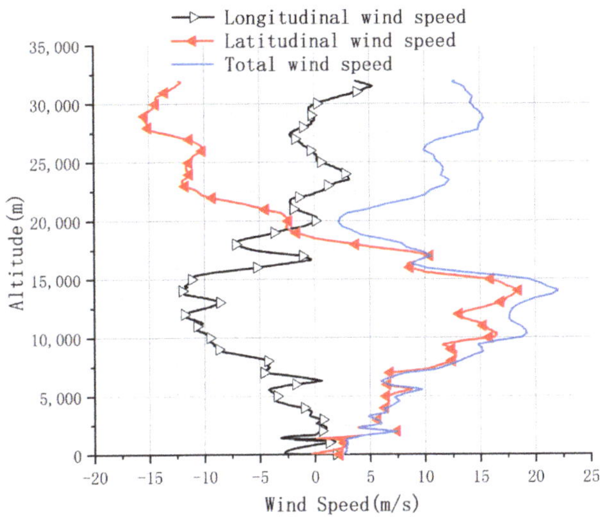

Figure 15. Wind speed profile vs. altitude on launch day.

Taking the floating mission time, horizontal motion, and landing speed into consideration, the results from the balloon flight system's trajectory simulation were as follows:

1. With the control strategy applied, the flight time and lateral drift distance were greatly reduced. For Mission A, the straight-line distance was 63.17 km. For Mission B, the straight-line distance was 62.28 km, and for Mission C, the straight-line distance was 70.46 km. For Mission D, the straight-line distance was 69.61 km. For all the flight missions, the flight time was less than 3 h. Furthermore, the landing velocity could be reduced to near zero with the ballast dropped at the designed ballast drop altitude. However, for the situations that did not use the control strategies, the trajectory of the balloon system could not be controlled from the ground after launch. This can be a cause of potential accidents.
2. The ballast flow rate mainly depended on the ballast drop altitude for each flight mission. The higher the ballast drop altitude, the smaller the ballast flow, which can help to design a ballast device. However, the higher the ballast drop altitude, the bigger the horizontal motion range, making recovery relatively more difficult.

These results can help with the overall design of a balloon flight system and planning a flight trajectory in advance. With the control strategies applied, this kind of flight mission can stay within the limited airspace, assisting the balloon system's rapid and safe recovery.

Table 2 shows the specifications of the missions. Figures 16–19 show the balloon system's change in vertical speed vs. time and change in altitude vs. time while the control strategies were applied (or not). Figure 20 shows the trajectories of the different missions.

Table 2. Flight mission specifications.

Flight Mission	Floating Time (min)	Ballast Drop Altitude (km)	Ballast Flow Rate (kg·s^{-1})
Mission A	30	10	1.023
Mission B	30	8	1.671
Mission C	60	10	1.024
Mission D	60	8	1.672

The launch took place at latitude 39° and longitude 105° on August 8, 2019 (UTC + 16). The initial helium mass was 626 kg.

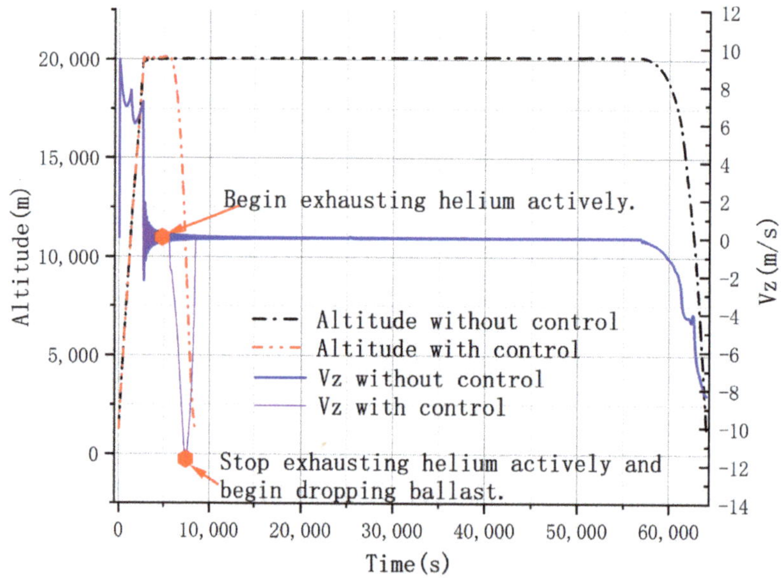

Figure 16. Simulation results of Mission A: balloon system's change in vertical speed vs. time and change in altitude vs. time while the control strategies are applied (or not).

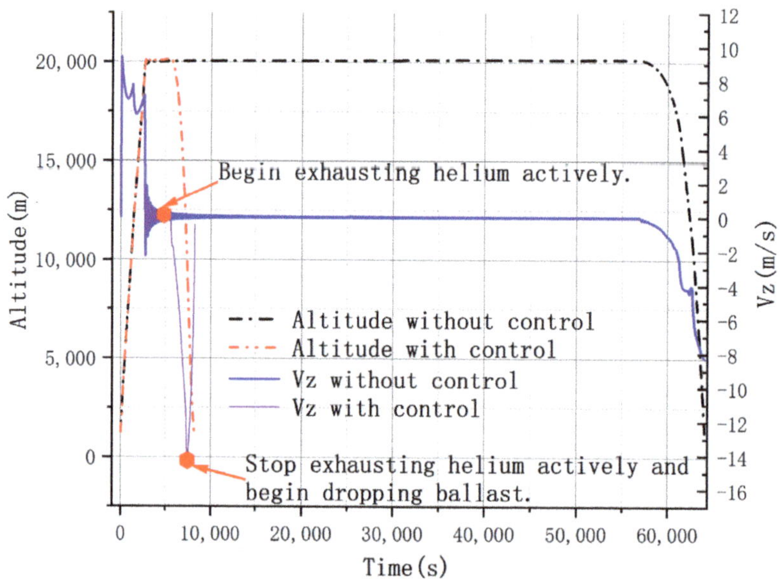

Figure 17. Simulation results of Mission B: balloon system's change in vertical speed vs. time and change in altitude vs. time while the control strategies are applied (or not).

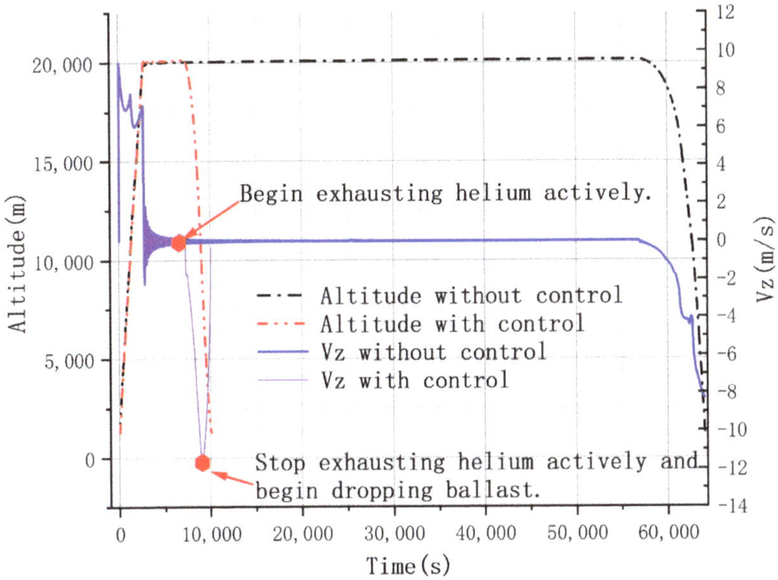

Figure 18. Simulation results of Mission C: balloon system's change in vertical speed vs. time and change in altitude vs. time while the control strategies are applied (or not).

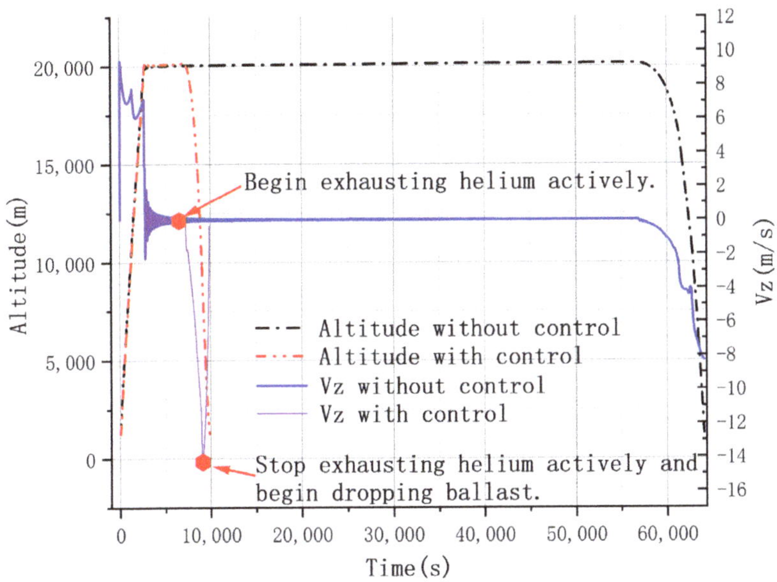

Figure 19. Simulation results of Mission D: balloon system's change in vertical speed vs. time and change in altitude vs. time while the control strategies are applied (or not).

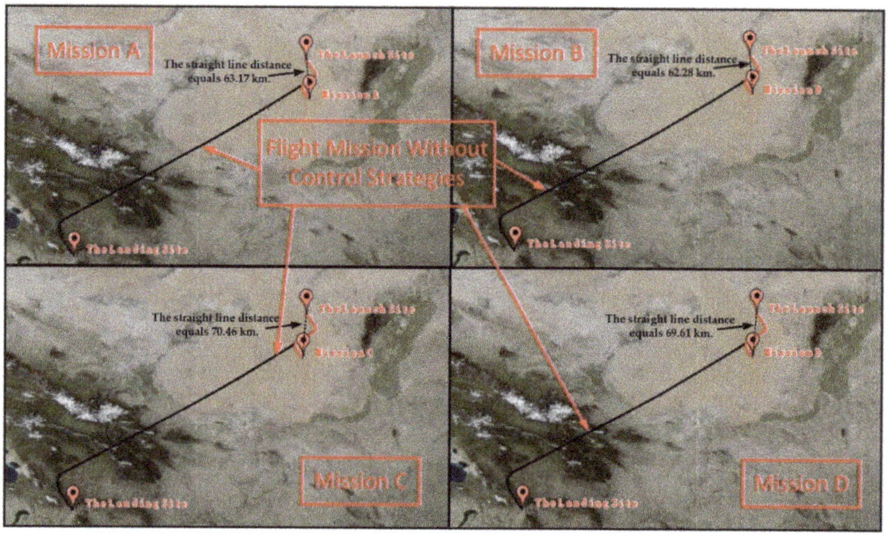

Figure 20. Trajectory of different flight missions discussed in the text.

5. Conclusions

A high-altitude zero-pressure balloon flight system flies with the wind in a horizontal direction, so it is necessary to predict its trajectory in advance to prevent the system from flying out of the required airspace and to avoid possible accidents. Most balloon flight systems need to fly within a limited airspace and accomplish rapid recovery to save on flight mission costs.

In the present work, a high-altitude zero-pressure balloon flight system with no parachute was simulated to predict trajectory, determine the impact of changing various parameters on balloon flight performance, and assist with rapid recovery. In order to further verify its accuracy, the simulation model needs to be validated by flight test data in the following research. The results can help in the planning of flight missions as well as design improvement for original equipment manufacturers of balloon flight systems.

The conclusions of the work are as follows:

1. During the ascent phase, the vertical velocity takes on a W shape. By the time the balloon flight system reaches the designed float altitude, the vertical velocity decreases sharply to 0 m/s and oscillates around 0 m/s until it finally drops gradually to 0 m/s.
2. During the ascent phase, the phenomenon of supercooling occurs, whereas the phenomenon of superheating occurs during the floating and descending phases. This information can help with balloon structure design.
3. The launch time plays an important role in trajectory prediction and has a significant impact on flight performance. The results show that the earlier the launch time, the longer the flight time and flight distance. However, the maximum flight time is less than 24 h. Furthermore, the launch time has no effect on the floating altitude and landing vertical velocity.
4. Maintaining the same launch time, changing the initial mass of helium has almost no effect on trajectory prediction. However, more helium filled on the ground will cause excessive vertical speed when ascending, leading to severe supercooling.
5. Based on realistic flight mission requirements, the floating mission time, horizontal motion range, and landing speed must be taken into consideration. The exhaust helium valve and ballast drop device can be used together to plan the trajectory. The simulation results show that with control strategies applied, flight missions can be

completed within limited airspace, and flight trajectories can be predicted to assist rapid and safe recovery and prevent harm from potential accidents.

6. The ballast flow rate can be determined by the ballast drop altitude for each flight mission by experimentally dropping the ballast device before the balloon flight mission.

Author Contributions: Conceptualization, J.T. and S.P.; methodology, S.P. and J.T.; software, S.P.; validation, J.T.; formal analysis, S.P.; investigation, P.Y. and W.X.; resources, S.P. and J.T.; data curation, Y.L.; writing—original draft preparation, P.Y. and B.H.; writing—review and editing, S.P. and J.T.; supervision, J.T.; project administration, J.T.; and funding acquisition, J.T. All authors have read and agreed to the published version of the manuscript.

Funding: This work is supported by the National Natural Science Foundation of China under grant number 51906141 and the Shanghai Sailing Program under grant number 19YF1421500.

Data Availability Statement: Not applicable.

Conflicts of Interest: The authors declare no conflict of interest.

References

1. Tang, J.; Duan, D.; Xie, W. Shape Exploration and Multidisciplinary Optimization Method of Semirigid Nearing Space Airships. *J. Aircr.* **2022**, *59*, 946–963. [CrossRef]
2. Tang, J.; Wang, X.; Duan, D.; Xie, W. Optimisation and analysis of efficiency for contra-rotating propellers for high-altitude airships. *Aeronaut. J.* **2019**, *123*, 706–726. [CrossRef]
3. Tang, J.; Xie, W.; Wang, X.; Chen, C. Simulation and Analysis of Fluid-Solid-Thermal Unidirectional Coupling of Near-Space Airship. *Aerospace* **2022**, *9*, 439. [CrossRef]
4. Yamada, K.; Abe, T.; Suzuki, K.; Honma, N.; Koyama, M.; Nagata, Y.; Abe, D.; Kimura, Y.; Hayashi, A.K.; Akita, D.; et al. Deployment and flight test of inflatable membrane aeroshell using large scientific balloon. In Proceedings of the 21st AIAA Aerodynamic Decelerator Systems Technology Conference and Seminar, Dublin, Ireland, 23–26 May 2011; p. 2579.
5. Jones, W.V. Evolution of scientific ballooning and its impact on astrophysics research. *Adv. Space Res.* **2014**, *53*, 1405–1414. [CrossRef]
6. Randolph, T.M.; Mullenax, R.; Schwantes, C.; Sell, S.W.; Ball, D.R.J. The first balloon flight of the Low Density Supersonic Decelerator technology demonstration mission. In Proceedings of the 2015 IEEE Aerospace Conference, Big Sky, MT, USA, 7–14 March 2015; pp. 1–12.
7. Qu, Z.; Li, C.; Hao, Y.; Yan, F.; Yang, Y. Project of high altitude balloon launched micro glider: Aircraft design, control and flight test. *Int. J. Micro Air Veh.* **2020**, *12*, 1756829320979955. [CrossRef]
8. Gemignani, M.; Marcuccio, S. Dynamic characterization of a high-altitude balloon during a flight campaign for the detection of ISM radio background in the stratosphere. *Aerospace* **2021**, *8*, 21. [CrossRef]
9. Farley, R. BalloonAscent: 3-D simulation tool for the ascent and float of high-altitude balloons. In Proceedings of the AIAA 5th ATIO and 16th Lighter-Than-Air System Technology and Balloon Systems Conferences, Arlington, VA, USA, 26–28 September 2005; p. 7412.
10. Dai, Q.; Fang, X.; Li, X.; Tian, L. Performance simulation of high altitude scientific balloons. *Adv. Space Res.* **2012**, *49*, 1045–1052. [CrossRef]
11. Sóbester, A.; Czerski, H.; Zapponi, N.; Castro, I. High-altitude gas balloon trajectory prediction: A Monte Carlo model. *AIAA J.* **2014**, *52*, 832–842. [CrossRef]
12. Zhang, Y.; Liu, D. Influences of initial launch conditions on flight performance of high altitude balloon ascending process. *Adv. Space Res.* **2015**, *56*, 605–618. [CrossRef]
13. Kayhan, Ö.; Yücel, Ö.; Hastaoğlu, M.A. Simulation and control of serviceable stratospheric balloons traversing a region via transport phenomena and PID. *Aerosp. Sci. Technol.* **2016**, *53*, 232–240. [CrossRef]
14. Lee, Y.; Yee, K. Numerical prediction of scientific balloon trajectories while considering various uncertainties. *J. Aircr.* **2017**, *54*, 768–782. [CrossRef]
15. Saleh, S.; He, W. Ascending performance analysis for high altitude zero pressure balloon. *Adv. Space Res.* **2017**, *59*, 2158–2172. [CrossRef]
16. Waghela, R.; Yoder, C.D.; Gopalarathnam, A.; Mazzoleni, A.P. Aerodynamic Sails for Passive Guidance of High-Altitude Balloons: Static-Stability and Equilibrium Performance. *J. Aircr.* **2019**, *56*, 1849–1857. [CrossRef]
17. Deng, X.; Yang, X.; Zhu, B.; Ma, Z.; Hou, Z. Analysis and simulation study of key technologies of intelligent stratospheric floatplane Loon. *J. Aeronaut.* **2022**, 1–18. Available online: http://kns.cnki.net/kcms/detail/11.1929.V.20220710.1648.034.html (accessed on 2 August 2022).
18. Atmosphere, US Standard. *US Standard Atmosphere*; National Oceanic and Atmospheric Administration: Washington, DC, USA, 1976.

19. Cathey, H.M. Advances in the thermal analysis of scientific balloons. In Proceedings of the 34th Aerospace Sciences Meeting an Exhibit, Reno, NV, USA, 15–18 January 1996.
20. Cheng, C.; Wang, X. Thermodynamic model of airship considering skin transmittance and its thermal characteristics. *J. Shangha Jiaotong Univ.* **2021**, *55*, 868–877.
21. Dai, Q. Study on Thermal Environment and Thermal Characteristics of Airships. Ph.D. Dissertation, Nanjing University o Aeronautics and Astronautics, Nanjing, China, 2014.
22. Li, X.; Fang, X.; Dai, Q. Research on thermal characteristics of photovoltaic array of stratospheric airship. *J. Aircr.* **2011**, *48* 1380–1386. [CrossRef]
23. Yang, X.; Hou, Z.; Ma, Z. Analysis of influential factors to supercooling phenomenon during ascent stage of stratospheric long duration balloons. *J. Natl. Univ. Def. Technol.* **2015**, *3i7*, 91–96.
24. Kreider, J.F. Mathematical modeling of high altitude balloon performance. In Proceedings of the 5th Aerodynamic Deceleratio Systems Conference, Albuquerque, NM, USA, 17–19 November 1975; p. 1385.
25. Rogers, R.M. *Applied Mathematics in Integrated Navigation Systems*; American Institute of Aeronautics and Astronautics, Inc Reston, VA, USA, 2000; ISBN 1-56347-397-6.
26. Sum of Sine Models. Available online: https://ww2.mathworks.cn/help/curvefit/sum-of-sine.html (accessed on 2 August 2022
27. Robyr, J.-L.; Bourquin, V.; Goetschi, D.; Schroeter, N.; Baltensperger, R. Modeling the vertical motion of a zero pressure ga balloon. *J. Aircr.* **2020**, *57*, 991–994. [CrossRef]
28. Lv, M.; Wu, Z. Thermodynamic model and numerical simulation of high altitude balloon ascending process. *J. Beijing Univ Aeronaut. Astronaut.* **2011**, *37*, 505–509. (In Chinese)
29. Farrell, J.A.; Barth, M. *The Global Positioning System & Inertial Navigation*; McGraw Hill: New York, NY, USA, 1999; ISBN 0-07-022045-X.
30. Yajima, N.; Izutsu, N.; Imamura, T.; Abe, T. *Scientific Ballooning: Technology and Applications of Exploration Balloons Floating in the Stratosphere and the Atmospheres of Other Planets*; Springer Science & Business Media: Berlin/Heidelberg, Germany, 2009 Volume 112.
31. Garde, G. Comparison of two balloon flight simulation programs. In Proceedings of the AIAA 5th ATIO and16th Lighter-Than-Ai System Technology and Balloon Systems Conferences, Arlington, VA, USA, 26–28 September 2005; p. 7413.
32. Palumbo, R. A Simulation Model for Trajectory Forecast, Performance Analysis and Aerospace Mission Planning with High Altitude Zero Pressure Balloons. Ph.D. Dissertation, University of Naples "Federico II", Naples, Italy, 2008.

aerospace

Article

Optimization Design and Experimental Verification for the Mixed-Flow Fan of a Stratospheric Airship

Wei Qu, Wentao Gong, Chen Chen, Taihua Zhang * and Zeqing He

Aerospace Information Research Institute, Chinese Academy of Sciences, Beijing 100094, China
* Correspondence: zth@aoe.ac.cn

Abstract: Large-flow, high-pressure fans are required to improve the shape-keeping and flight-height-adjusting ability of stratospheric airships. This study optimizes and analyzes a fan MIX-130 suitable for a stratospheric airship. Five design parameters are selected to optimize the fan's static pressure rise and efficiency: impeller outlet installation angle, installation angle increment, blade thickness, diffuser tilt angle, and $L_{16}(4^5)$ orthogonal test for optimization research. Based on the optimization results, the fan is processed, a fan test bench is built to verify the accuracy of the fan numerical analysis method, and the fan's performance curve in the stratosphere environment is given. The results demonstrate that after optimization, the static pressure rise in the MIX-130 fan increases by 47.5%, and the efficiency increases by 8%. The performance test data of the MIX-130 fan are consistent with the numerical analysis results. Furthermore, the flow pressure curve is significantly improved compared with the existing fan, satisfying the requirements of airship flight missions. The fan structure optimization and testing methods presented in this manuscript can provide a reference for designing and testing stratospheric airship fans.

Keywords: near-space; stratospheric airship; orthogonal optimization; fan test; similarity principle

1. Introduction

A stratospheric airship, an aerostat that can stay in near space (18–22 km) for an extended duration, can carry several loads to achieve communication, navigation, observation, early warning, and other functions [1–3]. It has a high cost-effectiveness ratio, broad coverage, and strong load capacity, demonstrating broad application potential [4–6].

It is a unique control feature of stratospheric airships, overpressure balloons, and other types of aerostats to use the main and auxiliary airbags for height adjustment [7]. The fan is used to compress the air to the auxiliary airbag, which can reduce the airship's standing altitude. The valve is used to exhaust the air in the auxiliary airbag, and the airship's standing altitude can be raised [8,9]. By changing the airship height, the wind layer at different altitudes is used to adjust the flight path, achieve specific trajectory control objectives, and complete the designated flight mission [10,11].

Mature aviation fans are used on stratospheric airships, such as MAX45016 and MAX80001 from AMETEK [12,13]. With the continuous development and progress of stratospheric airship technology, the demands of cross-day and night flight, wide-range cruise, fixed point hover, and other tasks are increasing, and the performance of the existing fans cannot meet the task requirements for three reasons. First, at the stratospheric airship standing altitude, the pressure head of the aviation fan is low, which cannot overcome the airship's internal pressure to fill with air [14,15]; thus, the airship cannot adjust its height through the auxiliary airbag during the day. Second, at the stratospheric airship standing altitude, the flow rate of the aviation fan is low. The pressure will decrease sharply if the airship encounters cold clouds [16]. The flow rate of the aviation fan is insufficient to quickly fill the air to maintain the aerodynamic shape of the stratospheric airship. Third, the fans of stratospheric airships are usually used with valves. Because of the limitation of the shape of aviation fans, the valves can only be installed at the outlet of the fans, which will significantly reduce fan efficiency [17].

Compared with traditional fans, stratospheric fans have several design difficulties.

(a) Harsh operating conditions: severe changes in atmospheric density, pressure, and temperature from the ground to the stratosphere, the large temperature difference between day and night at the standing altitude of the airship, low Reynolds number, and large drag coefficient [18].
(b) Power consumption limitations: The energy of stratospheric airships comes from onboard lithium batteries and solar arrays [19], and the energy supply is limited, so the power consumption of fans is limited.
(c) Weight constraints: the stratospheric airship platform itself requires a lightweight design, so the fan must be as lightweight as possible.
(d) High-performance requirements: in the stratospheric, low-temperature, low-pressure environment, the fan must have sufficient pressure head and flow rate but also high efficiency to save energy consumption. Satisfying these requirements necessitates relevant design and research for a stratospheric airship fan.

Sun et al. [20] designed centrifugal and axial fans for stratospheric conditions. The variation rule of fan characteristics of airship during descent is compared. Wei et al. [21] proposed a selection method for a centrifugal fan for an aerostat, which can provide a reference for the selection of aerostat fans. Zhao et al. [22] studied the operational characteristics of centrifugal fans at different altitudes by considering the changes in air density, pressure, and temperature in high-altitude environments. Yang et al. [23] studied the flow field characteristics of a high-flow axial fan in stratospheric environments using computational fluid dynamics (CFD). The results demonstrated that selecting a guide vane and diffuser would significantly affect fan efficiency. Zhang et al. [24] studied the relationship between the altitude, the pressure difference of the aerostat capsule, the number of fans used, and the fan inflation efficiency. The results demonstrated that it was feasible to use multiple fans in series or parallel for different aerostat flight tasks. In general, there is a lack of research in the field of stratospheric airship fans.

The core component of the fan is the impeller, which is related to the performance of the entire fan. It is necessary to optimize the impeller design to produce an efficient fan. Scholars have performed extensive research in this field. Fan et al. [25] obtained the optimal combination of impellers through optimization design, combined with range analysis and variance analysis, and significantly improved the aerodynamic performance of the fan. Wang et al. [26] used the number of blades and the blade outlet angle as variables to optimize the fan impeller parameters through the least squares method and obtained promising results. Xu et al. [27] adopted the orthogonal optimization method to optimize the design of a centrifugal pump, significantly improving its efficiency and head. Wang et al. [28] adopted an artificial intelligence optimization algorithm to carry out a multi-objective optimization design of a centrifugal pump, and the performance of the centrifugal pump has been significantly improved. Jung et al. [29] studied the influence of structural parameters of a mixed-flow fan on fan performance and analyzed the influence of a single parameter on performance using CFD. The research found that the diffuser structure would significantly affect fan performance. Wang et al. [30] adopted the combinatorial optimization system to improve the performance of mixed-flow pumps and achieved promising results.

Previous studies on stratospheric fans and an optimization method for fan impellers are relevant but have limitations, focused primarily on the theoretical derivation and simulation analysis. The weight and power consumption of the fan involved do not apply to stratospheric airships. Furthermore, the main body of research in most papers includes components that are not consistent with the use of the stratospheric airship environment, such as a water pump or industrial fan.

In this study, a stratospheric mixed-flow fan [31] is used as the research object, and the improvement of fan efficiency and static pressure rise is used as the optimization objectives.

The $L_{16}(4^5)$ orthogonal test is adopted to conduct the multi-parameter and multi-objective optimization study on the fan. Finally, the accuracy of the numerical analysis method and the rationality of the fan design are verified by combining the performance tests. The fan structure optimization and testing methods presented in this manuscript can provide a reference for designing and testing stratospheric airship fans.

The rest of the manuscript is organized as follows: Section 2 describes the research object of the MIX-130 fan and the numerical analysis method and boundary conditions adopted in this manuscript. In Section 3, the optimal design of the MIX-130 fan is carried out by the orthogonal optimization method, and the optimization results are analyzed. In Section 4, the performances of the MIX-130 fan are carried out to verify the accuracy of the numerical analysis method and the rationality of the fan design. Section 5 concludes the manuscript and provides future perspectives.

2. Numerical Methods and Boundary Conditions

2.1. Numerical Method

Traditional aviation fans, such as AMETEK's MAX45016 and MAX80001 [8,9], adopt an axial impeller design. Because of the saddle shape area on the performance curve of the axial fan, the available effective range is small and is unsuitable for the high-flow-rate, high-pressure head of stratospheric airships.

The mixed-flow impeller can consider the characteristics of the high-pressure head of a centrifugal fan and the large flow of an axial flow fan [32]. The vaneless diffuser has a wide flow range and simple structure [33], so the fan structure design with the mixed impeller and vaneless diffuser is more appropriate for the variable working conditions of the stratospheric airship fan. Figure 1 illustrates the MIX-130 fan, the object of this study.

Figure 1. MIX-130 fan.

The initial design parameters of the MIX-130 fan are listed in Table 1.

Table 1. Initial design parameters of the MIX-130 fan.

Description	Parameter	Value
Design flow rate	Q	1100 $m^3.h^{-1}$
Static pressure rise	P_R	700 Pa
Rated speed	n	25,000 Rpm
Power output	P	≯500 W
Blade number of impeller	Bn	18
Blade tip clearance	/	0.5 mm
Blade thickness	t	1.5 mm
Diameter of impeller in	D_2	80 mm
Diameter of impeller out	B_2	130 mm
Blade angle at leading edge	β_1	39–30°
Blade angle at trailing edge	β	45–49°
Diffuser tilt angle	ϕ	57°
Voltage	V	16–32 V
Total weight	/	≯2.5 kg

Based on the structure of the MIX-130 fan, its fluid domain can be extracted, as depicted in Figure 2. The fluid domain includes three parts: the inlet pipe, impeller, and diffuser.

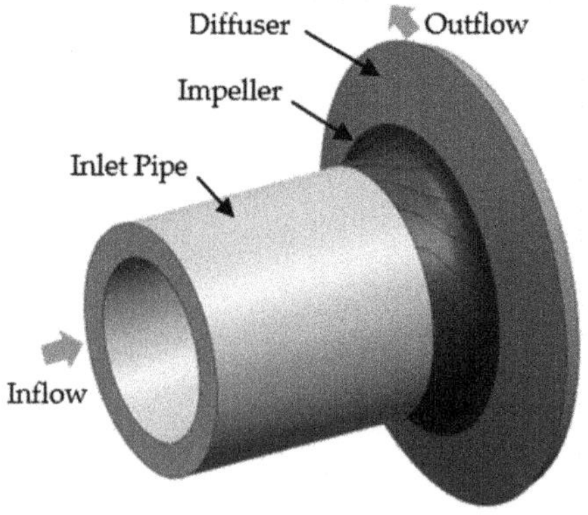

Figure 2. The fluid domain of MIX-130 fan.

In this study, fluid dynamics analysis software ANSYS CFX is used to analyze the fluid characteristics of the fan. Because the flow inside the fan is predominantly complex turbulent flow, an appropriate turbulence solution method must be selected to accurately simulate the flow of the air inside the fan [34]. Based on existing studies, the shear stress transport model (SST) [18,35] is more appropriate for analyzing a high altitude and a low Reynolds number. Therefore, the SST model is selected to solve the three-dimensional time-average Navier-Stokes equation.

Given the symmetrical characteristics of the fluid domain of the fan, the cyclic symmetric boundary is established, and a 1/18 flow passage is adopted for the numerical analysis to save calculation time, as depicted in Figure 3.

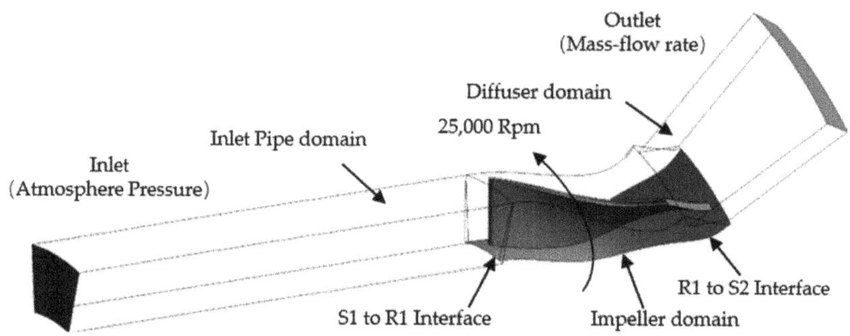

Figure 3. Computational domain.

The impeller is set as a rotating area with a speed of 25,000 Rpm, and the inlet tube and diffuser part are set as a stationary area. "Mixing-plane" technology is adopted to connect the interface between the rotary static interface, which axially averages the parameters of the interface of the upper-level components and then transmits them to the interface of the lower-level components [36]. All walls are in a slip-free condition. The calculation is considered convergent in steady-state computation when the root mean square residual is less than 10^{-5}.

2.2. Boundary Conditions

The boundary conditions of the MIX-130 fan simulation analysis should be consistent with the real-world environment. The specific value is related to the altitude of the airship target. The atmospheric temperature, pressure, and density in the high air can be calculated according to the ISA (International Standard Atmosphere) model [9] as follows:

$$T = \begin{cases} 288.15 - 0.0065h & (0 \leq h \leq 11000) \\ 216.65 & (11000 < h \leq 21000) \\ 288.15(0.682457 + h/288153.5) & (21000 < h \leq 32000) \end{cases} \quad (1)$$

$$P = \begin{cases} 101325(1 - 0.0065h/288.15)^{5.25588} & (0 \leq h \leq 11000) \\ 22631.8e^{1.73 - 0.000157 \times h} & (11000 < h \leq 21000) \\ 101325(0.988626 + h/198915)^{-34.16319} & (21000 < h \leq 32000) \end{cases} \quad (2)$$

$$\rho = \begin{cases} 1.22505(1 - 0.0065h/288.15)^{4.25588} & (0 \leq h \leq 11000) \\ 0.36392e^{1.73 - 0.000157 \times h} & (11000 < h \leq 21000) \\ 1.22505\ (0.988626 + h/201161)^{-35.16319} & (21000 < h \leq 32000) \end{cases} \quad (3)$$

As defined by Equations (1)–(3), the environmental parameters corresponding to the airship's standing altitude are as follows: atmospheric temperature of 216.65 K, the atmospheric pressure of 5500 Pa, and atmospheric density of 0.094 kg.m^{-3}.

The numerical analysis boundary conditions of the fan are presented in Table 2. The total pressure boundary pertains to the fan inlet, and the mass flow rate boundary pertains to the fan outlet.

Table 2. Numerical analysis of boundary conditions for MIX-130 fan.

Boundary	Setting	Value
S1 Inlet	Total Pressure	5500 Pa
S1 Inlet	Total Temperature	216.65 K
S2 Outlet	Mass Flow Rate	0.0288 Kg.s^{-1}
R1 Speed	/	25,000 Rpm
Reference Pressure	/	0 Pa
Residual	RMS	1×10^{-5}

2.3. Meshing and Independence

The fan inlet pipe and diffuser are meshed by ANSYS Mechanic. The mesh structure of the impeller is generated by ANSYS Turbo Mesh. As depicted in Figure 4, all the meshes are hexahedral.

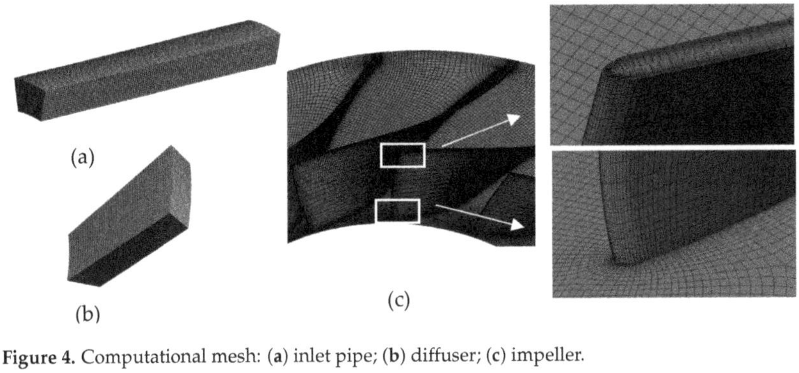

Figure 4. Computational mesh: (**a**) inlet pipe; (**b**) diffuser; (**c**) impeller.

The number of meshes will directly affect the accuracy of simulation results. Choosing a reasonable number of meshes can save calculation time, so it is critical to verify the mesh independence. In this study, six groups of different mesh numbers are selected for the mesh independence test. Table 3 presents the mesh distribution and independence test results.

Table 3. Mesh independence test.

Item	Mesh1	Mesh2	Mesh3	Mesh4	Mesh5	Mesh6
Inlet pipe	3520	16,744	26,400	26,400	26,400	10,080
Impeller	105,100	256,300	315,350	393,050	522,150	630,300
Diffuser	3059	19,008	10,080	10,080	10,080	10,080
Total mesh	111,679	292,052	351,830	429,530	558,630	650,460
PRi/PR1	1	1.015	1.016	1.017	1.019	1.018
$\eta i/\eta 1$	1	1.017	1.017	1.017	1.019	1.018

Based on the analysis results, when the inlet pipe mesh number is 26,400, the impeller mesh number is 522,150, the diffuser mesh number is 10,080, and the fan efficiency and static pressure ratio change rates are small. Therefore, in the subsequent calculation and analysis, it is reasonable to control the total mesh number at 558,630.

2.4. *Original Fan Performance*

Based on this analysis method, the efficiency of the MIX-130 fan is only 64%. The static pressure rise is only 511 Pa at the design condition point, which differs from the target design value and cannot meet the demand. Therefore, it is necessary to optimize the design and improve the fan structure.

3. Optimal Fan Design

3.1. *Optimization Parameters*

The impeller is the core energy conversion component of the fan. Its structural parameters directly affect the fan's pressure ratio and efficiency, so the optimization of the fan focuses primarily on the impeller.

The blade is the central part of the impeller. In terms of fluid, the thickness and number of blades directly determine the volume of the flow path in the impeller. If the number of blades is too small, secondary flow loss will occur in the impeller, reducing fan efficiency. If the number of blades is too large, the flow path of the impeller will be reduced, the flow loss and friction loss inside the impeller will be increased, and fan efficiency will be reduced [37]. In terms of structure, the thickness of the blade is too thick to increase the weight of the impeller, increasing the fan's power consumption. If the blade thickness is too thin, it will affect the strength of the impeller and increase the difficulty of processing. Therefore, blade thickness and blade number are important optimization parameters of the impeller. The blade outlet installation and variation angles directly determine the blade shape and significantly impact fan outlet pressure and efficiency [38]. Furthermore, the tilt angle of the diffuser in the MIX-130 fan is a special parameter. Its inclination angle will directly affect the structural size of the entire fan, so it is also selected as an optimization parameter—defined as the angle between the center line of the diffuser and the axial direction of the fan.

In this study, five parameters, including the number of blades, Bn, thickness, t, outlet installation angle, β, outlet installation angle increment, $\Delta\beta$, and the diffuser tilt angle, ϕ, are selected for optimization. In addition to the above parameters, other impeller parameters, such as impeller inlet and outlet diameter, width, and blade tip clearance [39], are limited by the overall fan structure design and remain the same as the original impeller without change.

3.2. *Orthogonal Test Analysis*

The orthogonal test method is a scientific method to study the multi-factor optimization test by applying the orthogonal principle of an orthogonal table and mathematical statistics analysis. It can optimize the combination of optimal parameters or conditions of each factor with the least number of tests [25] and is highly effective for optimizing the impeller design [38]. The orthogonal table consists of two parts: factor and factor level. The test individuals in the orthogonal table are symmetrical and evenly distributed to ensure that the results are representative of all factor combinations.

Factor levels were determined concerning the original impeller and design experience. In the five design factors selected in this study, each factor changes at four levels. For improving the static pressure rise and efficiency of the fan, the $L_{16}(4^5)$ orthogonal test was used to optimize the fan design. The factors and levels of the orthogonal test are presented in Table 4. A blade thickness t of 1~2 indicates that the blade thickness is 2 mm at the Hub and 1 mm at the shroud, and the thickness of the middle part changes linearly.

Table 4. Orthogonal test factors and levels.

Level/Factor	Bn	β/°	Δβ/°	t/mm	ϕ/°
1	17	35	1	1	57
2	18	40	2	1.5	60
3	19	45	3	2	63
4	20	50	4	1~2	68

As presented in Table 4, orthogonal Table 5 can be obtained with 16 groups of fan models with different factors and levels. The numerical method in Section 2 is used to analyze each group of fan models.

Table 5. Orthogonal table.

Level/Factor	Bn	β/°	Δβ/°	t/mm	ϕ/°
1	17	35	1	1	57
2	17	40	2	1.5	60
3	17	45	3	2	63
4	17	50	4	1~2	68
5	18	35	2	1~2	63
6	18	40	1	2	68
7	18	45	4	1.5	57
8	18	50	3	1	60
9	19	35	3	1.5	68
10	19	40	4	1	63
11	19	45	1	1~2	60
12	19	50	2	2	57
13	20	35	4	2	60
14	20	40	3	1~2	57
15	20	45	2	1	68
16	20	50	1	1.5	63

3.3. Optimization Results and Discussion

The static pressure rise and efficiency of 16 groups of different fans at the design point are presented in Table 6.

The static pressure rise in the fan is defined as:

$$P_R = P_{outlet} - P_{inlet} \tag{4}$$

P_R is the static pressure rise, P_{outlet} is the static pressure at the fan outlet, and P_{inlet} is the static pressure at the fan inlet.

The fan efficiency is defined as:

$$\eta = P_R \cdot Q / T \cdot W \tag{5}$$

P_R is the static pressure rise, Q is the flow rate, T is the impeller torque, and W is the impeller speed.

Table 6. Static pressure rise and fan efficiency.

NO.	Bn	β/°	Δβ/°	t/mm	ϕ/°	P_R	η
1	17	35	1	1	57	348	57.7%
2	17	40	2	1.5	60	386	60.4%
3	17	45	3	2	63	410	60.7%
4	17	50	4	1~2	68	707	69.1%
5	18	35	2	1~2	63	211	46.5%
6	18	40	1	2	68	273	48.6%
7	18	45	4	1.5	57	511	64.0%
8	18	50	3	1	60	733	71.5%
9	19	35	3	1.5	68	195	44.9%
10	19	40	4	1	63	486	64.4%
11	19	45	1	1~2	60	476	61.6%
12	19	50	2	2	57	474	61.6%
13	20	35	4	2	60	210	53.6%
14	20	40	3	1~2	57	314	54.7%
15	20	45	2	1	68	607	67.0%
16	20	50	1	1.5	63	574	64.5%

In the orthogonal method, k_i and R are used to evaluate the influence of specific factors on indicators at their level, defined as follows:

$$k_i = \sum_{1}^{n} \eta_i \tag{6}$$

$$R = max(k_1, k_2, k_3, k_4) - min(k_1, k_2, k_3, k_4) \tag{7}$$

Table 7 can be obtained from Equations (6) and (7). According to the R-value in Table 8, the primary and secondary order of the factors affecting the static pressure rise in the fan is β > t > Δβ > Bn > ϕ. The impeller outlet installation angle has the greatest influence, and the diffuser tilt angle has the smallest influence. Based on the k_i values, the optimal combination of factors is Bn = 17, β = 50°, Δβ = 4°, t = 1 mm, and ϕ = 60°.

Table 7. Range analysis of MIX-130 fan pressure rise.

No.	Bn	β/°	Δβ/°	t/mm	ϕ/°
K_1	1851	964	1671	2174	1647
K_2	1728	1459	1678	1666	1805
K_3	1631	2004	1652	1367	1681
K_4	1705	2488	1914	1708	1782
R	220	1524	262	807	158

Table 8. Range analysis of the MIX-130 fan Efficiency.

No.	Bn	β/°	Δβ/°	t/mm	ϕ/°
K_1	2.478	2.028	2.324	2.606	2.380
K_2	2.366	2.281	2.355	2.534	2.471
K_3	2.248	2.533	2.317	2.245	2.361
K_4	2.282	2.666	2.512	2.318	2.296
R	0.231	0.638	0.194	0.361	0.175

Table 8 can also be obtained from Equations (6) and (7). According to the R value in Table 7, the primary and secondary order of factors affecting fan efficiency is β > t > Bn > Δβ > ϕ. The impeller outlet installation angle has the greatest influence, and the diffuser tilt angle has the smallest influence. Based on the k_i values, the optimal combination of factors is Bn = 17, β = 50°, Δβ = 4°, t = 1 mm, and ϕ = 60°.

The combination of factors and levels that give the fan the highest efficiency and hydrostatic boost is the number of blades Bn = 17, outlet installation angle β = 50°, outlet installation angle increment Δβ = 4°, blade thickness t = 1 mm, and diffuser tilt angle ϕ = 60°.

Based on the optimization results, the MIX-130 fan is redesigned with the optimal parameters, and the performance of the optimized fan is analyzed using the numerical method (Section 2).

As depicted in Figure 5a, the static pressure rise and efficiency of the original and optimized fans change with the flow rate. The static pressure rise and efficiency of the optimized fan are higher than that of the original fan.

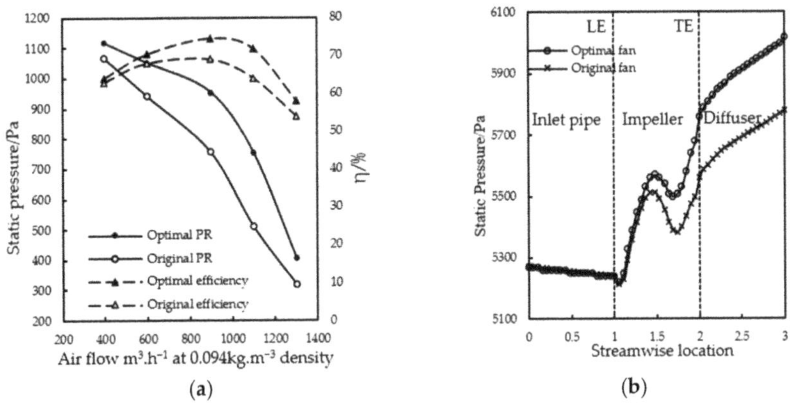

Figure 5. (**a**) Comparison of static pressure rise and fan efficiency before and after optimization; (**b**) Fan static pressure distribution along the streamwise.

Under the design flow rate, the static pressure rise in the original fan is 511 Pa, and the static pressure rise in the optimized fan is 754 Pa, which increases by 47.5%. The efficiency of the original impeller is 64%, and that of the optimized fan is 72%, an increase in 8%.

Figure 5b compares the streamwise static-pressure distribution between the optimized and original fans. After the inlet pipe, the static pressure in the optimized fan is higher than that in the original fan. In the impeller section, the static pressure of the optimized fan is increased to 10%, and that of the original fan is increased to 6%. In the diffuser section, the static pressure boost of the optimized fan is 4.5%, and that of the original fan is 3.9%. These findings reveal that the optimization of the impeller and diffuser is highly effective.

Figure 6 illustrates the static pressure distribution of the impeller of the original and optimized fans at a flow rate of 0.8–1.2 Q. The static pressure on the impeller increases continuously from the leading edge to the trailing edge of the blade, indicating that the booster effect of the impeller is ideal. The static pressure of the optimized impeller is higher than that of the original impeller under the same flow rate. After optimization, the static pressure distribution on the suction surface of the impeller is more uniform, the pressure gradient decreases, and the airflow stability increases, so flow separation does not readily occur.

Figure 7 illustrates the aerodynamic distribution on the blade. The force distribution uniformity on the optimized impeller increases, and maximum aerodynamic force is exerted on the suction surface of the blade. There are relatively concentrated aerodynamic forces at the front edge of the pressure and the trailing edge of the suction surface. However, the value is small, consistent with expectations. However, the aerodynamic distribution of the blade before the optimization is irregular, indicating that the airflow near the blade is relatively chaotic.

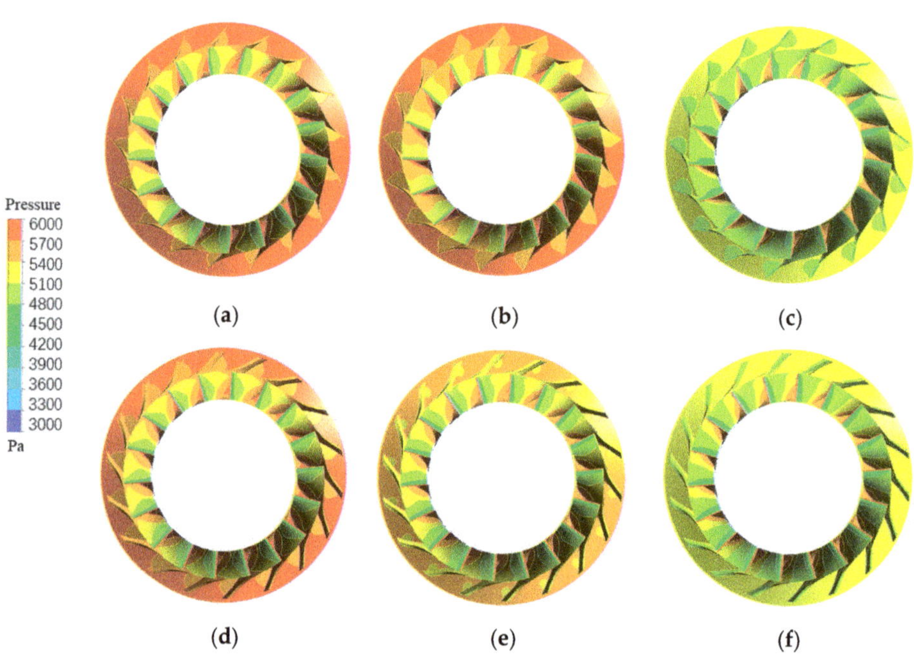

Figure 6. Static pressure distributions in impellers of original and optimized fans: (**a**) optimize impeller 0.8 Q; (**b**) optimize impeller Q; (**c**) optimize impeller 1.2 Q; (**d**) original impeller 0.8 Q; (**e**) original impeller Q; (**f**) original impeller 1.2 Q.

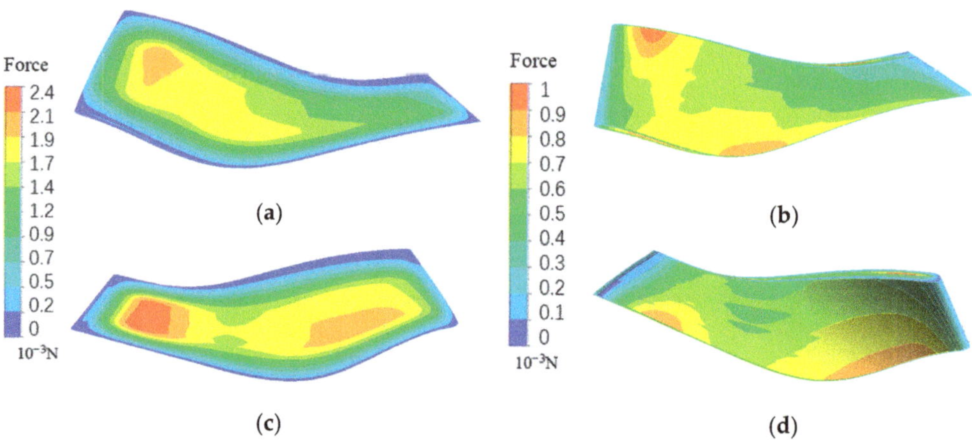

Figure 7. Blade force distribution (**a**) optimize blade pressure surface; (**b**) original blade pressure surface; (**c**) optimize blade suction surface; (**d**) original blade suction surface.

Figure 8 illustrates the static pressure distribution in the middle surface of the diffuser at a flow rate of 0.8–1.2 Q for the original and optimized fans. The static pressure in the fan increases with the increase in the diffuser diameter, exhibiting a good distribution, indicating that the diffuser has a high static-pressure transition ability. Under the same flow, the static pressure of the optimized diffuser is greater than that of the original diffuser, revealing that the optimization of the diffuser inclination angle is highly effective.

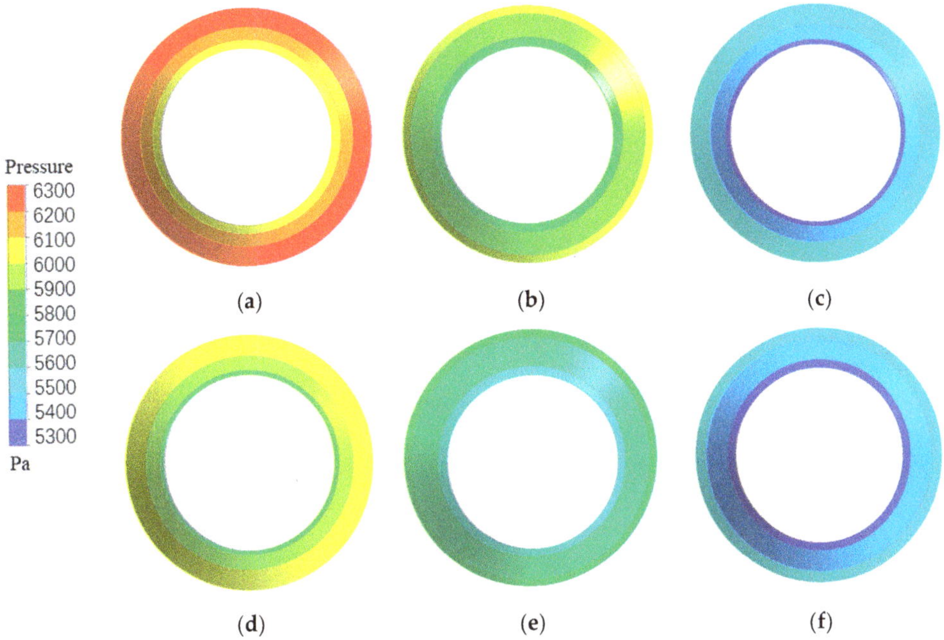

Figure 8. Static pressure distributions in the diffuser of original and optimized fans: (**a**) optimize diffuser 0.8 Q; (**b**) optimize diffuser Q; (**c**) optimize diffuser 1.2 Q; (**d**) original diffuser 0.8 Q; (**e**) original diffuser Q; (**f**) original diffuser 1.2 Q.

4. Fan Test

Figure 9 illustrates the optimized MIX-130 fan, including the overall structure of the fan and the structure of each component.

Figure 9. Optimized MIX-130 fan.

4.1. Test Method and Test Bench

Several mature methods and devices are available for testing fan performance at normal temperatures and pressure. However, measuring fan performance in a stratospheric, low-temperature, low-pressure environment requires a large enough environmental test chamber and specially customized test equipment. The cost is high, and the testing conditions are not currently available. This study proposes a method for testing and verifying fan performance at a high altitude. The fan's performance curve at a high altitude is obtained by measuring it at normal temperature and pressure, combined with the law of similarity conversion and numerical analysis. The accuracy of the performance curve obtained by the conversion is verified by measuring the maximum static pressure value of the fan at a high altitude.

Because the design point of the MIX-130 fan is the stratospheric, low-temperature, low-pressure condition, the full-speed test condition is not available when testing under the ground's normal temperature and pressure conditions, so the speed should be reduced for testing to ensure that the fan power is consistent with the high-altitude condition. The similarity principle converts fan performance under different working conditions [40,41]. The similarity principle is defined by Equations (8)–(10):

$$Q = Q_d n/n_d = NQ_d \tag{8}$$

$$p = p_d(n/n_d)^2(\rho/\rho_d) = N^2 p_d\left(\frac{\rho}{\rho_d}\right) \tag{9}$$

$$P = P_d(n/n_d)^3(\rho/\rho_d) = N^3 P_d\left(\frac{\rho}{\rho_d}\right) \tag{10}$$

where Q_d, p_d, P_d, n_d and ρ_d are the flow rate, pressure, power, speed, and medium density of the fan at the design point, and Q, p, P, n, and ρ are the operation values of the fan at other working conditions.

Based on Equation (10), the rated power P_d of the fan can be reached when the fan speed reaches 40% of the maximum speed under normal temperature and pressure on the ground.

The performance test of the MIX-130 fan at room temperature and pressure was conducted on the fan test bench by the AMCA (Air Movement and Control Association) specification, as depicted in Figure 10a, which was used to measure and record the fan performance curve. The test was conducted under an ambient temperature of 298 K, pressure of 0.1 atm, and relative humidity of 50%. The fan speed was set at 40% maximum rotation.

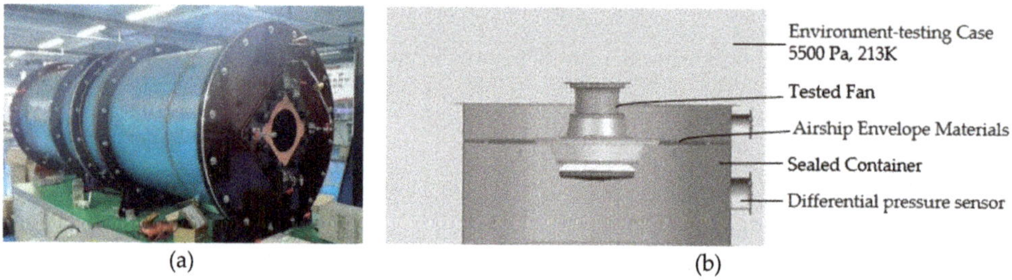

Figure 10. (**a**) Fan performance test bench for the ground condition; (**b**) Fan performance test bench for stratospheric condition.

The performance test of the MIX-130 fan at low temperature and low pressure is conducted on the test bench, as depicted in Figure 10b. Through this test bench [42], the maximum static pressure of the fan can be measured under any working condition at a high altitude. The test was conducted under a temperature of 213 K and pressure of 5500 Pa. The fan speed was set to the maximum speed.

4.2. Comparison of Test and Simulation Results

The curve of static pressure variation with the flow of the MIX-130 fan at normal temperature and pressure (P-Q curve for short) can be obtained using the numerical analysis model and method in Section 2 and changing the boundary conditions into the ground test environment. As depicted in Figure 11a, the two P-Q curves represent the numerical simulation and test results. The MIX-130 fan has a maximum static pressure of 2550 Pa and a maximum flow rate of 600 m^3/h at 40% of the maximum speed on the ground, and there is no saddle-shaped area in the entire P-Q curve. Based on the analysis, the average error between the numerical simulation and the experimental test is 3%, confirming the accuracy of the numerical analysis.

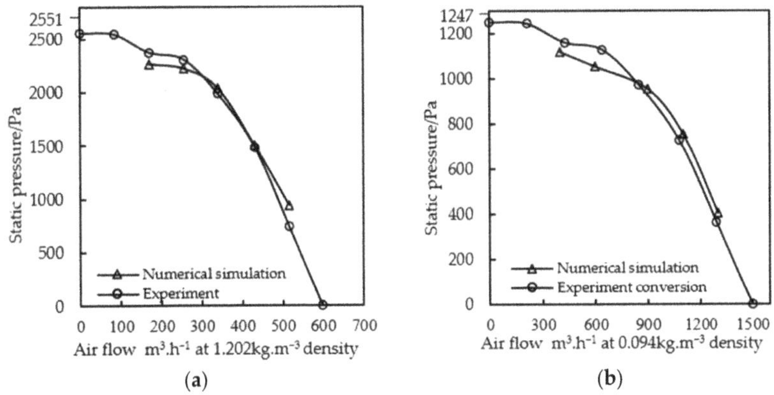

Figure 11. (**a**) The ground performance curve of the MIX-130 fan; (**b**) Stratospheric performance curve of the MIX-130 fan.

Based on Equations (8) and (9), the measured P-Q curve on the ground of the MIX-130 fan can be converted into the P-Q curve of stratospheric working conditions. As depicted in Figure 11b, the two curves are the stratospheric P-Q curves obtained from the numerical analysis in Section 2, and the stratospheric P-Q curves are measured and converted. Based on the analysis, the average error between the two is 4.4%, confirming that the similarity law is more accurate for converting fan performance in different working conditions.

The maximum static pressure of the MIX-130 and MAX45016 fans under stratospheric conditions was tested when the environment was stable at a temperature of 213 K and a pressure of 5500 Pa using the test bench in Figure 10b. As depicted in Figure 12, the test was completed in the Environmental Laboratory of Aerospace Information Research Institute, Chinese Academy of Sciences. The maximum static pressure of the MAX45016 and MIX-130 fans are 503 and 1210 Pa.

(a) (b)

Figure 12. (**a**) MIX-130 fan test status; (**b**) MAX45016 fan test status.

Figure 13 illustrates the P-Q curve of the MAX45016 and MIX-130 fans under stratospheric conditions. The P-Q curve of the stratospheric working condition of the MAX45016 fan can be converted from the P-Q curve at normal temperature and pressure using Equations (8) and (9), with a maximum static pressure of 506 Pa. The stratospheric P-Q curve of the MIX-130 fan is obtained by numerical simulation in Section 2, and its maximum static pressure is 1245 Pa. The error between the maximum static pressure obtained from the test and the data obtained from the numerical analysis is small, confirming the accuracy of the numerical method.

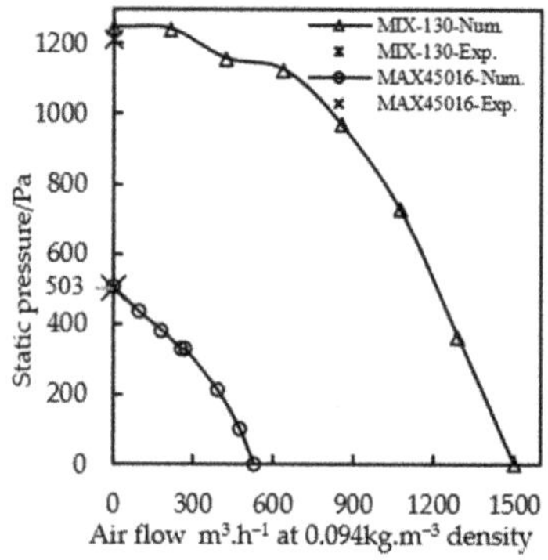

Figure 13. Performance comparison of fans in stratospheric conditions.

In stratospheric conditions, the maximum flow rate of the MIX-130 fan increases by 184%, and the maximum static pressure increases by 146% compared with the existing MAX45016 aviation fan. The performance of the MIX-130 fan can satisfy the requirements of a stratospheric airship in the first stage.

5. Conclusions

In this study, the $L_{16}(4^5)$ orthogonal test with five factors and four levels was used to optimize the performance of the mixed-flow fan used for a stratospheric airship, and a high optimization effect was obtained. The performance curves of the fan underground and stratospheric conditions were obtained using two different fan performance test stands, and the accuracy of the numerical calculation was verified by comparison.

The conclusions of this manuscript are as follows.

(1) The optimal fan design based on an orthogonal test has a significant effect. The results demonstrate that the static pressure lift of the optimized fan increases by 47.5%, and the efficiency increases by 8%.
(2) The average error between the performance curves obtained by testing the MIX-130 fan underground conditions and the numerical calculations is 3%, which proves the high accuracy of the CFD method and test means of this manuscript in obtaining the performance curves of the mixed-flow fan.
(3) The ground fan's performance curve is converted to the stratospheric performance curve by the law of similarity. Numerical and experimental methods confirm the accuracy of this conversion. This method can provide a reference for determining the performance curves of the fan at different altitudes.
(4) The MIX-130 fan can produce a flow rate of 1100 $m^3.h^{-1}$ and a static pressure lift of 754 Pa at the design point, which can meet the current requirements of stratospheric airships. The fan structure design with a mixed-flow impeller and blade-less diffuser is highly effective. The fan structure optimization and testing methods presented in this manuscript can provide a reference for designing and testing stratospheric airship fans.

For future works, In terms of fan optimization, we will carry out fan parameter optimization based on an algorithm [22,28,43] to continuously optimize fan performance; Practical application of fan: It is expected that in the summer of 2023, the MIX-130 fan will be mounted on stratospheric airships, overpressure balloons and other aerostats to carry out several studies in stratospheric environment, including the study on fan performance in low temperature and low-pressure environment, and the study on the sensitivity of fan parameters to the height regulation of overpressure balloons.

Author Contributions: Conceptualization, W.Q. and T.Z.; methodology, W.Q.; software, W.Q.; validation, W.G. and C.C.; investigation, Z.H.; data curation, W.Q.; writing—original draft preparation, W.Q.; writing—review and editing, T.Z. and Z.H.; visualization, C.C. All authors have read and agreed to the published version of the manuscript.

Funding: This research was funded by "CAS strategic leading science and technology project (Class A), XDA17000000"; CAS science and disruptive technology research pilot fund, E1Z220010F.

Data Availability Statement: Not applicable.

Acknowledgments: We thank Jiangsu Jiu Gao Tech Co., Ltd. for providing testing support for this study.

Conflicts of Interest: The authors declare no conflict of interest.

References

1. Jing, Y.; Wu, Y.; Tang, J.; Zhou, P.; Duan, D. Receding horizon trajectory generation of stratospheric airship in low-altitude return phase. *Aerospace* **2022**, *9*, 670. [CrossRef]
2. Liu, Y.; Xu, Z.; Du, H.; Lv, M. Multidisciplinary optimization of thermal insulation layer for stratospheric airship with a solar array. *Aerospace* **2022**, *9*, 83. [CrossRef]
3. Riboldi, C.E.D.; Rolando, A. Layout Analysis and Optimization of Airships with Thrust-Based Stability Augmentation. *Aerospace* **2022**, *9*, 393. [CrossRef]
4. Fu, C.; Zhu, M.; Ding, Y.; Tian, P.; Huang, M.; Liu, D.; Zhao, D. Influence of ice accretion on stratospheric airship in the non-forming ascending process. *Aerospace* **2022**, *9*, 536. [CrossRef]
5. Tang, J.; Xie, W.; Wang, X.; Chen, C. Simulation and analysis of fluid–solid–thermal unidirectional coupling of near-space airship. *Aerospace* **2022**, *9*, 439. [CrossRef]
6. Chaabani, S.; Azouz, N. Modelling and stabilisation of an unconventional airship: A Polytopic Approach. *Aerospace* **2022**, *9*, 252. [CrossRef]
7. Dumas, A.; Angel, D.; Trancossi, M. High altitude airship cabin sizing, pressurization and air conditioning. *Energy Procedia* **2014**, *45*, 977–986. [CrossRef]
8. Xiao, M.; Chen, X.; Yang, Y. A method for altitude control of stratospheric airships. *Electron. Opt. Control* **2014**, *21*, 86–89.
9. Zhang, J.; Yang, X.; Deng, X.; Guo, Z.; Zhai, J. Altitude control of stratospheric aerostat based on deep reinforcement learning. *J. Beijing Univ. Aeronaut. Astronaut.* 2021, *in press*. [CrossRef]
10. Lin, K.; Ma, Y.; Zheng, Z.; Wu, Z. Height control of stratospheric aerostat based on secondary airbag. *J. Beijing Univ. Aeronaut. Astronaut.* **2022**, *5*, 48.
11. Silva Marton, A.; Azinheira, J.R.; Fioravanti, A.R. Filtering and Estimation of State and Wind Disturbances Aiming Airship Control and Guidance. *Aerospace* **2022**, *9*, 470. [CrossRef]
12. Sun, P. Stratospheric Airship Blower Design and Its Characteristics Research. Master's Thesis, Shanghai Jiao Tong University, Shanghai, China, 2018.
13. Yan, L. Research on Finite Set Model Predictive Control of Drive Motor for High-Altitude Ventilator. Ph.D. Thesis, Northwestern Polytechnical University, Xi'an, China, 2018.
14. Zhong, H. The Mathematic Modeling and Simulation of Pressure Control of Stratospheric Airship. Master's Thesis, Shanghai Jiao Tong University, Shanghai, China, 2018.
15. Deng, X.; Yang, X.; Zhu, B. Simulation research and key technologies analysis of intelligent stratospheric aerostat Loon. *Acta Aeronaut. Astronaut. Sin.* 2022. Available online: https://kns.cnki.net/kcms/detail/11.1929.V.20220710.1648.034.html(accessed on 2 December 2022).
16. Zhu, R.; Wang, S.; Yang, Y.; Liu, Q. Analysis of day–night thermal properties of super-pressure balloon during cruising flight. *Comput. Simul.* **2020**, *37*, 54–59.
17. Liu, S. Research on flow variation of aerostat valve based on CFD. *J. Xi'an Aeronaut. Univ.* **2021**, *39*, 14–19, 25.
18. Niu, Z. Effects and Mechanisms of Low Reynolds Number on Compressors' Flow Stability. Master's Thesis, Tsinghua University, Beijing, China, 2021.
19. Gao, Y.; Xu, G.; Wang, S.; Li, Z.; Cai, R. Modeling and small signal stability analysis of stratospheric airship energy system. *Acta Energ. Sol. Sin.* **2022**, *43*, 50–57.
20. Sun, P.; Wang, X.; Xie, W. Centrifugal blower of stratospheric airship. *IEEE Access* **2018**, *6*, 10520–10529. [CrossRef]
21. Wei, K.; Wu, Q.; Zhang, Y.; Li, X. Aerodynamic performance calculation of centrifugal compressors for pressurized balloons based on similarity theory. *J. Xi'an Aeronaut. Univ.* **2019**, *03*, 28–32.
22. Zhao, D.; Hua, Z.; Dou, M.; Huang, F. Control oriented modeling and analysis of centrifugal compressor working characteristic at variable altitude. *Aerosp. Sci. Technol.* **2018**, *72*, 174–182. [CrossRef]
23. Yang, X.; Liu, Z.; Zhang, S.; Song, C. Aerodynamic design of key components of the large-discharge axial fan at high altitude. In Proceedings of the 31st Congress of the International Council of the Aeronautical Sciences, Belo Horizonte, Brazil, 14 September 2018.

24. Zhang, Q.; Miao, J.; Li, J.; Zhao, C.; Li, R. Characteristics research of fans in series and parallel on super-pressure balloon. In Proceedings of the 8th Annual High Resolution Earth Observation Academic Conference, Beijing, China, 1 May 2022.
25. Fan, H.; Zhang, J.; Zhang, W.; Liu, B. Multiparameter and multiobjective optimization design based on orthogonal method fo mixed flow fan. *Energies* **2020**, *13*, 2819. [CrossRef]
26. Wang, S.; Zhang, L.; Wu, Z.; Qian, H. Optimization research of centrifugal fan with different blade number and outlet blade angl In Proceedings of the 2009 Asia-Pacific Power and Energy Engineering Conference, Wuhan, China, 28–31 March 2009; pp. 1–4.
27. Wang, J.; Nakata, T.; Liu, H. Development of mixed flow fans with bio-inspired grooves. *Biomimetics* **2019**, *4*, 72. [CrossRef]
28. Wang, C.; Yang, F.; Thanh, V.; Nguyen, T. CFD Analysis and optimum design for a centrifugal pump using an effectively artificia intelligent algorithm. *Micromachines* **2022**, *13*, 1208. [CrossRef]
29. Jung, U.; Kim, J.; Kim, S.; Kim, J.; Choi, Y. Analyzing the shape parameter effects on the performance of the mixed-flow fan usin CFD & factorial design. *J. Mech. Sci. Technol.* **2016**, *30*, 1149–1161.
30. Wang, M.; Li, Y.; Yuan, J.; Meng, F.; Appiah, D.; Chen, J. Comprehensive improvement of mixed-flow pump impeller based o multi-objective optimization. *Processes* **2020**, *8*, 905. [CrossRef]
31. *CN202221392244*; A Mixed Flow Fan for the Stratosphere, Aerostat Control System; Aerospace Information Research Institute Beijing, China, 2022.
32. Thamsen, P.; Treder, A.; Hammer, S.; Tunsi, K. Mixed flow fans—Comparative investigations of blade design methods. Turbo Expo: Power for Land, Sea, and Air. *Am. Soc. Mech. Eng.* **2014**, *45578*, V01AT10A030.
33. Wang, Y.; Dai, J.; Wang, Z. Design of vaneless diffuser for centrifugal compressor. *J. Eng. Therm. Energy Power* **2022**, *37*, 15–21.
34. Wang, S. Design of Helium Centrifugal Compressor and Optimization of High Pressure Ratio. Master's Thesis, Harbin Engineering University, Harbin, China, 2021.
35. Han, F.; Chen, X.; Yang, Y. Numerical and experimental study on the effect of rotor–stator distance on rotor–stator interaction strength within mixed-flow centrifugal pumps. *J. Mar. Sci. Eng.* **2022**, *10*, 1114. [CrossRef]
36. Wang, S.; Liu, B.; Yue, G.; Jiang, Y. Research on high pressor ratio design of helium centrifugal compressor. *J. Eng. Therm. Energy Power* **2021**. [CrossRef]
37. Chen, Z.; Qing, D.; Tang, D. Study on the effects of blade number and setting angle on the performance of helium compressor *Mech. Eng.* **2019**, *12*, 26–28, 31.
38. Xu, Y.; Tan, L.; Cao, S.; Qu, W. Multiparameter and multiobjective optimization design of centrifugal pump based on orthogona method. *J. Mech. Eng. Sci.* **2017**, *231*, 2569–2579. [CrossRef]
39. Moghadam, S.M.A.; Meinke, M.; Schroeder, W. Analysis of tip-leakage flow in an axial fan at varying tip-gap sizes and operating conditions. *Comput. Fluids* **2019**, *183*, 107–129. [CrossRef]
40. Xie, H.; Zhang, S.; Yang, B. Research on the application of similarity principle on the centrifugal compressor. *J. Eng. Thermophys* **2021**, *36*, 147–154.
41. Wang, L.; Gu, B.; Wang, T. Fast calculation model of fan performance based on similarity theory. *Fluid Mach.* **2012**, *7*, 24–28.
42. *CN2020102600000*; An Air Tightness Detection System of Aerostat Valve and Test Method; Aerospace Information Research Institute: Beijing, China, 2022.
43. Nguyen, V.T.T.; Vo, T.M.N. Centrifugal pump design: An optimization. *Eurasia Proc. Sci. Technol. Eng. Math.* **2022**, *17*, 136–151 [CrossRef]

aerospace

MDPI

Article

Design of an Airship On-Board Crane

Fatma Guesmi [1,2], Naoufel Azouz [1,*] and Jamel Neji [2]

1 LMEE, UnivEvry, UFR Sciences and Technologies, Université Paris-Saclay, 91025 Evry, France
2 Lamoed, National Engineer School of Tunis, Civil Engineering Department, University of Tunis El Manar, Tunis 1068, Tunisia
* Correspondence: naoufel.azouz@univ-evry.fr

Abstract: This paper presents the design and mathematical model of an innovative smart crane, CHAYA-SC, based on the principle of a cable-driven parallel manipulator, as well as its stabilization. This crane is mounted on the airship hold and intended for handling at altitude. Our objective is to design a precise light crane that can be used for container loading or unloading, particularly in deep-sea ports. Thus, the model developed includes the oscillations as well as the transverse and longitudinal vibrations of the heavy cable supporting the load to be handled. The highly nonlinear partial differential equations (PDE) and ordinary derivative equations (ODE) that govern the motion of the system are obtained via the Lagrange equations coupled with a modal synthesis. So that the mathematical model of the system is compatible with control and real time, we developed a simplified dynamic model which proved to be equivalent to the complete model. As a first validation of the modelling, a simple control vector is applied to stabilize the airship and its load under the effect of a squall. Numerical simulations are presented at the end of the paper to show the relevance of the design.

Keywords: smart crane; large capacity airships; cable-driven parallel manipulator; freight transportation; modelling; stabilization

1. Introduction

Large capacity airships (LCAs) bring new perspectives for airships that hibernated for more than half a century after the Hindenburg drama. In fact, for more than two decades, researchers have studied the formidable potential of airships in terms of freight transport. These airships will be able to transport blades of wind turbines or logs from or to areas very difficult to access, load or unload container ships on the high seas in the absence of adequate ports (as seen in Figure 1), and install field hospitals in disaster areas. Other examples of the use of airships in land and maritime multimodal logistics can be found in [1,2]. These various examples prove, if necessary, the enormous potential available to these flying machines. However, we should mention that in order to be able to lift 1 kg of load, 1 m^3 of Helium must be provided within the hull of the airship. The airships targeted for the aforementioned examples will therefore have enormous volumes, and all the obstacles inherent in these large volumes have not yet been overcome. Among these obstacles, we note on the one hand, the high sensitivity of airships to gusts of wind. Various studies have looked into the stabilization of airships, particularly in hovering flight or for tracking trajectories (see, for example, [3–7]). On the other hand, the non-standard dimensions of these devices make the development of landing or handling areas very problematic or costly. This has left the designers of these machines to consider handling operations at altitude. It is in this context that our study takes place. We propose for this task a smart crane capable not only of loading and unloading but also of stabilizing the load during a gust of wind and of arranging the hold of the airship. This crane will be based on a cable-driven parallel manipulator (CDPM).

Figure 1. Prototype of a large capacity airship in handling.

Today, CDPMs are set to have a big impact on many aspects of modern life, from industrial manufacturing to healthcare, transportation, and the exploration of space and the seabed. Cable-driven parallel manipulators are a special class of parallel manipulators in which the effector is directly actuated by cables. By considering the number of cables and the degrees of freedom of the effector, the CDPM can be overconstrained, in particular as in [8,9], or underconstrained.

These manipulators offer a variety of potential advantages over traditional parallel robots. Because of their unique configurations, CDPMs are characterized by light structures and a large working space due to the location of the actuators at the fixed base of the structure, thus reducing the mass and inertia of the mobile platform. They can be produced on a very large scale at acceptable cost, which makes them very suitable for high velocities and high performance [10,11]. Another major advantage is that the CDPMs are reconfigurable, which allows them to be used for different tasks by moving the attachment points of cables as in [12].

In the literature, different approaches have been proposed to find the tension distribution of cables. The set of optimal tension chosen is usually performed using an optimization method, for example, linear programming [13], quadratic programming [14], and convex optimization for the minimization of L1 norm [15] and p norm [16]. However, neither of these methods can provide a continuous solution and cover the entire workspace as in [17]. Another disadvantage of cable parallel robots is the possibility of cables colliding with each other, but this is limited only to space redundant systems as in [12].

Given the advantages and unique characteristics of CDPMs, their use has spread to several types of applications. The idea of cable robots appeared from the studies of Albus et al. [18], Landsberger [19], and Higuchi [20]. Inspired by Stewart platform, the NIST-Robot Crane was developed by Albus and his team [18] to overcome the drawbacks of the motions underlying conventional cranes. Because of their large workspace and high velocities, the CDPMs have been used in sports recording, as in the case of Skycam [21]. Another application for this type of manipulator is medical rehabilitation as can be seen in [22–24].

We were therefore inspired by the CDPM principle to propose our smart crane on board the airship, the design of which we will present in the next section.

A model of a multibody system formed by the airship, the crane, and the suspended load will be presented, and the stabilization of the system subjected to a gust of wind will be proposed.

2. Problem Statement and Description of the Smart Crane

As part of the development of tools essential to the development of large capacity airships, we set out to design an embedded smart crane (the CHAYA-SC) that is as light as possible, can both handle the containers and stabilize the load during the handling, and finally ensure the precise arrangement of the containers within the hold.

The large airships that we are targeting have a large volume, greater than 100,000 m^3, and it would be difficult to provide them with landing and loading infrastructure. So that these means of transport could be used everywhere, it is essential to provide a means of handling at altitude. This will allow them to be used universally but must be accompanied by various measures, in particular robust control of the airship and a crane adapted to this configuration. It is in this context that we present our smart crane.

Presentation of the Smart Crane

The smart crane called CHAYA-SC is presented in Figure 2. It consists of a cuboid trunk (that we could see in details in Figure 3) supported by eight cables suspended from the four upper corners of the airship hold. These eight cables are arranged to kinematically constrain the trunk.

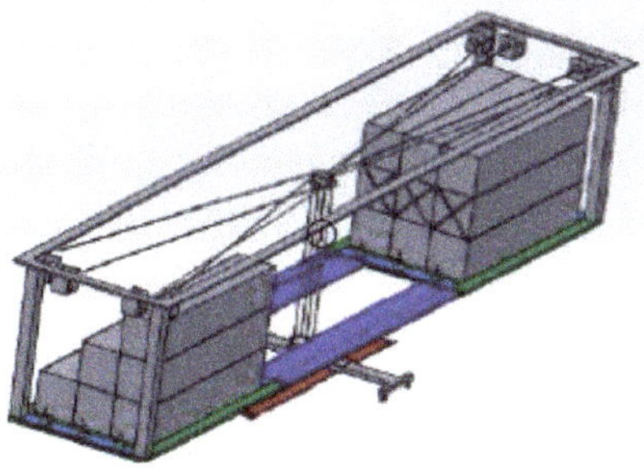

Figure 2. The smart crane: CHAYA.

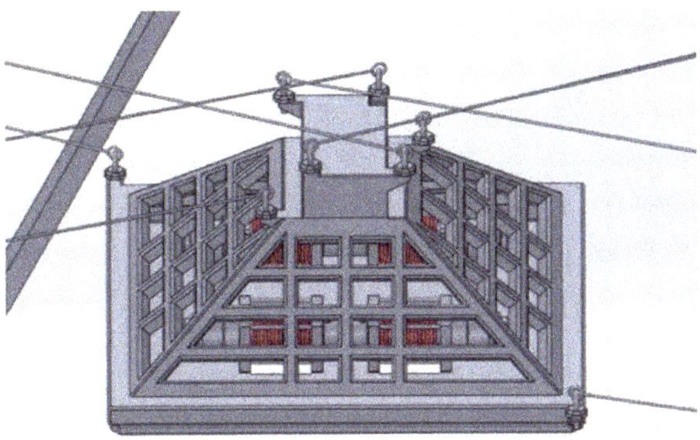

Figure 3. Cuboid details.

Eight winches have the role of controlling these eight cables. These are controlled and coordinated by a computer. The motion of the platform is controlled by an operator through a six-axis joystick with six DOFs. The operator can handle and manipulate the trunk and any load attached to it over a large volume of work inside the hold.

The trunk consists of a rigid structure overhung by four electrical engines. These engines pull a flexible cable connected to a container hook. The multiplication of electrical engines obeys safety constraints.

This crane has the following missions:

1. A classic container loading and unloading mission enabled by the winches integrated into the cuboid.
2. Stabilize the suspended containers during a sudden motion of the airship under the effect of a gust of wind by creating accelerations according to six possible degrees of freedom of the cuboid, depending on the nature of the oscillation of the load.
3. Drop off, collect, and store the containers in the airship hold (as shown in Figure 4).

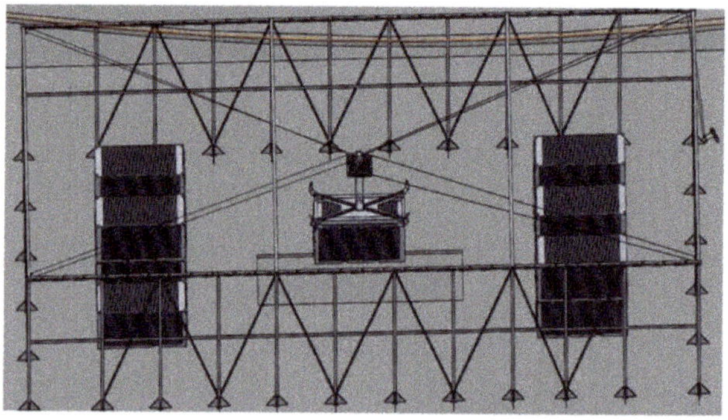

Figure 4. Diagram of the arrangement of the containers in the hold.

3. Dynamic Modelling

3.1. Modeling of the Cable-Driven Parallel Manipulator

3.1.1. Classical Case: Industrial CDPM

Let us consider a CDPM with m cables having the geometry illustrated in Figure 5.

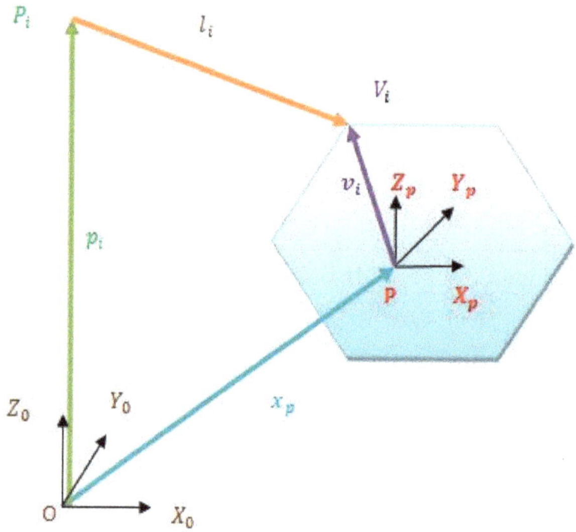

Figure 5. General Architecture of a CDPM.

The cables are attached to the fixed base at the exit points noted $P_i \in \mathbb{R}^3$ whose coordinates are defined in the earth fixed frame $R_0 = (O, X_0, Y_0, Z_0)$ by the vectors p_i, $i \in \{1, 2, \ldots, m\}$. The cable attachment points on the platform noted $V_i \in \mathbb{R}^3$ are expressed in the mobile frame attached to the suspended platform $R_p = (P, X_P, Y_P, Z_P)$ by v_i, and l_i is the length of the cable connecting these two points.

Unlike series-type robots, the inverse geometric and inverse kinematic models of parallel robots are easy to establish. These results are used to find a relation between the operational space with n dimensions of the spatial coordinates of the platform and the space with m dimensions of the articular coordinates.

We denote by $q = (q_1, q_2, \ldots, q_m)^t \in \mathbb{R}^m$ the column matrix of the articular coordinates of the CDPM and by $X_p = (X_{p_1}, X_{p_2}, \ldots, X_{p_n})^t \in \mathbb{R}^n$ the operational coordinates of the platform.

The inverse kinematics problem seeks to determine the lengths of the cables taking into account the position and orientation of the suspended platform. Inverse kinematics for parallel structures is easier to calculate than direct kinematics. The Jacobian matrix J of a cable-driven parallel robot is defined as a relationship between the velocity vector of the platform $\dot{X}_p = (\dot{x}_p, \omega_p)^t$ and the linear velocity of the cables $\dot{l} = \left(\dot{l}_1, \dot{l}_2, \ldots, \dot{l}_m\right)$:

$$\dot{l} = J(X_p)\dot{X}_p \tag{1}$$

The dynamic equations which describe the motion of the CDPM are obtained using the Lagrange equations that we will recall later in this paper. The general dynamic model can be divided into two expressions corresponding to the dynamics of the platform and the dynamics of the winches. The cables are considered in deformable and of negligible mass. The cables are controlled by motorized winches. These winches are equipped with a drum around which a cable is wound. Each winch (Figure 6) consists of asynchronous servomotor coupled to a planetary gearbox which is connected to a drum. In fact, the motor torques τ_i drives the cylindrical drum in rotation at an angle q_i about its axis of symmetry. This generates a tensile force t_i at the exit point of the cable P_i.

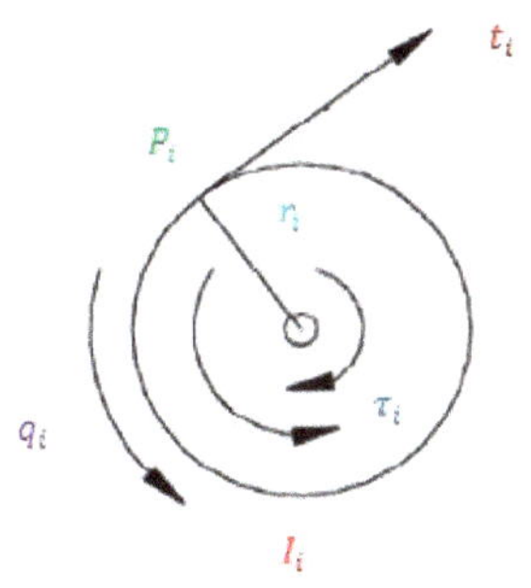

Figure 6. Diagram of an actuator.

The dynamic model of the actuators is given by the following equation:

$$\tau = I\ddot{q} + RT \tag{2}$$

$\tau = (\tau_1, \tau_2, \cdots, \tau_m)^t$ is the input vector of the torques exerted by the motors which control synchronously the cable lengths as a function of the tension $T = (t_1, t_2, \ldots, t_m)^t$ in order to provide the required movement of the platform in the Cartesian space by acting on the winches, and I is the diagonal inertia matrix of the moments of the drums, such as $diag(I) = \{I_1, I_2, \ldots, I_m\}$.

By applying the Lagrange equations, the dynamic equation of the lower base will be given, as in [25], by:

$$M(X_p)\ddot{X}_p + C\left(X_p, \dot{X}_p\right)\dot{X}_p = W(X_p).T + F_g \quad (3)$$

F_g represents the external forces, such as the weight of the platform and its loading or the force exerted by a gust of wind, $W(X_p) = -J^t(X_p)$, M is the mass matrix of the platform, C is the matrix of centrifugal and Coriolis forces. The latter two are defined in detail in Appendix C.

3.1.2. Case of the CDPM on Board an Airship

The CDPM which represents the upper part of our smart crane and which we present in this work is inspired by those used in industry from a technological and configuration point of view. We were inspired more precisely by the robot CoGiRo developed by Tecnalia presented in [26], which corresponds to our handling needs. In their study, Lamaury et al. defined the appropriate number of cables for the optimal control of the movement of the cuboid according to the six degrees of freedom. Their study concluded that this would require the use of eight cables. We followed this recommendation and used eight cables to manipulate the cuboid of our smart crane.

The main difference between these two CDPMs (apart from the dimensional aspect) is that ours is embarked on a mobile system (the airship) and is therefore subject to the turbulence and acceleration of the latter. The so-called "fixed" base actually follows the motion of the airship.

The mobile platform is embodied by the cuboid in our smart crane. Its motion depends not only on the motion of the eight cables that support it but also on the motion of the airship.

The CDPM is suspended by eight cables for six DOFs. The cables are connected on the other side to the hold of the airship at the top of the pillars (as seen in Figure 7). Each point of attachment of the cables is therefore directly deduced from the motion of the center of gravity G of the airship.

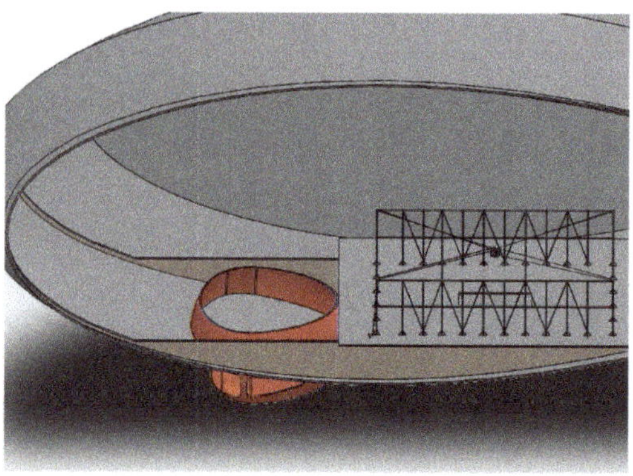

Figure 7. Positioning of the CDPM in the hold of the airship.

The pulleys guiding the cables (as shown in Figure 8) are connected to controlled motors which represent the actuators of the CDPM.

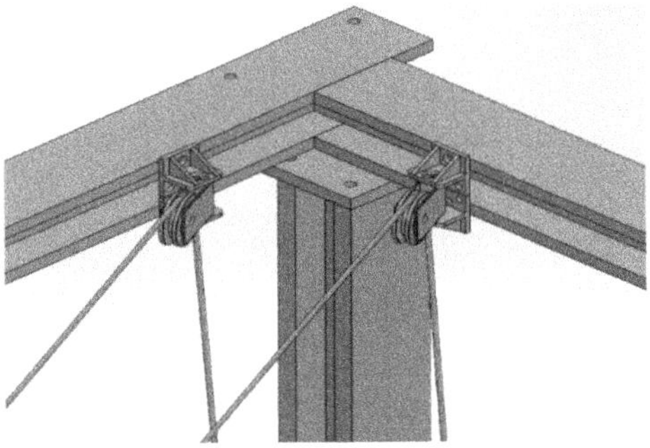

Figure 8. Arrangement of pulleys on the cargo hold structure.

The set of forces exerted on the platform (cuboid) is given by:

$$F_C = W(X_P)T \tag{4}$$

Compared to the equations developed previously and neglecting the term $I\ddot{q}$, we will have to use the kinetic energy of the airship to obtain the equation of motion of the CDPM:

$$M(X_P)\ddot{X}_P + C\left(X_P, \dot{X}_P\right)\dot{X}_P - F_g = F_C \tag{5}$$

More details concerning the obtaining of this equation will be developed in the particular case of the plane motion of the multibody system.

3.2. Dynamic Modeling of the Multibody System in a Plane Motion

As an application, we are interested in modeling the system assuming that the airship as well as the cuboid move in XZ space. They follow unidirectional motion along axis (GX_1) and axis $(G'X_2)$, respectively. The system to be modeled is shown in Figure 9. It is composed of an airship, a cuboid, and a load suspended by a cable. The latter, which holds the container, is thick and massive. Unlike the cables that support the cuboid, we will consider here its flexibility as well as its elongation. The cable has a length l_c, a mass per unit of length ρ, a modulus of Young E, and a moment of inertia I_{y3}. The load has mass m_L and matrix of inertia I_L. The distance between its center of mass and the end of the cable is represented by r_L, and the deformation is denoted by ω.

As part of the modelling of this system, this assumption will be retained: the rotational inertia of the pulleys is neglected in front of the other inertias.

The analysis of the global system motion is made with respect to six reference frames, namely:

i. An earth-fixed frame $R_0(O, X_0, Y_0, Z_0)$.
ii. A local frame fixed to the airship $R_1(G, X_1, Y_1, Z_1)$, having as origin the inertia center of the airship.
iii. A local frame fixed to the cuboid $R_2(G', X_2, Y_2, Z_2)$, having as origin the inertia center of the cuboid G'.
iv. A reference frame $R_3(G', X_3, Y_3, Z_3)$, linked to the cable in rotation at an angle θ with respect to R_2.
v. A reference frame $R_4(P_c, X_4, Y_4, Z_4)$, attached to any section of the cable located at a distance z_3 from the z_3 axis.
vi. A reference frame $R_5(P_L, X_5, Y_5, Z_5)$, at the end of the cable to describe the motion of the load.

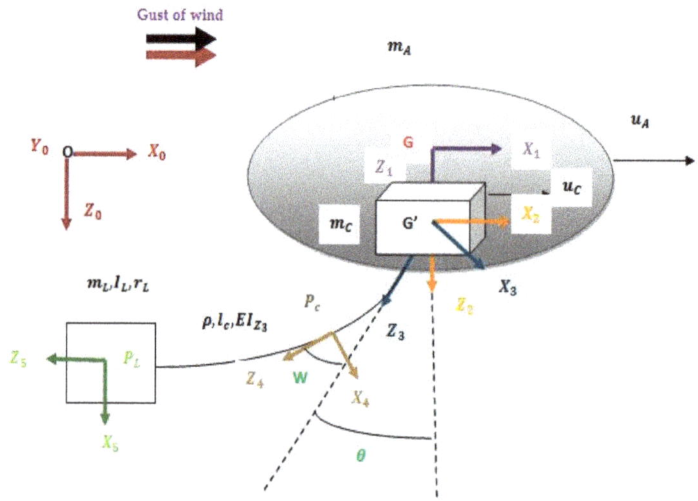

Figure 9. Coordinates of the different frame.

To facilitate the writing of the equations of motion, we adopt the notations below: $w = w(z_3, t)$,$w_{z_3} = \frac{\partial w(z_3,t)}{\partial z_3}$, $w_{z_3z_3} = \frac{\partial^2 w(z_3,t)}{\partial^2 z_3}$, $\dot{w} = \frac{dw(z_3,t)}{dt}$ represent the deformation of the cable as well as its derivatives at a point z_3 of the cable.

When point z_3 is equal to l_c, the deformation and its first and second derivatives are noted by:

$$w_{l_c} = w(l_c, t),\ w_{z_3,l_c} = \left(\frac{\partial w(z_3, t)}{\partial z_3}\right)_{|z_3=l_c},\ w_{z_3z_3,l_c} = \left(\frac{\partial^2 w(z_3, t)}{\partial^2 z_3}\right)_{|z_3=l_c}$$

In order to apply the Lagrange equations, we need to calculate the Lagrangian of the system. This Lagrangian can be defined as follows as in [27]:

$$L = T_{tot} - V_{tot} + F_R \tag{6}$$

With T_{tot} is the total kinetic energy of the system, $V_{tot'}$ is the total potential energy, and F_R is the Rayleigh dissipation function.

3.2.1. Kinetic Energy of the System

It should be noted that the kinetic energy of the whole system is made up of that of the airship (T_A), the cuboid (T_C), the payload (T_L), and the flexible cable.

$$T_{tot} = T_A + T_C + T_L + \int_0^{l_c} T_f \tag{7}$$

We will denote by u_A and u_C the speeds of the airship and of the cuboid projected onto the mobile reference frames linked to each of them. In our particular example of a plane case, we have: $u_A = \dot{x}_A\ and\ u_C = \dot{x}_C$.

The kinetic energy of the airship is simply expressed as:

$$T_A = \frac{1}{2} m_A\, u_A^2 \tag{8}$$

T_C is the kinetic energy of the cuboid:

$$T_C = \frac{1}{2} m_C \, u_C^2, \tag{9}$$

where m_C is the mass of the cuboid, and m_A is the mass of the airship including the added mass. For more details concerning the added masses one could consult [28]. The kinetic energy of the load has two terms: the first term which takes into consideration the linear motion of the body with respect to the frame R_3 and the second term which takes into consideration the rotation of the load. The expression of the kinetic energy of the load is:

$$\begin{aligned}
T_L &= \tfrac{1}{2} I_L \left(\dot{\theta}^2 + \dot{w}_{z_3,l_c}^2 + 2\dot{\theta}\, \dot{w}_{z_3,l_c} \right) \\
&+ \tfrac{1}{2} m_L \big(u_A^2 + u_C^2 + 2u_A u_C + \dot{w}_{l_c}^2 + \dot{w}_{z_3,l_c}^2 r_L^2 + 2\dot{w}_{l_c}(u_A + u_C) c\theta \\
&+ \dot{\theta}^2 (l_c + r_L)^2 + 2r_L \, \dot{w}_{z_3,l_c}(u_A + u_C) c\theta + 2(u_A + u_C)\dot{\theta} c\theta (l_c + r_L) \\
&- r_L (u_A + u_C)\, \dot{\theta} c\theta w_{z_3,l_c}^2 - \dot{\theta}(u_A + u_C) c\theta \int_0^{l_c} w_{z_3}^2 dz_3 + 2r_L \, \dot{w}_{z_3,l_c} \dot{w}_{l_c} \\
&+ 2\dot{\theta}\dot{w}_{l_c}(l_c + r_L) + \dot{\theta}^2 w_{l_c}^2 + r_L^2 \dot{\theta}^2 w_{z_3,l_c}^2 + 2r_L (l + r_L)\dot{\theta}\dot{w}_{z_3,l_c} \\
&- 2(u_A + u_C) s\theta \int_0^{l_c} w_{z_3} \dot{w}_{z_3} dz_3 - 2r_L (u_A + u_C) s\theta \dot{w}_{z_3,l_c} w_{z_3,l_c} \\
&- 2(u_A + u_C)\dot{\theta} s\theta w_{l_c} + 2r_L (u_A + u_C) s\theta w_{z_3,l_c} + o(3) \big)
\end{aligned} \tag{10}$$

$o(3)$ are small terms of the third order that can be neglected later.

On the other hand, the kinetic energy of the cable is computed by a summation over the entire length of the latter. We can see its expression in detail in Appendix A.

By developing the equations of the kinetic energy of the cable and of the load, we will keep the expressions of the position and the speed exact to the second order. All other quantities can be linearized. Finally, to obtain the expression for the linearized total kinetic energy of the system, it would suffice to add and rearrange the corresponding terms. The developed expression of the total kinetic energy of the system can be seen in Appendix A.

3.2.2. Potential Energy

The potential energy will be computed using the same reasoning as that of the kinetic energy: in theory, the potential energy of the multibody system is the sum of the potential energies of the different bodies.

However, for the case mentioned in Section 3.2.1, and which will be our illustrative example in this paper, the airship as well as the cuboid will have horizontal rigid body motions and will therefore not affect the potential energy. For these reasons, the total potential energy can be written as:

$$V_{tot} = V_L + \int_0^{l_c} V_f \tag{11}$$

The potential energy of the load is given by:

$$V_L = m_L \, g \, s\theta (w_{l_c} - r_L w_{z_3,l_c}) - m_L \, g \, c\theta \left(l_c + r_L - \frac{1}{2} r_L w_{z_3,l_c}^2 - \frac{1}{2} \int_0^{l_c} w_{z_3}^2 dz_3 \right) \tag{12}$$

On the other hand, the potential energy of the cable will not only have the contribution of gravity but also a contribution given by its elasticity, which we could write as follows:

$$V_f = \frac{1}{2} E I_{z_3} w_{z_3 z_3}^2 dz_3 + g\rho\, w(z_3, t) s\theta\, dz_3 \;\; - \rho g\, c\theta \left(z_3 - \frac{1}{2} \int_0^{z_3} w_{z_3}^2(s,t)\, ds \right) dz_3 \qquad (13)$$

Hence the expression of the total potential energy becomes:

$$\begin{aligned} V_{tot} &= \tfrac{1}{2} E I_{z_3} \int_0^{l_c} w_{z_3 z_3}^2 dz_3 + g\rho\, s\theta \int_0^{l_c} w(z_3, t) dz_3 \\ &-\rho\, g\, c\theta \int_0^{l_c} \left(z_3 - \tfrac{1}{2} \int_0^{z_3} w_{z_3}^2(s,t) ds \right) dz_3 + m_L g\, s\theta \left(w_{l_c} - r_L w_{z_3, l_c} \right) \\ &- m_L g\, c\theta \left(l_c + r_L - \tfrac{1}{2} r_L w_{z_3, l_c}^2 - \tfrac{1}{2} \int_0^{l_c} w_{z_3}^2 dz_3 \right) \end{aligned} \qquad (14)$$

3.2.3. Dissipation Function

Within a flexible metal cable, there is an internal energy dissipative damping. To model this characteristic, the Rayleigh dissipation function was used. Its expression is:

$$F_R = \frac{1}{2} k_e\, E I_{z_3} \int_0^{l_c} \dot{w}_{z_3 z_3}^2\, dz_3 \qquad (15)$$

In order to obtain a dynamic model having a reasonable number of degrees of freedom, modal synthesis will be used to discretize the energies calculated previously. This procedure is usual in the vibratory problems and allows the number of degrees of freedom to be minimized in our precise case by minimizing the modes retained. This will have the advantage of facilitating the implementation of control laws.

3.2.4. Modal Synthesis

The vibration of cables is extensively investigated in the literature. Recently, vibration of axially moving materials has received a great deal of attention in the literature (see, for example, [29]). These survey papers present a picture of the state of the art in the vibration and dynamic stability of axially moving strings or beams. Little research on the vibration behavior of overhead cranes with flexible cables has been studied in [30,31]. In the study of [32], the authors used the Rayleigh–Ritz discretization method. Meirovitch [33] describes the behavior of the deformation of the cable as an ordinary differential equation model. Another approach named the Galerkin method has been used by a number of researchers to solve problems of deformation of cables of fixed length [34]. Vibration problems of materials whose effective lengths vary with time have also been the subject of recent research activities where the system is discretized via a modified finite element technique [35,36], and finally a modified Galerkin method was used by Fung [37].

In the present work, we have used the assumed modes method. This approach consists in representing the deformation as a weighted sum of the shape functions. The solution is approximated by a series of finite superimposed functions multiplied by indeterminate coefficients. In other words, this deformation is decomposed into two functions as in [38]: a spatial function over the length of the cable $\Phi_j(z_3)$, and a time varying function $\delta_j(t)$ as follows:

$$w(z_3, t) = \sum_{j=1}^{m} \Phi_j(z_3)\, \delta_j(t), \qquad (16)$$

where $\Phi = (\Phi_1, \Phi_2, \ldots, \Phi_m)^t$ is the vector of shape functions. This vector gives the general configuration of the cable, $\delta = (\delta_1, \delta_2, \ldots, \delta_m)^t$ is the vector of the generalized coordinates relating to the flexible modes. This vector represents the nature of the motion made by the cable and m being the number of modes retained in the series.

By replacing the expression of the deformation defined above in that of kinetic energy, we obtain a new developed expression of the kinetic energy (see Appendix B).

This expression can be written in the following compact form:

$$T_{tot} = \frac{1}{2}\begin{pmatrix} u_A & u_C & \dot{\theta} & \dot{\delta}^t \end{pmatrix}^T \overline{M} \begin{pmatrix} u_A & u_C & \dot{\theta} & \dot{\delta}^t \end{pmatrix} \tag{17}$$

Let us denote by:

- $\overline{M}(X)$ the mass matrix;
- $\overline{B}$ the damping matrix;
- $\overline{K}$ the stiffness matrix;
- $\overline{G}(X)$ the vector of forces due to the gravity;
- $\overline{C}\left(X, \dot{X}\right)$ the vector of Coriolis and centrifugal forces;
- $\tau = (F_A, F_C, 0, 0)^t$ the vector of the forces produced by the actuators and applied on the airship and on the cuboid. (The details of these matrices are given in Appendixes B–D).

Using the same procedure as that developed in the discretization of kinetic energy, the expression of the discretized potential energy is given by:

$$\begin{aligned} V_{tot} = {} & \tfrac{1}{2}\delta^t EI_{z_3} \int_0^{l_c} \Phi_{z_3 z_3} \Phi^t_{z_3 z_3} dz_3\, \delta + \left(-\rho \tfrac{l_c^2}{2} - m_L(l_c + r_L)\right) g.c\theta + \\ & \tfrac{1}{2}\delta^t \left(\rho \int_0^{l_c}\int_0^{z_3} \Phi_{z_3}\Phi^t_{z_3} ds dz_3 + m_L\left(\tfrac{1}{2} r_L \Phi_{z_3,l_c}\Phi^t_{z_3,l_c} + \tfrac{1}{2}\int_0^{l_c} \Phi_{z_3}\Phi^t_{z_3} dz_3\right)\right) \\ & g\delta c\theta + s\theta.g\left(\rho \int_0^{l_c} \Phi^t dz_3 + m_L\left(\Phi^t_{l_c} - r_L \Phi^t_{z_3,l_c}\right)\right)\delta \end{aligned} \tag{18}$$

The discretized potential energy of the system can be rewritten in a more compact form such as:

$$V_{tot} = \frac{1}{2}\delta^t \overline{K}_{ff}\, \delta + \overline{G}_{rr}\, c\theta + \frac{1}{2}\delta^t \overline{G}_{rf}\, \delta\, c\theta + s\theta\, \overline{G}_{ff}\, \delta \tag{19}$$

The discretized Rayleigh function can be written in the form:

$$F_R = \frac{1}{2}\dot{\delta}^t k_e\, EI_{z_3} \int_0^{l_c} \Phi_{z_3 z_3}\Phi^t_{z_3 z_3} dz_3\, \dot{\delta} \tag{20}$$

In matrix form, it will be written as follows:

$$F_R = \begin{pmatrix} u_A & u_C & \dot{\theta} & \dot{\delta}^t \end{pmatrix}^T \overline{B} \begin{pmatrix} u_A & u_C & \dot{\theta} & \dot{\delta}^t \end{pmatrix} \tag{21}$$

3.2.5. Equations of Motion

Once the energies have been discretized as well as the Rayleigh function we obtain a discretized Lagrangian L (Equation (6)) to which we apply the Lagrange equations:

$$\frac{d}{dt}\frac{\partial L}{\partial \dot{X}_i} - \frac{\partial L}{\partial X_i} = F_{R_i} \tag{22}$$

The dynamics of the system can be written in compact form in terms of the generalized coordinates $X = \begin{pmatrix} x_A & x_C & \theta & \delta \end{pmatrix}^t$:

$$\overline{M}(X)\ddot{X} + \overline{B}\dot{X} + \overline{K}X + \overline{G}(X) + \overline{C}\left(X, \dot{X}\right) = \tau \tag{23}$$

3.2.6. Simplified Model

The model of the multibody system developed above corresponds to a nonlinear system fully coupled to (m + 2) states. In order to remedy the complexity of the model, we have made some simplifications and arrived at a reduced and simplified model.

(a) The first simplification concerns the shape functions: we have chosen the shape functions $w_i(z_3) = z_3^2, \quad i \in \{1,2,\ldots,m\}$ and we have taken only one mode (m = 1), the expression of the deformation then becomes:

$$w(z_3, t) = z_3^2 \delta(t) \tag{24}$$

(b) With the aim of simplifying the dynamic model in order to be able to apply control laws to it, we have analyzed the different terms of the matrix $\overline{M}(X)$ and succeeded in highlighting the terms whose calculation is complex, but whose absolute value can be neglected compared to the other terms of the matrix. The mass matrix $\overline{M}(X)$ can thus be subdivided as the sum of two matrices: main matrix noted $\widetilde{M}(X)$ and complementary matrix $\overline{dM}(X)$:

$$\overline{M}(X) = \widetilde{M}(X) + \overline{dM}(X) \tag{25}$$

The expressions of matrices $\widetilde{M}$ and $\overline{dM}$ are defined in Appendix D. We note that matrix $\overline{dM}$ consists of small terms compared to those of the main matrix. These terms will be neglected in continuation and consequently, we can suppose that $\overline{M} \cong \widetilde{M}$.

(c) A third simplification is carried out for the Coriolis vector. We replace $\overline{M}$ with $\widetilde{M}$ in Equation (A24) (see Appendix D) and thus obtain the following expression of $\overline{C}\left(X, \dot{X}\right)$:

$$\overline{C}\left(X, \dot{X}\right) = \dot{\widetilde{M}}\,\dot{X} - \frac{1}{2}\frac{\partial}{\partial X}\left(\dot{X}^t\,\widetilde{M}\,X\right) \tag{26}$$

The new Coriolis vector will be:

$$\overline{C}(X, \dot{X}) \cong \begin{pmatrix} -\frac{1}{2}a_5\,\dot{\theta}\,s\theta\,\dot{\delta} \\ -\frac{1}{2}a_5\,\dot{\theta}\,s\theta\,\dot{\delta} \\ \frac{1}{2}a_5\,s\theta\,(u_A + u_C)\,\dot{\delta} \\ -\frac{1}{2}a_5\,s\theta\,\dot{\theta}\,(u_A + u_C) \end{pmatrix} \tag{27}$$

with

Remark 1. *It should be specified here that this study is established for a given length of cable lc. During the loading or unloading phase, this length will vary. Because of the sensors at the level of the pulleys, the controller intended to stabilize the load will be able to know this length at time t and will take this information into account in real time.*

3.3. *Stabilization*

The main objective of the design of this smart crane is to develop a control vector based on the various actuators of the system to stabilize the airship and its load during the loading or unloading phase. As a first approach, we present a classic system based on the Proportional-Derivative (PD) technique. Stabilization by the qLPV method and by model-free control will be presented in future work.

The proposed PD control law is based on measurable variables for the airship and the cuboid, namely the displacement and speed of the airship and the cuboid, the oscillation angle and its angular speed, and the deformation of the cable and its speed.

Different scenarios were considered and simulated numerically in order to validate our mathematical model.

4. Simulation Results

The numerical simulations presented in this section concern the most feared scenario of a gust of wind impacting the airship during the loading or unloading phase. A controller is proposed to drive the system to a stabilized position and at the same time minimize the cable oscillations in order to preserve the products loaded in the container. The main control objective is to use the cuboid acceleration to stabilize the motion of the cable while damping the cable vibrations. The various computer developments were carried out within the Matlab software R2022a.

The actuators available are, on the one side, the motorized pulleys which pull the eight upper cables in order to control the motion of the cuboid and, on the other side, the thrusters which act on a longer timescale in order to stabilize the airship.

In these simulations, the characteristics of the different elements of the system (airship-cuboid-load and flexible cable) are listed in Table 1:

Table 1. Characteristics of the different elements of the system.

Parameter	Symbol	Value
Airship mass	m_A	500 kg
Cuboid mass	m_C	50 kg
Load mass	m_L	40 kg
Load radius	r_L	0.5 m
Cable length	l_c	15 m
Linear mass	ρ	1.5 kg/m
Young's modulus	E	2.10^8 Pa
Moment of inertia of the load	I_L	$2\, m_L\, r_L^2/5\, (\text{kg.m}^2)$
Quadratic moment of the cable	I_{z3}	10^{-4} m^4
Damping coefficient	k_e	0.1

We have assumed that the airship is subjected to a gust of wind which will be represented by a Ricker wavelet along the X_1 axis as we can see in Figure 10:

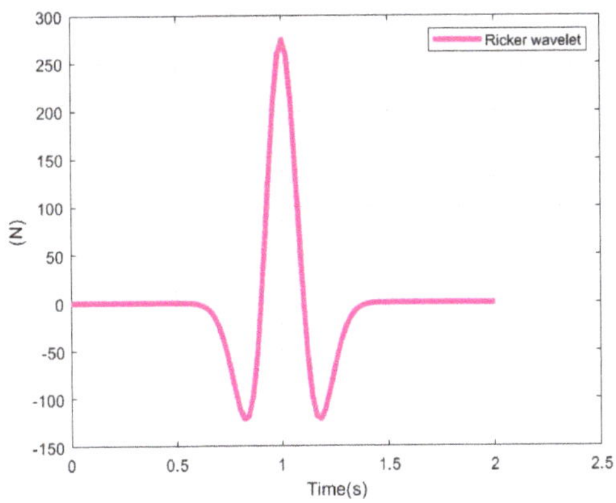

Figure 10. Representation of the gust of wind.

During our simulations, we found that the difference in response between the complete model and the reduced model is less than 1%. The different curves are superimposed perfectly, which justifies the reduction of the model that we have made and which allows a notable reduction of time in the simulation in comparison with the complete model. We have therefore chosen to present the curves of the different variables calculated from the reduced model.

We notice also that under the effect of this impulse caused by the gust of wind, the system oscillate due to the inertial coupling between the various generalized coordinates (displacement of the airship x_A, displacement of the cuboid x_C, oscillation and deformation of the cable θ and δ).

We apply a control vector which acts, on the one hand on the airship, through force F_A developed by the thrusters along axis X_1 in order to stabilize the latter around a desired position x_{Ad} and, on the other hand, on the cuboid by force F_C along axis X_1 produced by the eight winches which drive the cuboid, such as:

$$\begin{array}{c} F_A = -k_{pA}(x_A - x_{Ad}) - k_{vA}\,\dot{x}_A \\ F_C = k_{p\theta}\theta + k_{v\theta}\dot{\theta} + k_{p\delta}\delta + k_{v\delta}\dot{\delta} - k_{pC}(x_C - x_{Cd}) - k_{vC}\dot{x}_C \end{array} \tag{28}$$

The gains are chosen as:

$$\begin{array}{c} k_{pA} = k_{vA} = k_{p\theta} = k_{p\delta} = 200; \\ k_{v\delta} = k_{v\theta} = 150; \quad k_{pC} = k_{vC} = 100 \end{array} \tag{29}$$

Unlike agile flying machines such as helicopters, the dynamics of the airship are slow dynamics. This is what motivated our choice to stabilize the suspended load by the acceleration of the cuboid, which benefits from the advantages of the CDPM and therefore has a quick dynamic.

According to Figure 11, the airship reaches slowly the desired position (origin). The control vector succeeds in stabilizing the system. The motion of the cuboid, the oscillation angle θ, and the generalized deformation of the cable δ tend to zero.(Figures 12–14).

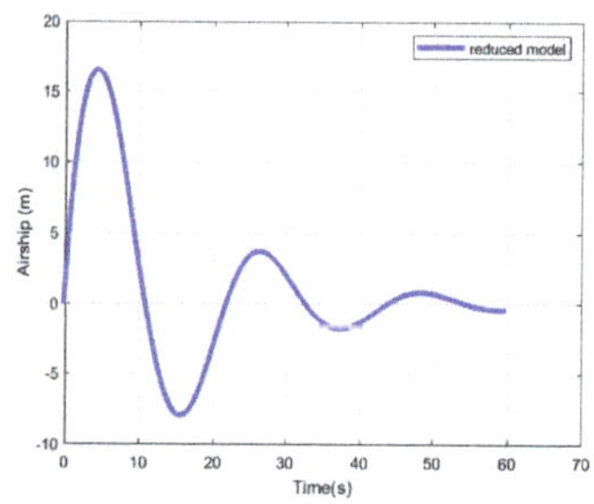

Figure 11. Position of the airship.

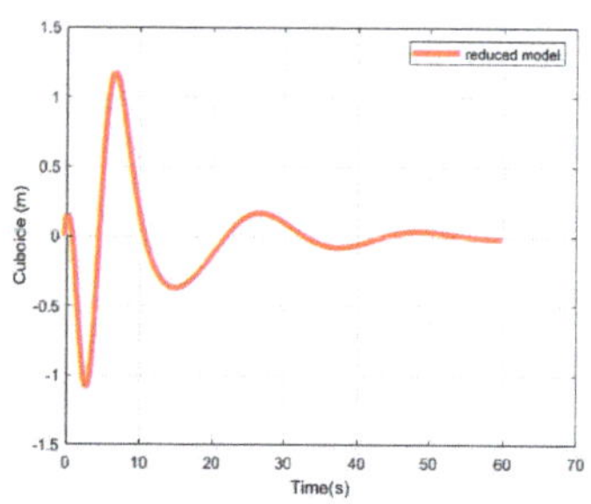

Figure 12. Position of the cuboid.

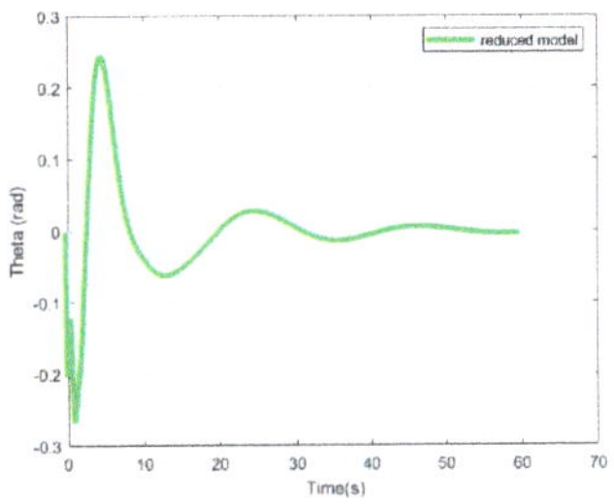

Figure 13. Oscillation angle θ.

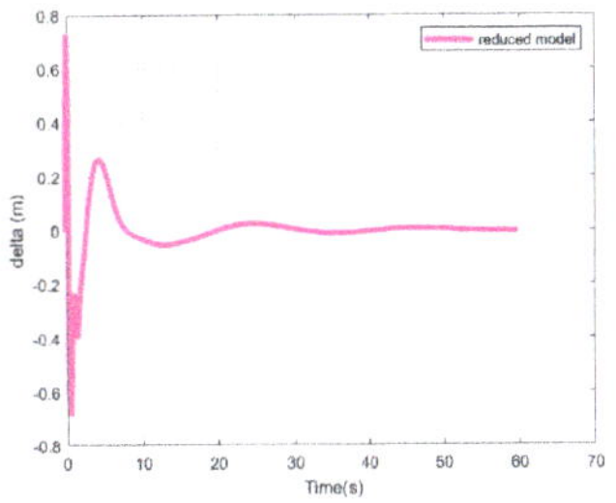

Figure 14. Deformation of the cable δ.

We can see that the proposed PD-type closed loop control system reasonably suppresses the load oscillation, and it is also efficient at damping cable vibrations. In addition, the final positions of the airship and the cuboid are reached in a relatively reasonable time.

The performance of the proposed control scheme is good. Our goal thereafter is to highlight the behavior of the eight winches driving the cuboid. We show in Figure 15 the detail of all the simulation of the accelerations of each winch.

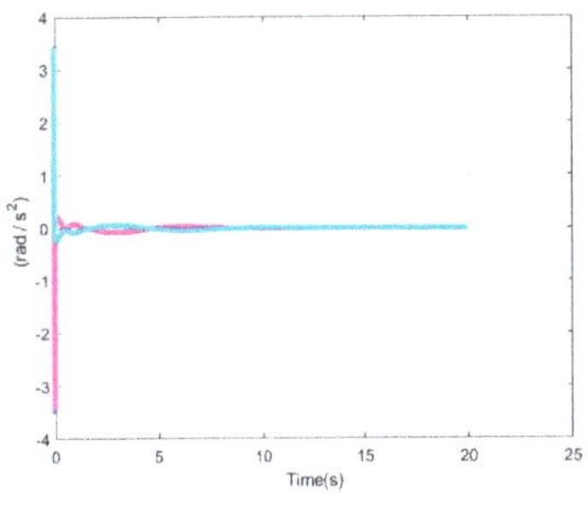

Figure 15. Angular accelerations of winches.

We recall that the expression which links the cuboid accelerations with the angular accelerations $\ddot{q}_i$ of the eight actuators is given by:

$$\ddot{q} = R^{-1}\left(J\dot{u}_C + \dot{J}u_C\right), \tag{30}$$

where J is the Jacobian matrix of the robot.

The coordinates of the exit points and the attachment points used in this example are taken from a reference example given by the Tecnalia Company for an industrial CoGiRo robot. They are listed in Table 2:

Table 2. Coordinates of exit points and attachment points.

p_i	x_{pi} (m)	y_{pi} (m)	z_{pi} (m)
p_1	0.500	−0.507	0.555
p_2	−0.488	0.361	0.554
p_3	−0.500	−0.260	0.555
p_4	0.503	0.342	0.548
p_5	−0.500	0.507	0.555
p_6	0.497	−0.353	0.554
p_7	0.499	0.260	0.549
p_8	0.499	0.260	0.549
vi	x_{vi} (m)	y_{vi} (m)	z_{vi} (m)
v_1	−7.224	−5.359	−5.468
v_2	−7.435	−5.058	5.477
v_3	−7.425	5.196	5.486
v_4	−7.210	5.497	5.495
v_5	7.139	5.463	5.481
v_6	7.440	5.158	5.494
v_7	7.415	−5.089	5.481
v_8	7.113	−5.388	5.492

The diagonal matrix $R = diag\{1,1,1,1,1,1,1,1\}$. The following figure shows the acceleration behavior of these eight actuators.

Figures 15 and 16 show the angular accelerations and speeds of the motorized pulleys which drive the cuboid. With the movement of the latter being rectilinear along axis X_1, it is noted that the accelerations of the "front" pulleys (colored in blue) are identical and opposite to those placed at the rear of the cuboid (colored in magenta).

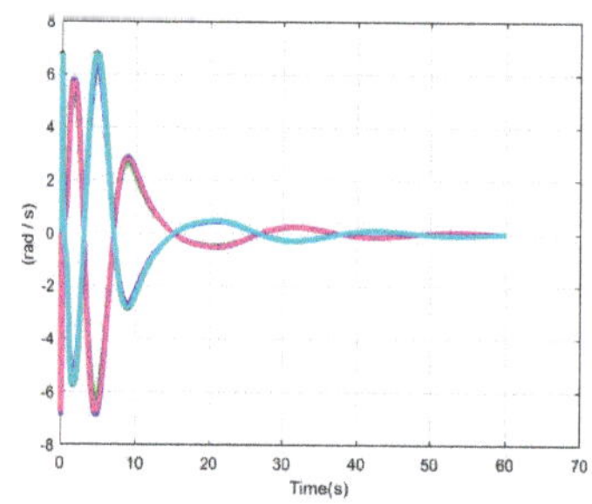

Figure 16. Angular speeds of winches.

5. Discussion

In this study we wanted to validate a concept. It involved equipping a Large Capacity Airship with a smart on-board crane capable of stabilizing the load, in addition to its usual functions of lifting the load and arranging the containers in the hold of the airship. By design, this innovative crane can stabilize the load regardless of the direction of oscillation. We have conducted an extensive study on the dynamics of the multibody system including the flexible heavy cable. However, in this study, we made a strong hypothesis and confined ourselves to a disturbance of the wind along the x axis alone. The results obtained are encouraging and allow us to validate the concept. We remain aware that the dynamic model used for the airship is relatively simplistic. Our first objective was to present the smart crane and its interaction with a particular movement of the airship.

Generalization to external disturbances along various directions and taking into account the complete dynamic model of the airship [3] is being studied and will be the subject of future publications.

6. Conclusions

As part of the development of large capacity airships, we have presented in this paper the design and modelling of an on-board smart crane based on the CDPM principle. The main objective of this crane is to ensure safe handling at altitude, in particular for loading or unloading container ships on the high seas. For this objective, we have established a precise mathematical model of a multibody system including the airship, the crane, the flexible lifting cable, and the suspended load. For control requirements, we proposed a reduced dynamic model which proved to be very reliable and consistent with the complete model, while being very useful for control applications.

A classical control vector was applied to the system for its stabilization after a disturbance due to a gust of wind. The numerical results were conclusive and validated our model.

More elaborate control laws (qLPV control and modeless control) are under study and will be the subject of future publications.

Author Contributions: Conceptualization, F.G. and N.A.; methodology, all authors; software, F.G.; validation, all authors; investigation, all authors, writing—original draft preparation, all authors; writing—review and editing, F.G. and N.A.; supervision, J.N.; project administration, N.A. All authors have read and agreed to the published version of the manuscript.

Funding: This research received no external funding. All authors certify that they have no affiliations with or involvement in any organization or entity with any financial interest or non-financial interest in the subject matter or materials discussed in this manuscript.

Institutional Review Board Statement: Not applicable.

Informed Consent Statement: Not applicable.

Data Availability Statement: The data presented in this study are available on request from the corresponding author.

Conflicts of Interest: The authors declare that they have no conflicts of interest.

Nomenclature

$\|\| . \|\|$	is the norm of the vector (.).
$\dot{A}$	is the time derivative of A.
x	is the cross product.
$Diag(R)$	is the column matrix of the diagonal components of R.
A^t	is the transpose of the matrix A.
w_{z_3}	is the partial derivative of w with respect to z_3.
$o(3)$	are small terms of the third order

Appendix A

$$
\begin{aligned}
T_{tot} = \tfrac{1}{2}(& (m_A + m_L + \rho\, l_c)\, u_A^2 + (m_C + m_L + \rho\, l_c) u_C^2 \\
& + \left(I_L + m_L\,(l_c + r_L)^2 + \rho \tfrac{l_c^3}{3}\right) \dot{\theta}^2 + \left(-\rho \int_0^{l_c} z_3 \int_0^{z_3} w_{z_3}^2\, ds \right. \\
& \left. dz_3 + \rho \int_0^{l_c} w^2\, dz_3 + m_L\, w_{l_c}^2 + r_L^2\, m_L\, w_{z_3,l_c}^2 \right) \dot{\theta}^2 + (2\rho \\
& \int_0^{l_c} \dot{w}\, dz_3 + 2\, m_L\, \dot{w}_{l_c} + 2\, m_L\, r_L\, \dot{w}_{z_3,l_c}) c\theta\, u_A + \left(2\rho \int_0^{l_c} \dot{w} \right. \\
& dz_3 + 2\, m_L\, \dot{w}_{l_c} + 2\, m_L\, r_L\, \dot{w}_{z_3,l_c})\, c\theta\, u_C + (-2\, m_L\, r_L\, \dot{w}_{z_3,l_c} \\
& \left. -2\, m_L \int_0^{l_c} w_{z_3} \dot{w}_{z_3,l_c} dz_3 + 2\rho \int_0^{l_c}\int_0^{z_3} w_{z_3} \dot{w}_{z_3} ds\, dz_3 \right) s\theta\, u_A + \\
& \left(-2\, m_L\, r_L\, \dot{w}_{z_3,l_c} - 2\, m_L \int_0^{l_c} w_{z_3} \dot{w}_{z_3,l_c} dz_3 + 2\rho \int_0^{l_c}\int_0^{z_3} w_{z_3} \dot{w}_{z_3} \right. \\
& ds\, dz_3)\, s\theta\, u_C + \left(2\rho \int_0^{l_c} z_3 w\, dz_3 + 2\, I_L \dot{w}_{z_3,l_c} + 2\, m_L \right. \\
& (l_c + r_L)\, \dot{w}_{l_c} + 2\, m_L\, r_L (l_c + r_L)\, \dot{w}_{z_3,l_c}) \dot{\theta} + (l_c^2 \rho + 2\, m_L \\
& \left. (l_c + r_L) - m_L\, r_L\, w_{z_3,l_c}^2 - m_L \int_0^{l_c} w_{z_3}^2\, dz_3 - \rho \int_0^{l_c}\int_0^{z_3} w_{z_3}^2\, ds\, dz_3 \right) \\
& c\theta\, u_A\, \dot{\theta} + (l_c^2 \rho + 2\, m_L (l_c + r_L) - m_L\, r_L\, w_{z_3,l_c}^2 - m_L \int_0^{l_c} w_{z_3}^2\, dz_3 \\
& -\rho \int_0^{l_c}\int_0^{z_3} w_{z_3}^2\, ds\, dz_3)\, c\theta\, u_C\, \dot{\theta} + \left(2\rho \int_0^{l_c} w\, dz_3 + 2\, m_L\, r_L \right. \\
& w_{z_3,l_c} - 2\, m_L\, w_{l_c})\, s\theta\, u_A\, \dot{\theta} + \left(2\rho \int_0^{l_c} w\, dz_3 + 2\, m_L\, r_L\, w_{z_3,l_c} \right. \\
& -2\, m_L\, w_{l_c})\, s\theta\, u_C\, \dot{\theta} + \rho \int_0^{l_c} \dot{w}^2\, dz_3 + I_L\, \dot{w}_{z_3,l_c}^2 + m_L \left(\dot{w}_{l_c}^2 + \right. \\
& \left. \dot{w}_{z_3,l_c}^2 r_L^2 + 2\, r_L\, \dot{w}_{z_3,l_c} \dot{w}_{l_c}\right) + 2\,(\rho + m_L)\, u_A\, u_C + o(3))
\end{aligned}
\tag{A1}
$$

$$
\begin{aligned}
\int_0^{l_c} T_f = \tfrac{1}{2}\rho(& l_c (u_A + u_C)^2 + \int_0^{l_c} \dot{w}^2 dz_3 + \tfrac{l_c^3}{3} \dot{\theta}^2 + 2\, u_A\, c\theta \int_0^{l_c} \dot{w}\, dz_3 \\
& + 2\, u_C\, c\theta \int_0^{l_c} \dot{w}\, dz_3 + l_c^2\, u_A\, \dot{\theta}\, c\theta + l_c^2\, u_C\, \dot{\theta}\, c\theta - \dot{\theta}\, u_A\, c\theta \int_0^{l_c}\int_0^{z_3} w_{z_3}^2 \\
& ds\, dz_3 - \dot{\theta}\, u_C\, c\theta \int_0^{l_c}\int_0^{z_3} w_{z_3}^2\, ds\, dz_3 + 2\dot{\theta} \int_0^{l_c} z_3\, \dot{w}\, dz_3 - \dot{\theta}^2 \int_0^{l_c} z_3 \int_0^{z_3} w_{z_3}^2 \\
& ds\, dz_3 + \dot{\theta}^2 \int_0^{l_c} w^2\, dz_3 + 2\, u_A\, \dot{\theta}\, s\theta \int_0^{l_c} w\, dz_3 + 2\, u_C\, \dot{\theta}\, s\theta \int_0^{l_c} w\, dz_3 + \\
& 2\, u_A\, s\theta \int_0^{l_c}\int_0^{z_3} w_{z_3}\, \dot{w}_{z_3}\, ds\, dz_3 + 2\, u_C\, s\theta \int_0^{l_c}\int_0^{z_3} w_{z_3}\, \dot{w}_{z_3}\, ds\, dz_3 + o(3))
\end{aligned}
\tag{A2}
$$

Appendix B

$$
\begin{aligned}
& T_{tot} = \tfrac{1}{2}((m_A + m_L + \rho\, l_c)\, u_A^2 + (m_C + m_L + \rho\, l_c) u_C^2 \\
& + \Big(I_L + m_L (l_c + r_L)^2 + \rho\, \tfrac{l_c^3}{3} \Big) \dot{\theta}^2 + \delta^t \Bigg(-\rho \int_0^{l_c} z_3 \int_0^{z_3} \Phi_{z_3} \\
& \Phi_{z_3}^t\, ds\, dz_3 + \rho \int_0^{l_c} \Phi\, \Phi^t dz_3 + m_L\, \Phi_{l_c}\, \Phi_{l_c}^t + r_L^2\, m_L\, \Phi_{z_3,l_c} \\
& \Phi_{z_3,l_c}^t \Bigg)\, \delta\, \dot{\theta}^2 + \Bigg(2\rho \int_0^{l_c} \Phi^t\, dz_3 + 2\, m_L\, \Phi_{l_c}^t + 2\, m_L\, r_L\, \Phi_{z_3,l_c}^t \Bigg) \\
& \dot{\delta}\, c\theta\, u_A + \Bigg(2\rho \int_0^{l_c} \Phi^t\, dz_3 + 2\, m_L\, \Phi_{l_c}^t + 2\, m_L\, r_L\, \Phi_{z_3,l_c}^t \Bigg) \dot{\delta} \\
& c\theta\, u_C + \dot{\delta}^t\, (-2\, m_L\, r_L\, \Phi'_{l_c}\, \Phi_{z_3,l_c}^t - 2\, m_L \int_0^{l_c} \Phi_{z_3,l_c}\, \Phi_{z_3} dz_3 \\
& + 2\rho \int_0^{l_c} \int_0^{z_3} \Phi_{z_3} \Phi_{z_3}^t dz_3 \Bigg)\, s\theta\, u_A\, \dot{\delta} + \dot{\delta}^t\, (-2\, m_L\, r_L\, \Phi'_{l_c}\, \Phi_{z_3,l_c}^t \\
& -2\, m_L \int_0^{l_c} \Phi_{z_3,l_c}\, \Phi_{z_3} dz_3 + 2\rho \int_0^{l_c} \int_0^{z_3} \Phi_{z_3} \Phi_{z_3}^t dz_3 \Bigg)\, s\theta\, u_C\, \dot{\delta} \\
& + \Bigg(2\rho \int_0^{l_c} z_3\, \Phi^t dz_3 + 2\, I_L\, \Phi_{z_3,l_c}^t + 2\, m_L\, (l_c + r_L)\, \Phi_{l_c}^t + \\
& 2\, m_L\, r_L\, (l_c + r_L)\, \Phi_{z_3,l_c}^t \Bigg)\, \dot{\delta}\, \dot{\theta} + (l_c^2 \rho + 2\, m_L\, (l_c + r_L) + \\
& \delta^t (-m_L\, r_L\, \Phi_{z_3,l_c}\, \Phi_{z_3,l_c}^t - m_L \int_0^{l_c} \Phi_{z_3} \Phi_{z_3}^t dz_3 - \rho \int_0^{l_c} \int_0^{z_3} \Phi_{z_3} \\
& \Phi_{z_3}^t ds\, dz_3) \delta) c\theta\, u_A \dot{\theta} + (l_c^2 \rho + 2\, m_L\, (l_c + r_L) + \delta^t \\
& (-m_L\, r_L\, \Phi_{z_3,l_c}\, \Phi_{z_3,l_c}^t - m_L \int_0^{l_c} \Phi_{z_3} \Phi_{z_3}^t dz_3 - \rho \int_0^{l_c} \int_0^{z_3} \Phi_{z_3} \\
& \Phi_{z_3}^t ds\, dz_3) \delta) c\theta\, u_C \dot{\theta} + \Bigg(2\rho \int_0^{l_c} \Phi^t dz_3 + 2\, m_L\, r_L\, \Phi_{z_3,l_c}^t \\
& -2\, m_L \Phi_{l_c}^t \Bigg) \delta\, s\theta\, u_A \dot{\theta} + \Bigg(2\rho \int_0^{l_c} \Phi^t dz_3 + 2\, m_L\, r_L\, \Phi_{z_3,l_c}^t \\
& -2\, m_L \Phi_{l_c}^t \Bigg) \delta\, s\theta\, u_C \dot{\theta} + \dot{\delta}^t \Bigg(\rho \int_0^{l_c} \Phi\, \Phi^t dz_3 + I_L \Phi_{z_3,l_c}\, \Phi_{z_3,l_c}^t \\
& + m_L \Big(\Phi_{l_c}\, \Phi_{l_c}^t + \Phi_{z_3,l_c}\, \Phi_{z_3,l_c}^t\, r_L^2 + 2\, r_L\, \Phi_{l_c}\, \Phi_{z_3,l_c}^t \Big)\, \dot{\delta} + \\
& 2\, (\rho + m_L)\, u_A\, u_C)
\end{aligned}
\tag{A3}
$$

Appendix C

$M(X_p) = \begin{pmatrix} m_p\, I_{3x3} & 0_{3x3} \\ 0_{3x3} & R_p^0\, I_p R_p^0 \end{pmatrix}$ is the classical mass matrix, $C\big(X_p, \dot{X}_p\big)$ is the matrix of centrifugal and Coriolis forces defined by:

$$
C\big(X_p, \dot{X}_p\big)\dot{X}_p = \begin{pmatrix} 0_{3x1} \\ \omega_p \times R_0^p\, I_p R_p^0\, \omega_p \end{pmatrix}
\tag{A4}
$$

m_p is the mass of the platform, $I_{3\times 3}$ is the identity matrix, and I_p is the inertia matrix of the platform.

Appendix D

$$\overline{M} = \begin{pmatrix} a_0 & \frac{a_2}{2} & \frac{(a_8+\delta^t a_9\,\delta\, c\theta+a_{10}\,\delta\, s\theta)}{2} & \frac{(a_5\, c\theta+\delta^t a_6 s\theta)}{2} \\ \frac{a_2}{2} & a_1 & \frac{a_8}{2} & \frac{(a_5\, c\theta+\delta^t a_6 s\theta)}{2} \\ \frac{(a_8+\delta^t a_9\,\delta\, c\theta+a_{10}\,\delta\, s\theta)}{2} & \frac{(a_8+\delta^t a_9\,\delta\, c\theta+a_{10}\,\delta\, s\theta)}{2} & a_3+\delta^t a_4\,\delta & \frac{a_7}{2} \\ \frac{(a_5\, c\theta+\delta^t a_6 s\theta)}{2} & \frac{(a_5\, c\theta+\delta^t a_6 s\theta)}{2} & \frac{a_7}{2} & a_{11} \end{pmatrix} \quad \text{(A5)}$$

$$\overline{dM} = \begin{pmatrix} 0 & 0 & \frac{(\delta a_9 \delta c\theta+a_{10}\delta s\theta)}{2} & \frac{\delta a_6 s\theta}{2} \\ 0 & 0 & \frac{(\delta a_9 \delta c\theta+a_{10}\delta s\theta)}{2} & \frac{\delta a_6 s\theta}{2} \\ \frac{(\delta a_9 \delta c\theta+a_{10}\delta s\theta)}{2} & \frac{(\delta a_9 \delta c\theta+a_{10}\delta s\theta)}{2} & \delta a_4 \delta & 0 \\ \frac{\delta a_6 s\theta}{2} & \frac{\delta a_6 s\theta}{2} & 0 & 0 \end{pmatrix} \quad \text{(A6)}$$

$$\widetilde{M} = \begin{pmatrix} a_0 & \frac{a_2}{2} & \frac{a_8}{2} & \frac{a_5 c\theta}{2} \\ \frac{a_2}{2} & a_1 & \frac{a_8}{2} & \frac{a_5 c\theta}{2} \\ \frac{a_8}{2} & \frac{a_8}{2} & a_3 & \frac{a_7}{2} \\ \frac{a_5 c\theta}{2} & \frac{a_5 c\theta}{2} & \frac{a_7}{2} & a_{11} \end{pmatrix} \quad \text{(A7)}$$

$$a_0 = m_A + m_L + \rho\, l_c, \;\; a_1 = m_C + m_L + \rho\, l_c\,, \;\; a_2 = 2\,(\rho + m_L) \quad \text{(A8)}$$

$$a_3 = I_L + m_L\,(l_c + r_L)^2 + \rho\,\frac{l_c^3}{3} \quad \text{(A9)}$$

$$a_4 = m_L\,\Phi_{l_c}\,\Phi_{l_c}^t + \rho\int_0^{l_c} \Phi\,\Phi^t dz_3 + m_L\, r_L^2\,\Phi_{z_3,l_c}\,\Phi_{z_3,l_c}^t - \rho\int_0^{l_c} z_3\int_0^{z_3} \Phi_{z_3}\Phi_{z_3}^t ds\, dz_3 \quad \text{(A10)}$$

$$a_5 = 2\,\rho\int_0^{l_c} \Phi^t dz_3 + 2\,m_L\Phi^t + 2\,m_L\, r_L\,\Phi_{z_3,l_c}^t \quad \text{(A11)}$$

$$a_6 = -2\,m_L\, r_L\,\Phi'_{l_c}\,\Phi_{z_3,l_c}^t - 2\,m_L\int_0^{l_c} \Phi_{z_3,l_c}\,\Phi_{z_3} dz_3 + 2\,\rho\int_0^{l_c}\int_0^{z_3} \Phi_{z_3}\Phi_{z_3}^t ds\, dz_3 \quad \text{(A12)}$$

$$a_7 = 2\,\rho\int_0^{l_c} z_3\Phi^t dz_3 + 2\,m_L\,(l_c + r_L)\Phi_{l_c}^t + 2\,m_L\, r_L\,(l_c + r_L)\Phi_{z_3,l_c}^t + 2\,I_L\Phi_{z_3,l_c}^t \quad \text{(A13)}$$

$$a_8 = l_c^2\rho + 2\,(l_c + r_L) \quad \text{(A14)}$$

$$a_9 = -\,m_L\, r_L\,\Phi_{z_3,l_c}\,\Phi_{z_3,l_c}^t - m_L\int_0^{l_c} \Phi_{z_3}\,\Phi_{z_3}^t dz_3 - \rho\int_0^{l_c}\int_0^{z_3} \Phi_{z_3}\Phi_{z_3}^t ds\, dz_3 \quad \text{(A15)}$$

$$a_{10} = 2\,\rho\int_0^{l_c} \Phi^t dz_3 - 2\,m_L\Phi_{l_c}^t + 2\,m_L\, r_L\,\Phi_{z_3,l_c}^t \quad \text{(A16)}$$

$$a_{11} = I_L \Phi_{z_3,l_c} \Phi^t_{z_3,l_c} + \rho \int_0^{l_c} \Phi \Phi^t dz_3 + m_L \left(\Phi_{l_c} \Phi^t_{l_c} + r_L^2 \Phi_{z_3,l_c} \Phi^t_{z_3,l_c} + 2 r_L \Phi_{z_3,l_c} \Phi^t_{z_3,l_c} + 2 r_L \Phi_{l_c} \Phi^t_{l_c} \right) \quad \text{(A17)}$$

$$\overline{B} = \begin{pmatrix} 0 & 0 & 0 & 0 \\ 0 & 0 & 0 & 0 \\ 0 & 0 & 0 & 0 \\ 0 & 0 & 0 & k_e \overline{K}_{ff} \end{pmatrix} \overline{K} = \begin{pmatrix} 0 & 0 & 0 & 0 \\ 0 & 0 & 0 & 0 \\ 0 & 0 & 0 & 0 \\ 0 & 0 & 0 & \overline{K}_{ff} \end{pmatrix} \quad \text{(A18)}$$

$$\overline{K}_{ff} = E I_{y_3} \int_0^{l_c} \Phi_{z_3 z_3} \Phi^t_{z_3 z_3} dz_3 \quad \text{(A19)}$$

$$\overline{G}(X) = \begin{pmatrix} 0 \\ 0 \\ -\left(\overline{G}_{rr} - \frac{1}{2} \delta^t \overline{G}_{rf} \delta\right) s\theta + \overline{G}_{ff} c\theta\, \delta \\ \overline{G}_{ff} s\theta + \overline{G}_{rf} c\theta\, \delta \end{pmatrix} \quad \text{(A20)}$$

$$\overline{G}_{ff} = \left(\rho \int_0^{l_c} \Phi^t dz_3 + m_L \left(\Phi^t_{l_c} - r_L \Phi^t_{z_3,l_c} \right) \right) g \quad \text{(A21)}$$

$$\overline{G}_{rf} = \left(\int_0^{l_c} \int_0^{z_3} \Phi_{z_3} \Phi^t_{z_3} ds\, dz_3 + \frac{1}{2} m_L \left(r_L \Phi_{z_3,l_c} \Phi^t_{z_3,l_c} + \int_0^{l_c} \Phi_{z_3} \Phi^t_{z_3} dz_3 \right) \right) g \quad \text{(A22)}$$

$$\overline{G}_{rr} = \left(-\frac{l_c^2}{2} - m_L \left(l_c + r_L \right) \right) g \quad \text{(A23)}$$

$$\overline{C}\left(X, \dot{X}\right) = \left(\overline{C}_{11} \quad \overline{C}_{21} \quad \overline{C}_{31} \quad \overline{C}_{41} \right)^T \overline{C}_{21} = \overline{C}_{11} \quad \text{(A24)}$$

$$\overline{C}_{11} = \frac{1}{2} \left(-a_5 \dot{\theta}\, s\theta + \delta^t a_6 \dot{\theta}\, c\theta + \dot{\delta}^t a_6\, s\theta \right) \dot{\delta} + \left(-\delta^t a_9\, \delta\, \dot{\theta}\, s\theta + 2 \delta^t a_9\, \dot{\delta} c\theta + a_{10}\, \delta\, \dot{\theta}\, c\theta + a_{10}\, \dot{\delta}\, s\theta \right) \dot{\theta} \quad \text{(A25)}$$

$$\overline{C}_{31} = 2 \delta^t a_4\, \dot{\delta}\, \dot{\theta} - \frac{1}{2} \left(-a_5\, s\theta + \delta^t a_6\, c\theta \right) u_A\, \dot{\delta} - \frac{1}{2} \left(-a_5\, s\theta + \delta^t a_6\, c\theta \right) u_C\, \dot{\delta} - \frac{1}{2} \left(-\delta^t a_9\, \delta\, s\theta + a_{10}\, \delta\, c\theta \right) u_A \dot{\theta} - \frac{1}{2} \left(-\delta^t a_9\, \delta\, s\theta + a_{10}\, \delta\, c\theta \right) u_C \dot{\theta} + \frac{1}{2} \left(-\delta^t a_9\, \delta\, \dot{\theta}\, s\theta + 2 \delta^t a_9\, \dot{\delta}\, c\theta + a_{10}\, \delta\, \dot{\theta}\, c\theta + a_{10}\, \dot{\delta}\, s\theta \right) u_A + \frac{1}{2} \left(-\delta^t a_9\, \delta\, \dot{\theta}\, s\theta + 2 \delta^t a_9\, \dot{\delta}\, c\theta + a_{10}\, \delta\, \dot{\theta}\, c\theta + a_{10}\, \dot{\delta}\, s\theta \right) u_C \quad \text{(A26)}$$

$$\overline{C}_{41} = \frac{1}{2} \left(-a_5\, \dot{\theta}\, s\theta + \delta^t a_6\, \dot{\theta}\, c\theta + \dot{\delta}^t a_6\, s\theta \right) u_A + \frac{1}{2} \left(-a_5\, \dot{\theta}\, s\theta + \delta^t a_6\, \dot{\theta}\, c\theta + \dot{\delta}^t a_6\, s\theta \right) u_C - a_4\, \delta\, \dot{\theta}^2 - \frac{1}{2} \left(2 a_9\, \delta\, c\theta + a_{10}\, s\theta \right) u_A\, \dot{\theta} - \frac{1}{2} \left(2 a_9\, \delta\, c\theta + a_{10}\, s\theta \right) u_C\, \dot{\theta} - \frac{1}{2} a_6\, s\theta\, u_A\, \dot{\delta} - \frac{1}{2} a_6\, s\theta\, u_C\, \dot{\delta} \quad \text{(A27)}$$

References

1. Liao, L.; Pasternak, I. A review of airship structural research and development. *Prog. Aerosp. Sci.* **2009**, *45*, 83–96. [CrossRef]
2. Kanoria, A.A.; Pant, R.S. Winged Aerostat Systems for Better Station Keeping for Aerial Surveillance. In Proceedings of the International Conference on Mechanical and Aerospace Engineering, New Delhi, India, 19–20 March 2011; pp. 273–277.

3. Azouz, N.; Khamlia, M.; Lerbet, J.; Abichou, A. Stabilization of an Unconventional Large Airship When Hovering. *Appl. Sci.* **2021**, *11*, 3551. [CrossRef]
4. Kulczycki, E.A.; Joshi, S.S.; Hess, R.A.; Elfes, A. Controller design for autonomous airships using SLC and LQR methods. In Proceedings of the AIAA Guidance, Navigation, and Control Conference and Exhibit, AIAA-6778, Keystone, CO, USA, 5–8 August 2006; pp. 21–24.
5. Azinheira, J.R.; Moutinho, A.; De Paiva, E.C. A backstepping controller for path-tracking of an underactuated autonomous airship. *Int. J. Robust Nonlinear Control.* **2009**, *19*, 418–441. [CrossRef]
6. Zhu, E.; Pang, J.; Sun, N.; Gao, H.; Sun, Q.; Chen, Z. Airship horizontal trajectory tracking control based on Active Disturbance Rejection Control (ADRC). *Nonlinear Dyn.* **2014**, *75*, 725–734. [CrossRef]
7. Beji, L.; Abichou, A. Tracking control of trim trajectories of a blimp for ascent and descent flight maneuvers. *Int. J. Control.* **2005**, *78*, 706–719. [CrossRef]
8. Kawamura, S.; Choe, W.; Tanaka, S.; Kino, H. Development of an Ultrahigh Speed Robot FALCON Using Parallel Wire Drive Systems. *J. Robot. Soc. Jpn.* **1997**, *15*, 82–89. [CrossRef]
9. Aref, M.M.; Taghirad, H.D. Geometrical workspace analysis of a cable-driven redundant parallel manipulator: Kntucdrpm. In Proceedings of the IEEE/RSJ, International Conference on Intelligent Robots and Systems, IEEE, Nice, France, 22–26 September 2008; pp. 1958–1963. [CrossRef]
10. Kawamura, S.; Choe, W.; Tanaka, S.; Pandian, S. Development of an ultrahigh speed robot FALCON using wire drive system. In Proceedings of the 1995 IEEE International Conference on Robotics and Automation, Nagoya, Japan, 21–27 May 1995; pp. 215–220. [CrossRef]
11. Rosati, G.; Zanotto, D.; Agrawal, S.K. On the Design of Adaptive Cable-Driven Systems. *J. Mech. Robot.* **2011**, *3*, 021004. [CrossRef]
12. Merlet, J.P. *Parallel Robots*; Springer Science & Business Media: Berlin/Heidelberg, Germany, 2006; Volume 128.
13. Fang, S.; Franitza, D.; Torlo, M.; Bekes, F.; Hiller, M. Motion Control of a Tendon-Based Parallel Manipulator Using Optimal Tension Distribution. *IEEE/ASME Trans. Mechatronics* **2004**, *9*, 561–568. [CrossRef]
14. Li, H.; Zhang, X.; Yao, R.; Sun, J.; Pan, G.; Zhu, W. Optimal Force Distribution Based on Slack Rope Model in the Incompletely Constrained Cable-Driven Parallel Mechanism of FAST Telescope. In *Cable-Driven Parallel Robots*; Springer: Berlin/Heidelberg, Germany, 2013; pp. 87–102. [CrossRef]
15. Snyman, J.A.; Hay, A.M. Analysis and Optimization of a Planar Tendon-Driven Parallel Manipulator. In *On Advances in Robot Kinematics*; Springer: Berlin/Heidelberg, Germany, 2004; pp. 303–312. [CrossRef]
16. Gosselin, C.; Grenier, M. On the determination of the force distribution in overconstrained cable-driven parallel mechanisms. *Meccanica* **2011**, *46*, 3–15. [CrossRef]
17. Pott, A. An Improved Force Distribution Algorithm for Over-Constrained Cable-Driven Parallel Robots. In *Computational Kinematics*; Springer: Berlin/Heidelberg, Germany, 2014; pp. 139–146. [CrossRef]
18. Albus, J.; Bostelman, R.; Dagalakis, N. The NIST RoboCrane. *J. Robot. Syst.* **1993**, *10*, 709–724. [CrossRef]
19. Landsberger, S.E.; Sheridan, T.B. A New Design for Parallel Link Manipulators. In Proceedings of the IEEE International Conference on Systems, Man and Cybernetics, Tucson, AZ, USA, 12–15 November 1985; pp. 812–814.
20. Higuchi, T. Application of multi-dimensional wire crane in construction. In Proceedings of the 5th International Symposium On Robotics in Construction, Tokyo, Japan, 6–8 June 1988; pp. 661–668.
21. Tanaka, M.; Seguchi, Y.; Shimada, Y.S. Kineto-statics of skycam-type wire transport system. In Proceedings of the USA-Japan Symposium on Flexible Automation, Crossing Bridges: Advances in Flexible Automation and Robotics, Minneapolis, MN, USA, 18–20 July 1988; pp. 689–694.
22. Rosati, G.; Gallina, P.; Masiero, S. Design, Implementation and Clinical Tests of a Wire-Based Robot for Neurorehabilitation. *IEEE Trans. Neural Syst. Rehabil. Eng.* **2007**, *15*, 560–569. [CrossRef] [PubMed]
23. Boukraa, Y. Modeling and Control of a Cable-Driven Parallel Robot for Lower Limb Rehabilitation. Ph.D. Thesis, Superior Technology School, University of Quebec, Québec City, QC, Canada, 2019.
24. Surdilovic, D.; Zhang, J.; Bernhardt, R. STRING-MAN: Wire-robot technology for safe, flexible and human-friendly gait rehabilitation. In Proceedings of the 2007 IEEE 10th International Conference on Rehabilitation Robotics, Noordwijk, The Netherlands, 13–15 June 2007; pp. 446–453. [CrossRef]
25. BenAbdallah, F.; Azouz, N.; Beji, L.; Abichou, A. Modeling and control of an aerial robocrane using a wire driven system. In Proceedings of the ASME IDETC-67798, Cleveland, OH, USA, 27–29 June 2018; pp. 508–513.
26. Lamaury, J.; Gouttefarde, M.; Michelin, M.; Tempier, O. Design and Control of a Redundant Suspended Cable-Driven Parallel Robots. In *Latest Advances in Robot Kinematics*; Lenarcic, J., Husty, M., Eds.; Springer: Dordrecht, The Netherlands, 2012; pp. 237–244. [CrossRef]
27. Azouz, N.; Khamlia, M.; Lerbet, J.; Abichou, A. Modelling and Stabilization of a load suspended by cable from an airship. *J. Multibody Syst.* **2022**, *55*, 399–431. [CrossRef]
28. Chaabani, S.; Azouz, N. Estimation of the virtual masses of a large unconventional airship based on purely analytical method to aid in the preliminary design. *Aircr. Eng. Aerosp. Technol.* **2022**, *94*, 531–540. [CrossRef]
29. Abrate, S. Vibrations of belts and belt drives. *Mech. Mach. Theory* **1992**, *27*, 645–659. [CrossRef]
30. Andréa-Novel, B.D.; Coron, J. Exponential stabilization of an overhead crane with flexible cable via a back-stepping approach. *Automatica* **2000**, *36*, 587–593. [CrossRef]

1. Ali, H.; Singh, T. Passive control of overhead cranes. In Proceedings of the 1998 IEEE International Conference on Control Applications, Trieste, Italy, 4 September 1998; pp. 1046–1050. [CrossRef]
2. Fatehi, M.H.; Eghtesad, M.; Necsulescu, D.S.; Fatehi, A.A. Tracking control design for a multi-degree underactuated flexible-cable overhead crane system with large swing angle based on singular perturbation method and an energy-shaping technique. *J. Vib. Control.* **2019**, *25*, 1752–1767. [CrossRef]
3. Meirovitch, L. *Fundamentals of Vibrations*; McGraw-Hill: New York, NY, USA, 2001.
4. Lanczos, C. *The Variational Principles of Mechanics*; University of Toronto Press: Toronto, ON, Canada, 1996.
5. Chien, C.G.; Fung, R.F.; Tsal, C.L. Non-linear vibration analysis of the coupled textile/rotor system by finite element method. *J. Sound Vib.* **1999**, *221*, 67–84. [CrossRef]
6. Stylianou, M.; Tabarrok, B. Finite Element Analysis of an Axially Moving Beam, Part I: Time Integration. *J. Sound Vib.* **1994**, *178*, 433–453. [CrossRef]
7. Fung, R.F.; Lin, J.H. Vibration analysis and suppression control of an elevator string actuated by b PM synchronous servo motor. *J. Sound Vib.* **1997**, *206*, 399–423. [CrossRef]
8. Du, J.; Agrawal, S.K. Dynamic Modeling of Cable-Driven Parallel Manipulators with Distributed Mass Flexible Cables. *J. Vib. Acoust.* **2015**, *137*, 021020. [CrossRef]

aerospace

MDPI

Article

Analysis of Influence of Stratospheric Airship's Key Parameter Perturbation on Motion Mode

Jiwei Tang [1], Shilong Bai [2], Weicheng Xie [3], Junjie Wu [3], Hanjie Jiang [4,*] and Yuxuan Sun [1]

[1] School of Aeronautics and Astronautics, Shanghai Jiao Tong University, Shanghai 200240, China; tangjw@sjtu.edu.cn (J.T.); lien4051@sjtu.edu.cn (Y.S.)

[2] Chongqing Near Space Innovation R&D Center, Shanghai Jiao Tong University, Chongqing 401135, China; bsl2014@163.com

[3] Aerospace System Engineering Shanghai, Shanghai 201108, China; xieweicheng@asesspace.com (W.X.); wujunjie@asesspace.com (J.W.)

[4] School of Aerospace Engineering, Universiti Sains Malaysia, Nibong Tebal 14300, Malaysia

* Correspondence: jianghanjie@163.com

Abstract: The stratospheric airship is taken as the research object, and the motion mode analysis of the stratospheric airship is carried out. The influence of key parameters such as the center of mass, the center of buoyance, and the aerodynamic stability moment on the motion mode of stratospheric airship are analyzed and summarized in detail. According to the simulation and analysis results, unlike high-speed and high-dynamic aircrafts such as airplanes, the motion modes of the stratospheric airship are hardly affected by the perturbation of aerodynamic stability moment; the perturbations of the vertical center of mass and the vertical center of buoyancy have a great influence on the pitch pendulum motion modes, and their parameter perturbations affect the frequency of the pitch pendulum motion and also the stability of the pitch pendulum motion; the axial mass center location perturbation not only changes the damping of pitch pendulum motion but also affects the frequency of the yaw motion attitude motion mode to a certain extent.

Keywords: stratospheric airship; motion mode; key parameters; perturbation; analysis

1. Introduction

As a new type of aircraft, the stratospheric airship relies on buoyancy to balance the influence of gravity to realize the long-term flying and stationing in near space at the altitude of 20 km [1–4]. As a new type of Earth observation and surveillance platform, the stratospheric airship, compared with satellites and unmanned aerial vehicles (UAV), has many outstanding advantages, such as a large task load and long-term stationing in the air [5] The stratospheric airship is widely used in Earth observation, regional surveillance, navigation and positioning, comprehensive early warning detection, electronic reconnaissance, confrontation, communication, and relay and has high military use and civil value. Therefore, the development of the stratospheric airship has great strategic value and significance.

The motion mode can well reflect the disturbed motion mode of the aircraft, and modal analysis uses the corresponding means to analyze the disturbed motion characteristics of the aircraft [6]. Motion mode analysis plays an important role in aircraft design. It provides the corresponding theoretical basis for the overall optimization of the aircraft, flight quality analysis, and flight control system design, and it forms a set of mature analysis methods [7]. As a new type of aircraft, the stratospheric airship has obvious differences in its flight mechanism and flight characteristics compared to airplanes and other aircrafts. Therefore, it is necessary to conduct in-depth research on the motion mode of the stratospheric airship according to its characteristics.

Many scholars have carried out related technical research on the motion mode analysis of the stratospheric airship. Khoury [8] provided the motion mode analysis method of a low-altitude airship according to its motion characteristics; Ouyang et al. [9] took the "Zhiyuan 1" stratospheric verification airship as the research object, studied the disturbed motion of the airship in detail, and described the motion mode of the airship. Li et al. [10] analyzed the motion modes of a Skyship-500 airship through simulations and experiments and provided the characteristic root distribution at different flight speeds; Liu et al. [11] took the stratospheric airship with a volume of 500 m^3 as the research object, calculated the eigenvalues of the disturbed motion of the airship, and analyzed the influence of flight speed on the motion mode and kinetic characteristics of the airship; Yang et al. [12] systematically studied the motion modes of the stratospheric airship and provided the modal characteristics of longitudinal motion and lateral motion; Wang [13] revealed the influence of flight speed on the motion mode of an airship through analyzing the pole distribution and frequency characteristics of the airship at typical flight speeds. Liu [14] described the process of stratospheric airship mode calculation in detail and provided the correlation mechanism of flight speed change on airship motion mode; Miao [15] took the 1000 m^3 stratospheric verification airship as the research object, introduced the free motion characteristics of the airship at typical flight speeds, and simulated and analyzed the control response characteristics of the airship; Wang [16] introduced the motion mode characteristics of the V-shaped new concept stratospheric airship and calculated the motion mode of the airship under typical working conditions; Wang et al. [17] analyzed the characteristics of the longitudinal motion modes of the stratospheric airship from the perspective of stability and provided the modes sensitive to the longitudinal motion; Anshul et al. [18] conducted the trim and stability analysis of an airship with bifurcation techniques and provided the pole distribution diagram in different motion states; Wu et al. [19] analyzed and calculated the motion mode and dynamic characteristics of an airship driven by a new type of driving system with Matlab according to the characteristics of the "buoyant-slider" driven airship; Zhang [20] took GoodYear ZP4K as the research object, analyzed the characteristics of the longitudinal and lateral mode motion of the airship, and provided the generation mechanism of the pendulum motion mode; Liu [21] analyzed the influence of different motion models on the motion modes of airship under the conditions of straight and level flight at a constant speed and pointed out the root cause of the difference of motion modes. Yuan [22] proposed a control strategy combining model prediction, sliding mode control, and extended state observer for the space trajectory tracking of a stratospheric airship under state constraints, input saturation, and position disturbance and carried out a corresponding simulation analysis; Huang [23] analyzed the impact of external environmental changes on buoyant gas, airship mass, and internal and external pressure differences; Liu [24] proposed the corresponding dynamic model of a stratospheric airship according to the different description methods of aerodynamic parameters; Li [25] carried out dynamic modeling and a longitudinal stability analysis for a stratospheric airship with a double-hull configuration and gave the simulation results of pole distribution and handling characteristics; Gobiha [26] carried out dynamic modeling for an autonomous flying airship and used the established nonlinear 6 DOF model to simulate and evaluate the motion characteristics of the airship.

To summarize, the analysis of the current literature research on the motion modes of the stratospheric airship basically focuses on the determination of the overall scheme of the airship and the motion modes at typical flight speeds. In the actual engineering practices, except when the flight speed affects the motion mode of the stratospheric airship, the locations of the center of mass and the center of buoyance, aerodynamic characteristics, and other parameters of the airship will also have a greater impact on the motion mode of the stratospheric airship; even the impact is subversive. At the same time, compared with the airplane, the stratospheric airship has a large volume and a complex structure, so it is difficult to ensure that all the parameters of the airship in the design, production, and assembly process are testable and to ensure the closed-loop control of the airship technology in the full state and the whole process with the existing design means and process; under the influence of the external environment during the flight, there is great uncertainty between the actual flight state and the theoretical design state in terms of the related parameters of the airship. Therefore, it is necessary to carry out in-depth research on the influence of the perturbation of the key parameters such as the center of mass, the center of float, and the aerodynamic characteristics of the stratospheric airship on the motion mode. Such research can provide corresponding technical support for the overall optimization, the design of the flight control system, and the stable flight of the stratospheric airship. Therefore, the main goal of this paper is to focus on the uncertainty of key parameters such as the center of mass, the center of buoyancy, and the aerodynamic characteristics during the manufacturing and flight test of the stratospheric airship and analyze the influence of the perturbation of these key parameters on the motion mode. The effects of these key parameters on the motion stability, damping characteristics, and frequency change trend of each motion mode of the airship are investigated to optimize the overall design of the stratosphere airship flight control system.

The paper is organized as follows: in Section 2, the description of the nonlinear model and the linearized model of the disturbed motion of the stratospheric airship are given; in Section 3, the motion modes of the stratospheric airship at a typical flight speed is analyzed; in Section 4, the influence of the key parameters of the stratospheric airship on motion modes is analyzed in detail; finally, the influence of the key parameters of the stratospheric airship on motion modes is summarized.

2. Overall Layout and Structure Parameters of Stratospheric Airship

The stratospheric airship studied in this paper adopts the aerodynamic layout of an unstreamlined and X-shaped empennage. The empennage is mainly used to enhance the stability of the airship, as shown in Figure 1. R represents the aerodynamics force vector, B represents the buoyancy vector, G represents the gravity vector, C_V represents the center of aerodynamics force, C_B represents the center of buoyancy, C_G represents the center of gravity, C represents the center of body.

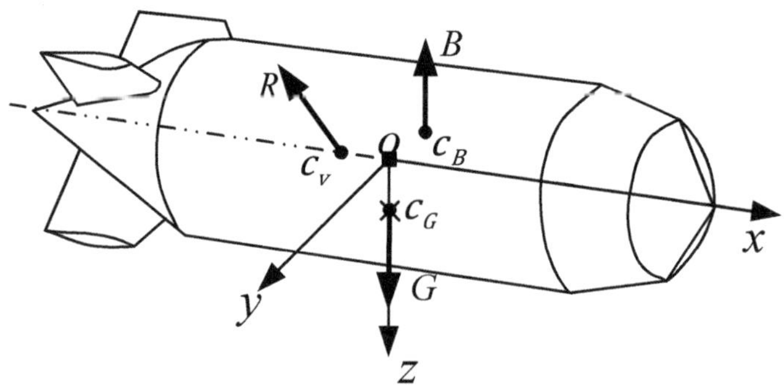

Figure 1. Aerodynamic Layout Diagram of Stratospheric Airship.

The relevant parameters of the stratospheric airship are shown in the Table 1.

Table 1. Relevant Parameters of Stratospheric Airship.

Parameter	Value
Mass [kg]	11,800
Length [m]	138
Diameter [m]	35
Volume [m^3]	134,037
Flight altitude [km]	20

3. Motion Model of Stratospheric Airship

3.1. 6 DOF Nonlinear Model

The kinetic model of the stratospheric airship is the basis and key to studying the spatial motion characteristics of the airship. According to the Newton–Euler method, the 6 DOF nonlinear model of the airship's spatial motion can be established as follows [27,28]:

$$\begin{bmatrix} mE+\lambda_A & -mr_G\times \\ mr_G\times & I+I_A \end{bmatrix}\begin{bmatrix} \dot{V}^b \\ \dot{\Omega}^b \end{bmatrix}+\begin{bmatrix} \Omega^b\times(mE+\lambda_A)V^b+m\Omega^b\times\left(\Omega^b\times r_G\right) \\ \Omega^b\times(I+I_A)\Omega^b+mr_G\times\left(\Omega^b\times V^b\right) \end{bmatrix}=\begin{bmatrix} F \\ M \end{bmatrix} \quad (1)$$

(1) Kinetic Equation

wherein m represents the airship mass; λ_A denotes the additional mass matrix of airship; I represents the moment of inertia matrix; I_A represents the additional moment of inertia matrix of the airship; E denotes the third-order identity matrix; r_G represents the projection of the center of mass of the airship in the hull coordinate system; V^b and Ω^b represent the projection of velocity vector and angular velocity vector in the hull coordinate system; F and M represent the resultant force and torque in the null coordinate system of the airship.

(2) Kinematical Equation

$$\dot{\Phi}=\begin{bmatrix} L_b^I & 0_{3\times3} \\ 0_{3\times3} & \Gamma_b^I \end{bmatrix}\begin{bmatrix} V^b \\ \Omega^b \end{bmatrix} \quad (2)$$

$$L_b^I=\begin{bmatrix} \cos\vartheta\cos\psi & \sin\phi\sin\vartheta\cos\psi-\cos\phi\sin\psi & \cos\phi\sin\vartheta\cos\psi+\sin\phi\sin\psi \\ \cos\vartheta\sin\psi & \sin\phi\sin\vartheta\sin\psi+\cos\phi\cos\psi & \cos\phi\sin\vartheta\sin\psi-\sin\phi\cos\psi \\ -\sin\vartheta & \sin\phi\cos\vartheta & \cos\phi\cos\vartheta \end{bmatrix} \quad (3)$$

$$\Gamma_b^I=\begin{bmatrix} 1 & \tan\vartheta\sin\phi & \tan\vartheta\cos\phi \\ 0 & \cos\phi & -\sin\phi \\ 0 & \sec\vartheta\sin]\phi & \sec\vartheta\cos\phi \end{bmatrix} \quad (4)$$

wherein ϑ, ψ, and ϕ denote the pitch, yaw, and roll attitude angles of the airship, respectively; $\dot{\Phi}$ represents the state variable composed of the velocity of the ground coordinate system and the angular velocity of the attitude.

3.2. Linear Processing of Nonlinear Model

The six-degrees-of-freedom model of aircraft belongs to an equation of nonlinear motion; a nonlinear model can only be solved by numerical method rather than analytical method, and therefore, it is not conducive to the analysis of the aircraft's motion mode and the design of the flight control system. For this reason, it is necessary to linearize the equation of the nonlinear motion of the aircraft [29].

The airship motion can be described by six DOFs (degrees of freedom) and 12-order state variables. From its state transition matrix, it can be seen that the motion mode is a multi-variable cross-coupled motion process. In order to better describe and control the airship motion, we can ignore the weak correlation items in the motion in each state according to the actual flight characteristics of the airship and retain the dominant factors of the system motion modal; in this way, the motion modes of the airship can be decoupled into longitudinal motion and lateral motion [30].

(1) Longitudinal Motion Model Linearization

To analyze the motion morphology of the airship's longitudinal motion channel, the motion of the horizontal and straight cruise flight on the longitudinal plane is selected as the reference motion, and trim the nonlinear model according to the speed and altitude of the cruise flight, and the nonlinear model is linearized under the trim state to obtain the linearized model of longitudinal motion [31].

$$\begin{bmatrix} m+m_{11} & 0 & mz_G & 0 \\ 0 & m+m_{33} & -mx_G & 0 \\ mz_G & -mx_G & I_y+m_{55} & 0 \\ 0 & 0 & 0 & 1 \end{bmatrix}\begin{bmatrix}\Delta\dot{u}\\ \Delta\dot{\alpha}\\ \Delta\dot{q}\\ \Delta\dot{\theta}\end{bmatrix} = \begin{bmatrix}\bar{X}^{u} & \bar{X}^{\alpha} & 0 & \bar{X}^{\theta}\\ 0 & \bar{Z}^{\alpha} & \bar{Z}^{q} & \bar{Z}^{\theta} \\ 0 & \bar{M}^{\alpha} & \bar{M}^{q} & \bar{M}^{\theta}\\ 0 & 0 & 1 & 0\end{bmatrix}\begin{bmatrix}\Delta u\\ \Delta\alpha\\ \Delta q\\ \Delta\theta\end{bmatrix} + \begin{bmatrix}0\\ \bar{Z}^{\delta_E}\\ \bar{M}^{\delta_E}\\ 0\end{bmatrix}\delta_E \quad (5)$$

wherein m_{11}, m_{33}, and m_{55} represent the additional mass and additional moment of inertia of the airship, respectively; I_y represents the moment of inertia of the airship around the *y*-axis; x_G and z_G represent the locations of axial and vertical mass centers of the airship, respectively; Δu and $\Delta\dot{u}$ represent the velocity deviation and its derivative of disturbed motion, respectively; $\Delta\alpha$ and $\Delta\dot{\alpha}$ respectively represent the angle of attack deviation of disturbed motion and its derivative; Δq and $\Delta\dot{q}$ respectively represent the pitch rate deviation of disturbed motion and its derivative; $\Delta\theta$ and $\Delta\dot{\theta}$ respectively represent the pitch angle deviation of disturbed motion and its derivative; $\bar{X}^{u}$, $\bar{X}^{w}$, and $\bar{X}^{\theta}$ respectively represent the derivatives of the axial force related to forward velocity u, vertical velocity w, and attitude θ; $\bar{Z}^{\alpha}$ and $\bar{M}^{\alpha}$ respectively represent the derivatives of normal aerodynamic force and pitch aerodynamic stability moment; $\bar{Z}^{q}$ and $\bar{M}^{q}$ respectively represent the derivatives of the coefficients related to pitch damping force and torque; $\bar{Z}^{\theta}$ and $\bar{M}^{\theta}$ respectively represent the derivatives of force and moment related to the pitch attitude angle; $\bar{Z}^{\delta_E}$ and $\bar{M}^{\delta_E}$ respectively represent the derivatives of pitch control force and torque; δ_E represents the angle of elevator deflection.

(2) Horizontal Motion Model Linearization

The motion of the horizontal and straight cruise flight is selected as the reference motion, the nonlinear model is trimmed according to the speed and altitude of cruise flight, and the nonlinear model is linearized under the trim state to obtain the linearized model of horizontal motion [31].

$$\begin{bmatrix} m+m_{22} & -mz_G & mx_G & 0 \\ -mz_G & I_x+m_{44} & 0 & 0 \\ mx_G & 0 & I_z+m_{66} & 0 \\ 0 & 0 & 0 & 1 \end{bmatrix}\begin{bmatrix}\Delta\dot{\beta}\\ \Delta\dot{p}\\ \Delta\dot{r}\\ \Delta\dot{\phi}\end{bmatrix} = \begin{bmatrix}\bar{Y}^{\beta} & 0 & \bar{Y}^{r} & \bar{Y}^{\phi}\\ \bar{L}^{\beta} & \bar{L}^{p} & 0 & \bar{L}^{\phi} \\ \bar{N}^{\beta} & 0 & \bar{N}^{r} & \bar{N}^{\phi}\\ 0 & 0 & 1 & 0\end{bmatrix}\begin{bmatrix}\Delta\beta\\ \Delta p\\ \Delta r\\ \Delta\phi\end{bmatrix} + \begin{bmatrix}\bar{Y}^{\delta_R}\\ 0\\ \bar{N}^{\delta_R}\\ 0\end{bmatrix}\delta_R \quad (6)$$

wherein m_{22}, m_{44}, and m_{66} represent the additional mass and additional moment of inertia of the airship, respectively; I_x and I_z respectively represent the moment of inertia of the airship around the *x*-axis and *z*-axis; $\Delta\beta$ and $\Delta\dot{\beta}$ represent the slideslip angle deviation and its derivative of disturbed motion, respectively; Δp and $\Delta\dot{p}$ represent the roll angle rate deviation and its derivative of disturbed motion, respectively; $\Delta\alpha$ and $\Delta\dot{\alpha}$ respectively represent the angle of attack deviation of disturbed motion and its derivative; Δr and $\Delta\dot{r}$ respectively represent the yaw angle deviation of disturbed motion and its derivative; $\Delta\phi$ and $\Delta\dot{\phi}$ respectively represent the roll angle deviation of disturbed motion and its derivative; $\bar{Y}^{\beta}$ and $\bar{N}^{\beta}$ respectively represent the derivatives of lateral aerodynamic force and yaw aerodynamic stability moment; $\bar{Y}^{r}$ and $\bar{N}^{r}$ respectively represent the derivatives of yaw damping force and damping force torque; $\bar{L}^{\beta}$ represent the derivative of the inclined blowing moment; $\bar{L}^{p}$ represent the derivative of the roll damping force moment; $\bar{Y}^{\phi}$, $\bar{L}^{\phi}$, and $\bar{N}^{\theta}$ respectively represent the derivatives of the lateral force, roll moment, and yaw moment related to the roll angle; $\bar{Y}^{\delta_R}$ and $\bar{N}^{\delta_R}$ respectively represent the derivatives of yaw control force and control force torque; δ_R represents the angle of rudder deflection.

4. Motion Mode of Stratospheric Airship

4.1. Description of Motion Mode

The so-called mode, namely the vibration mode of motion, is an inherent property of the time-invariant system. The mode is the most basic independent unit of the motion of the system, and the total motion of the system is a linear combination of the modes. The motion mode embodies the free motion characteristics of the airship after being disturbed, and it is the basic component of the disturbed motion of the airship. The motion mode reflects the law of motion variables changing with time and the amplitude ratio of variables and phase relationship between variables. The assignment of the motion parameters in a certain mode has a fixed proportional relationship, and the phase difference between motion parameters is also constant [32].

The motion of the airship is the linear combination of different eigenvalues in the corresponding motion modes, and the influence of eigenvectors and initial values is reflected in the weight of different motion modes. Under the condition of the same initial value in the same mode, the performance of the corresponding motion parameters is determined by the relative sizes of eigenvectors.

4.2. Characteristics of Motion Mode

According to the corresponding time scale of the six-DOF motion of the airship, the motion of the airship can be divided into long-term motion and short-term motion. The long-term motion is used to describe the translation of the center of mass of the airship, and the short-term motion is used to describe the attitude change process of the airship. The mode of the short-term motion is more representative for the description of the motion characteristics of the airship and plays a leading role in the influence of the flight quality of the airship. Therefore, we further analyze the short-term motion of the airship below.

(1) Pitching Channel

The characteristic equation of the pitching channel of the airship can be derived according to the linear model of the longitudinal motion:

$$\left|sI - m_L^{-1}A_L\right| \approx (s - Z^\alpha)\left(s^2 - M^q s - M^\theta\right) \tag{7}$$

wherein $A_L = \begin{bmatrix} \bar{Z}^\alpha & \bar{Z}^q & \bar{Z}^\theta \\ \bar{M}^\alpha & \bar{M}^q & \bar{M}^\theta \\ 0 & 1 & 0 \end{bmatrix}$, $m_L = \begin{bmatrix} (m + m_{33})s & -mx_G & 0 \\ -mx_G & (I_y + m_{55})s & 0 \\ 0 & 0 & s \end{bmatrix}$

wherein Z^α represents the derivative of normal aerodynamic coefficient; M^q represents the derivative of pitch damping moment coefficient; M^θ represents the derivative of pitch attitude moment coefficient.

From the characteristic equation, it can be seen that the motion mode of the pitching channel is composed of a monotone damping mode and a second-order oscillation convergence mode; the frequency of the second-order oscillation mode is determined by M^θ, and the damping is determined by M^q.

(2) Yaw Channel

The characteristic equation of the yaw channel of the airship can be derived according to the linear model of the lateral motion:

$$\left|sI - m_S^{-1}A_S\right| = s^2 - (Y^\beta + N^r)s - (Y^\beta N^r + N^\beta Y^r) \tag{8}$$

wherein $A_S = \begin{bmatrix} \bar{Y}^{\beta} & \bar{Y}^{r} \\ \bar{N}^{\beta} & \bar{N}^{r} \end{bmatrix}$, $m_S = \begin{bmatrix} m + m_{22} & mx_G \\ mx_G & I_z + m_{66} \end{bmatrix}$.

wherein Y^{β} represents the derivative of lateral aerodynamic coefficient; N^{β} represents the derivative of the yaw aerodynamic stabilization coefficient; Y^{r} and N^{r} respectively represent the derivatives of yaw damping force and damping force moment coefficients.

From the characteristic equation above, it can be seen that the motion mode of the yaw channel for the stratospheric airship is a second-order oscillation convergence mode; the change of Y^{β} and N^{r} affects the damping effect of the motion mode; the change of Y^{r}, N^{r}, Y^{β}, and N^{β} affects the frequency of the motion mode; upon the large damping of the mode, the second-order oscillation mode transforms into two one-order inertia damping modes.

(3) Analysis of Difference between Pitch Motion Mode and Yaw Motion Mode

From the above analysis of the pitch motion mode and yaw motion mode, it can be seen that the pitch motion mode is quite different from the yaw motion mode of the airship. The pitch attitude angle motion mode is composed of an inertial damping mode and a second-order oscillation convergence mode, while the yaw rate motion mode is composed of a second-order oscillation mode and adds an integration link on the basis of the second-order oscillation mode, thus resulting in unstable motion modes. According to the longitudinal motion linearized model, the internal state structure diagram of the pitch channel can be obtained in order to further analyze the roots, as shown in Figure 2.

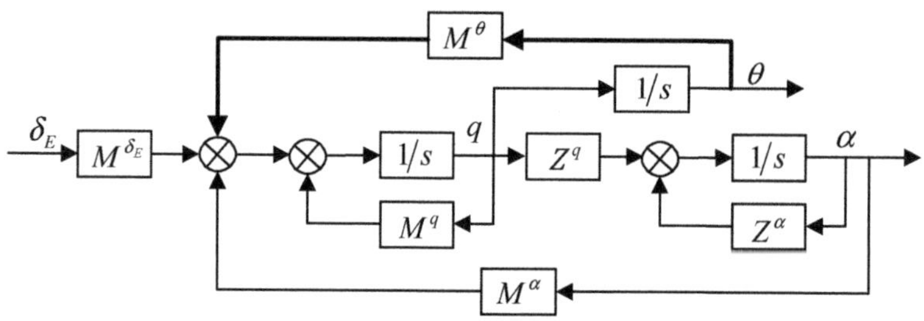

Figure 2. Internal State Structure Diagram of Airship's Pitch Channel.

The internal state structure diagram of the yaw channel of airship can be derived according to the linear model of the lateral motion, as shown in the Figure 3:

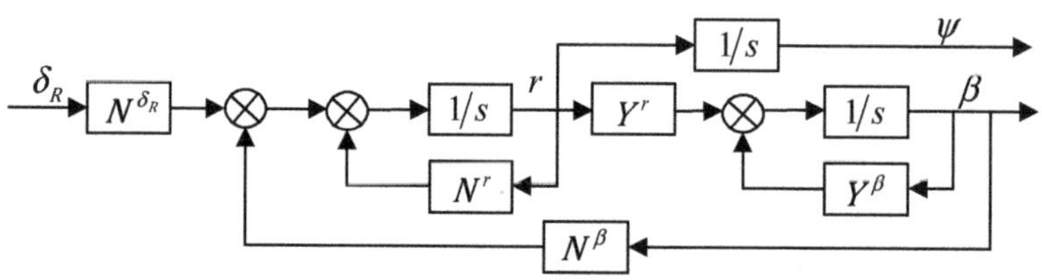

Figure 3. Internal State Structure Diagram of Airship's Yaw Channel.

Comparing the internal state structure diagrams of the airship's pitch channel and yaw channel, it can be found that the attitude angle forms a negative feedback loop through M^{θ} in the internal state structure diagram of the pitch channel, but there is no feedback loop of attitude angle in the internal state structure diagram of the yaw channel. The fundamental reason for this phenomenon is that the center of mass of the airship is located directly below the coordinate system of the airship hull and the gravity will produce a moment to feed the motion information of the pitch angle back to the system during the attitude motion, while the gravity will not produce any moment in the horizontal plane, so the motion information of the yaw angle cannot be fed back to the system. Similarly, when the center of buoyancy is above the body center and the attitude of the airship changes, the buoyance force will generate a torque through the center of buoyancy and feed the information of pitch attitude angle back to the system.

5. Influence of Key Parameter Perturbation on Motion Mode

5.1. Moment of Aerodynamic Stabilization

For high-speed and high-dynamic aircrafts such as missiles and airplanes, the moment of aerodynamic stability plays a crucial role in the motion modes of the pitch channel and yaw channel. The value and polarity of the moment determine the frequency of motion mode and the stability of the motion, and the parameter perturbation has a great influence on the kinetic characteristics of the system. Since the airship belongs to the category of low-speed and low-dynamic aircrafts, there is a big difference between the airship and a high-dynamic aircraft; therefore, it is necessary to study the influence of the aerodynamic stability moment perturbation on the disturbed motion mode of the airship.

(1) Pitching Channel

According to the state equation of the pitching channel, the characteristic equation of short-term motion of pitch channel can be obtained:

$$D(s) = s[(s - M^{q})(s - Z^{\alpha}) - M^{\alpha}Z^{q}] - M^{\theta}(s - Z^{\alpha}) \tag{9}$$

By factoring the equation above, the short-term motion modes of the pitch channel can be obtained and consist of a first-order inertial damping mode and a second-order oscillatory convergence mode. The corresponding characteristic equation is shown as follows:

$$D(s) = (s - \omega_{w})(s^{2} - 2\xi_{\theta}\omega_{\theta}s - \omega_{\theta}^{2}) \tag{10}$$

wherein ω_w represents the frequency of the first-order inertial damping mode; ξ_θ and ω_θ respectively denote the damping and frequency of the second-order oscillation convergence mode.

Then, the calculation formula of pitch aerodynamic stability moment M_α is shown as follows:

$$M_{\alpha} = k_{\alpha} \cdot \bar{M}^{\alpha} \cdot \alpha \tag{11}$$

wherein α represents the angle of attack; k_α represents the perturbation coefficient of pitch aerodynamic stability moment, $k_\alpha \in [-2, 2]$.

In the perturbation range of the selected pitch aerodynamic stability moment, the change curves of the key parameters of the pitching channel in different motion modes with the perturbation of the pitch aerodynamic stability moment can be calculated, respectively, as shown in the Figure 4.

Figure 4a depicts the frequency change curves of the first-order inertial damping mode under the perturbation of the pitch aerodynamic stability moment, while Figure 4b depicts the damping and frequency change curves of the second-order oscillation convergence mode under the perturbation of the pitch aerodynamic stability moment. As can be seen from the above figures, with the increase of the perturbation coefficient of the pitch aerodynamic stability moment, the frequency of the first-order inertial damping mode will gradually decrease, while the damping and frequency of the second-order oscillation convergence mode will increase. However, the change amplitude of the above data shows that the influence on the disturbed motion mode of the pitching channel is very small and can be ignored. Therefore, the influence of the pitch aerodynamic stability moment can be ignored in the analysis of the pitching channel disturbed motion mode.

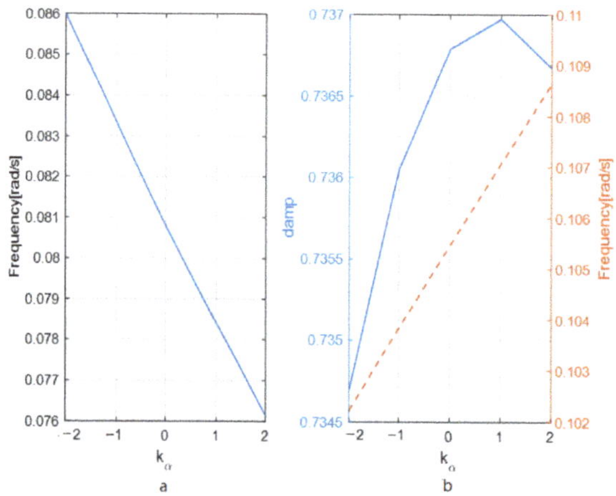

Figure 4. Curve of Influence of Moment of Aerodynamic Stabilization on Pitching Channel Motion Mode: (**a**) frequency change curves of the first-order inertial damping mode under the perturbation of the pitch aerodynamic stability moment (**b**) damping and frequency change curves of the second-order oscillation convergence mode under the perturbation of the pitch aerodynamic stability moment.

According to the above analysis conclusions, the characteristic equation of the pitch channel disturbed motion can be simplified as follows:

$$D(s) = (s - Z^{\alpha})(s^2 - M^q s - M^{\theta}) \tag{12}$$

To further illustrate the influence of the pitch aerodynamic stability moment perturbation on the pitching motion mode of the stratospheric airship and to verify the rationality and validity of the assumption that the influence of pitch aerodynamic stability moment can be ignored in the mode analysis, the Bode plot and deviation curve under the state of pitch aerodynamic stability moment perturbation are respectively presented in Figure 5.

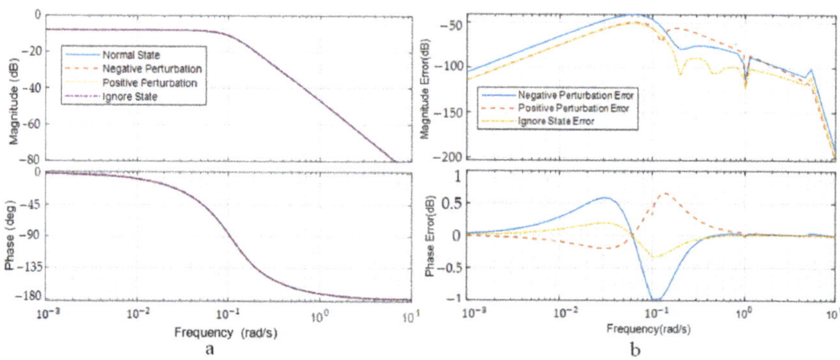

Figure 5. Bode Plot and Deviation Curve of Pitching Channel in Perturbation State of Moment of Pitching Aerodynamic Stabilization: (**a**) Bode plot of the pitching channel in the perturbation state of the pitch aerodynamic stability moment, (**b**) amplitude frequency and phase frequency deviation curves of the pitching channel in the perturbation state of the pitch aerodynamic stability moment.

Figure 5a depicts the Bode plot of the pitching channel in the perturbation state of the pitch aerodynamic stability moment, and Figure 5b depicts the amplitude frequency and phase frequency deviation curves of the pitching channel in the perturbation state of the pitch aerodynamic stability moment. The above figures show that the amplitude-frequency characteristics and phase-frequency characteristics of the corresponding system are basically unchanged in the perturbation state of the pitch aerodynamic stability moment; the amplitude-frequency and phase-frequency deviation in the pitch aerodynamic stability moment perturbation state are in a small range relative to the nominal state. Therefore, the influence of the pitch aerodynamic stability moment can be ignored during the analysis of the motion mode of the pitch channel. This is the huge difference between the stratospheric airship and the high-speed and high-dynamic aircrafts in terms of pitch disturbed motion.

(2) Yaw Channel

As mentioned above, the short-term motion mode of yaw channel can be described with a second-order oscillation mode, and the corresponding characteristic equation is shown as follows:

$$D(s) = (s^2 + 2\xi_r\omega_r s + \omega_r^2) \tag{13}$$

wherein ξ_r and ω_r denote the damping and frequency of the yaw channel short-term motion mode, respectively.

Then, the calculation formula of the yaw aerodynamic stability moment N_β is shown as follows:

$$N_\beta = k_\beta \cdot \overline{N}^\beta \cdot \beta \tag{14}$$

wherein β represents the angle of slideslip; k_β represents the perturbation coefficient of the yaw aerodynamic stability moment, $k_\beta \in [-2, 2]$.

In the perturbation range of the selected yaw aerodynamic stability moment, the change curves of the key parameters of the yaw channel in different motion modes with the perturbation of the yaw aerodynamic stability moment can be calculated respectively, as shown in Figure 6.

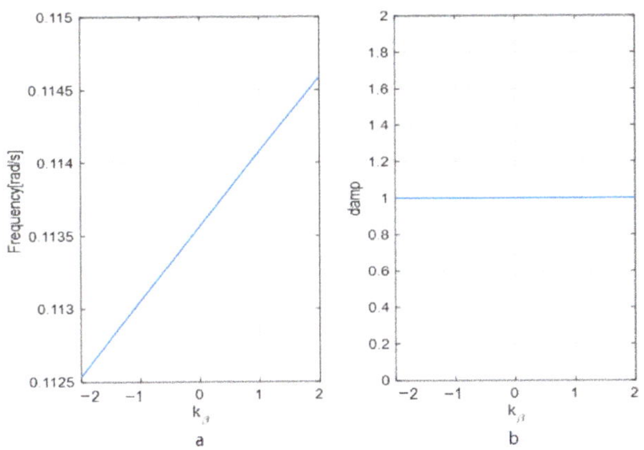

Figure 6. Curve of Influence of Moment Perturbation of Yaw Aerodynamic Stabilization on Pitching Channel Motion Mode: (**a**) frequency characteristic curve of the yaw channel motion mode under the perturbation of the yaw aerodynamic stability moment, (**b**) damping characteristic curve of the yaw channel motion mode under the perturbation of the yaw aerodynamic stability moment.

Figure 6a depicts the frequency characteristic curve of the yaw channel motion mode under the perturbation of the yaw aerodynamic stability moment, while Figure 6b depicts the damping characteristic curve of the yaw channel motion mode under the perturbation of the yaw aerodynamic stability moment. As can be seen from the above figures, the yaw disturbed motion is stable when the yaw aerodynamic stability moment is perturbed within the specified range; with the increase of the perturbation coefficient of the yaw aerodynamic stability moment, the frequency of the yaw disturbed motion mode will also increase; in the whole change process, however, the damping of the yaw disturbed motion is always one, indicating that the disturbed motion is overdamped, and the second-order oscillation convergence mode transforms into two inertia damping modes. The comparison result of the change amplitude of the above data shows that the frequency of the yaw disturbed motion mode increases with a small amplitude when the pitch aerodynamic stability moment is perturbed, and therefore, the influence of the yaw aerodynamic stability moment on the yaw disturbed motion is minimal and can be ignored.

According to the above analysis conclusions, the characteristic equation of the yaw channel disturbed motion can be simplified as follows:

$$D(s) \approx (s - Y^v)(s - N^r) \tag{15}$$

The above equation shows that the yaw motion mode can be simplified into two first-order inertial damping modes, whose characteristics are related to the derivative Y^v of the lateral aerodynamic coefficient and the derivative N^r of the yaw damping moment coefficient, respectively.

To further illustrate the influence of the yaw aerodynamic stability moment perturbation on the yaw motion mode of the stratospheric airship and to verify the rationality and validity of the assumption that the influence of yaw aerodynamic stability moment can be ignored in the mode analysis, the Bode plot and deviation curve under the state of yaw aerodynamic stability moment perturbation are respectively presented in Figure 7.

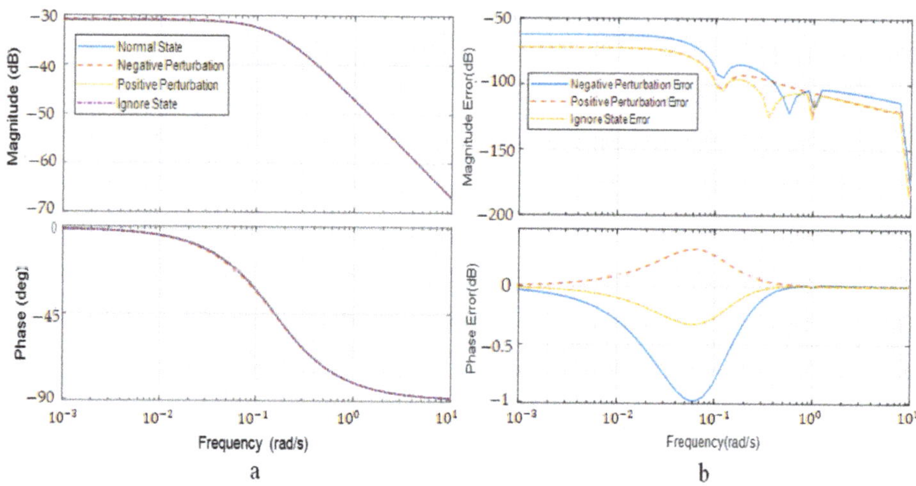

Figure 7. Bode Plot and Deviation Curve of Yaw Channel in Perturbation State of Moment of Yaw Aerodynamic Stabilization: (**a**) Bode plot of the yaw channel in the perturbation state of the yaw aerodynamic stability moment, (**b**) amplitude frequency and phase frequency deviation curves of the yaw channel in the perturbation state of the yaw aerodynamic stability moment.

Figure 7a depicts the Bode plot of the yaw channel in the perturbation state of the yaw aerodynamic stability moment, and Figure 7b depicts the amplitude frequency and phase frequency deviation curves of the yaw channel in the perturbation state of the yaw aerodynamic stability moment. The above figures show that the amplitude-frequency characteristics and phase-frequency characteristics of the corresponding system are basically unchanged in the perturbation state of the yaw aerodynamic stability moment; the amplitude-frequency and phase-frequency deviation in the yaw aerodynamic stability moment perturbation state are in a small range relative to the nominal state. Therefore, the influence of the yaw aerodynamic stability moment can be ignored during the analysis of the motion mode of the yaw channel. This is the huge difference between the stratospheric airship and the high-speed and high-dynamic aircrafts in terms of yaw disturbed motion.

5.2. Location of Mass Center

The mass centers of aircrafts such as missiles and airplanes are basically distributed uniformly inside the aircrafts around the carrier coordinate systems. Therefore, the mass center of the aircraft can be well-configured to the origin of the carrier coordinate system so that the gravity only has an effect on the motion of the mass center of the aircraft and will not have any effect on the attitude motion. Due to the special structure layout, the mass center of the airship does not overlap with its center of body; its gravity will have a great influence on the attitude motion of the airship with the change in the location of the mass center, resulting in the great differences of the airship's pitching motion mode from that of missiles and airplanes.

5.2.1. Axial Location

(1) Pitching Channel

According to the linearized model, the perturbation in the axial location of the mass center mainly affects $\overline{M}^{q}$ in the state transition matrix, as shown in the equation below:

$$\overline{M}^{q} = M_Y^q q_i S_r L_r - x_G m U_0 \tag{16}$$

wherein M_Y^q represents the derivative of the pitch aerodynamic damping moment coefficient; q_i represents the dynamic pressure; S_r represents the characteristic area; L_r represents the characteristic length; x_G represents the axial of the airship's mass center; U_0 represents the airship's trim axial velocity.

It can be seen from the above equation that when other parameters are unchanged and the axial location of the airship's mass center is perturbed in a positive way, the damping of the second-order oscillation convergence mode will increase; otherwise, the damping of the second-order oscillation convergence mode will decrease.

According to the requirements of the overall scheme, the change range of the axial location of mass center shall be from −10 m to 10 m; then, the Bode plot and deviation curve of the perturbation boundary of the airship's pitch angle motion in the axial location and that in nominal state are shown in the Figure 8.

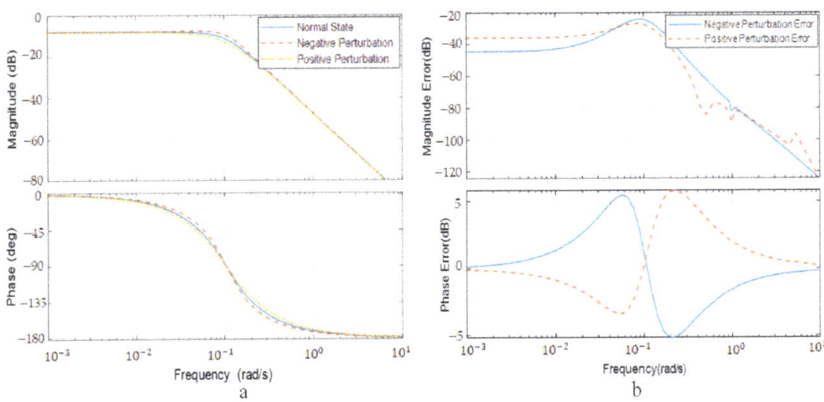

Figure 8. Bode Plot and Deviation Curve of Axial Mass Center Perturbation for Pitching Channel: (**a**) Bode diagram of the pitch channel under the state of perturbation of the pitch axial position, (**b**) amplitude frequency and phase frequency error curves of the pitch channel under the state of perturbation of the axial position.

The influence of the change in axial location of mass center on the pitching channel disturbed motion modes of the stratospheric airship is shown in Figure 9.

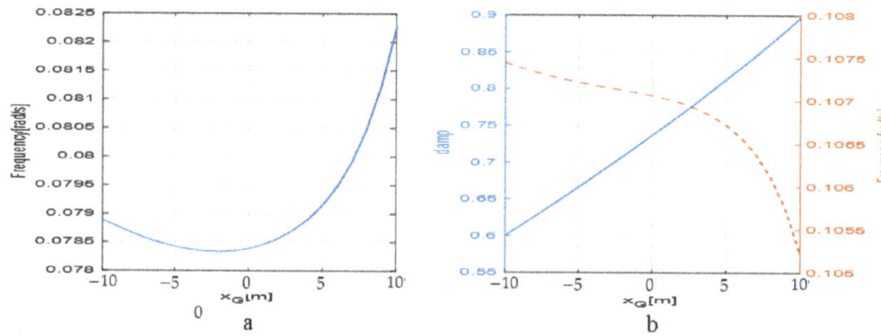

Figure 9. Curve of Influence of Axial Mass Center Perturbation on Pitching Channel Motion Mode: (**a**) Bode plot of the pitching channel with axial location perturbation of the mass center, (**b**) amplitude frequency and phase frequency deviation curves of the pitching channel with axial location perturbation of the mass center.

Figure 8a depicts the Bode plot of the pitching channel with axial location perturbation of the mass center; Figure 8b depicts the amplitude frequency and phase frequency deviation curves of the pitching channel with axial location perturbation of the mass center; Figure 9a depicts the curve of the influence of the axial location perturbation of the mass center on the frequency of the first-order inertia damping motion mode of the pitching channel; Figure 9b depicts the damping and frequency characteristic curves of the second-order oscillatory convergence mode of the pitching channel under the axial location perturbation of the mass center. It can be seen from Figures 8 and 9 that the axial location perturbation of the mass center has a great influence on the damping of the pendulum motion mode but has little influence on the frequency of the pendulum motion and the frequency of the first-order inertia damping mode. Moreover, the damping increases linearly with the axial location of the mass center.

(2) Yaw Channel

According to the yaw channel linear model, N^r can be described as:

$$N^r = M_Z^r q_i S_r L_r - x_G m U_0 \tag{17}$$

wherein M_Z^r represents the derivative of yaw aerodynamic damping moment.

It can be seen from the above equation that the axial location perturbation of mass center will affect N^r. When the axial location of the airship's mass center is perturbed in a positive way, N^r will increase; otherwise, N^r will decrease. As mentioned earlier, the yaw motion mode of the airship can be described as two first-order inertia damping modes, which determines the frequency of one of the modes so that the axial location perturbation of the mass center will directly affect the moment-dependent motion mode.

The Bode plot of the corresponding yaw motion in the perturbation boundary of the axial location of the mass center is shown in Figure 10.

It can be seen from the Bode plot that the influence of the axial location of the mass center on the motion mode of the yaw channel is consistent with the results of the previous analysis: the axial location of the mass center mainly affects the dominant motion mode of the yaw motion and has a relatively obvious influence on the motion characteristics of the yaw motion within the middle- and low-frequency bands.

5.2.2. Vertical Location

The main flight mode of the stratospheric airship is horizontal flight. Due to the inherent characteristics of gravity, the vertical location perturbation of the mass center mainly affects the motion mode of pitching channel in the disturbed motion. As mentioned earlier, the motion mode of the stratospheric airship's pitch channel can be composed of an inertia damping mode and an oscillation convergence mode, the latter of which is usually referred to as the "pendulum effect" mode. The fundamental cause of the stratospheric airship pendulum effect mode is that the mass center of the airship does not overlap with the body center, and the mass center of the airship that lies under the system so that the attitude disturbance of the airship in the pitch and roll direction can quickly converge.

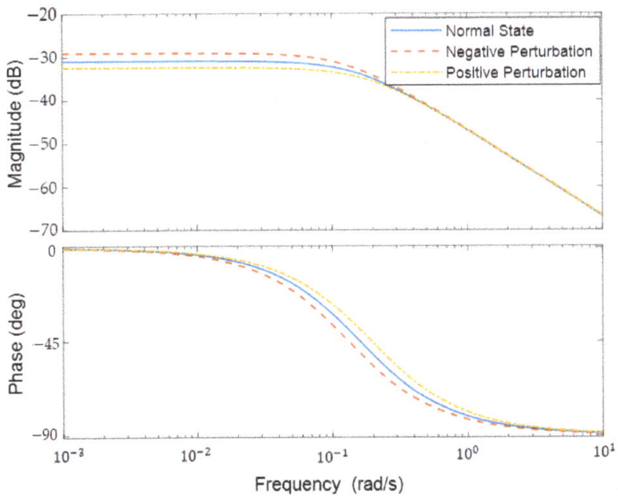

Figure 10. Yaw Channel Bode Plot of Axial Mass Center Perturbation.

When the overall aerodynamic layout is determined, the key parameter affecting the pendulum motion mode is M^{θ}, and such a parameter not only determines the frequency of the pendulum motion but also has a certain impact on the damping of the pendulum motion. According to the linearized model, we can see that:

$$M^{\theta} = z_B B - z_G G \quad (18)$$

wherein z_B and z_G respectively represent the vertical projections of the center of buoyancy and the center of mass of the airship; B and G respectively denote the buoyancy and gravity of the airship.

In order to better analyze the influence of the center of mass on the pendulum motion, it can be assumed that the center of buoyancy coincides with the body center; in this case, both the weight and the mass center location of the airship will not only affect the frequency of the pendulum motion, but the mass center location of the airship will also affect the stability of the pendulum motion. In order to ensure the stability of the pitch motion mode, the center of mass of the airship shall be generally required to be located below the body center.

The vertical location of mass center is selected to vary in the range of 0 m to 10 m. The influence of the change in the vertical location of mass center on the disturbed motion mode of the stratospheric airship's pitch channel is shown in Figure 11.

Figure 11a depicts the frequency characteristic curve of the first-order inertia damping mode of the pitching channel affected by the vertical perturbation of the mass center; Figure 11b depicts the damping and frequency characteristic curves of the second-order oscillation convergence mode of the pitching channel affected by the vertical perturbation of the mass center. The left figure above depicts the frequency change curve of the first-order inertia damping mode of the pitching channel with the vertical location perturbation, and the right figure above depicts the parameter change curve of the second-order oscillation convergence mode of the pitching channel with the vertical location perturbation. As can be seen from the left figure, when the vertical location of the mass center is less than 3 m, the frequency of the first-order inertia damping mode increases linearly with the increase of the vertical location; after the vertical location is larger than 3 m, the frequency of the first-order inertia damping mode remains essentially the same. As can be seen from the right figure, when the vertical location is larger than 3 m, the second-order oscillation convergence mode will linearly increase with the increase of the vertical location, while the damping will decrease with the increase of vertical location.

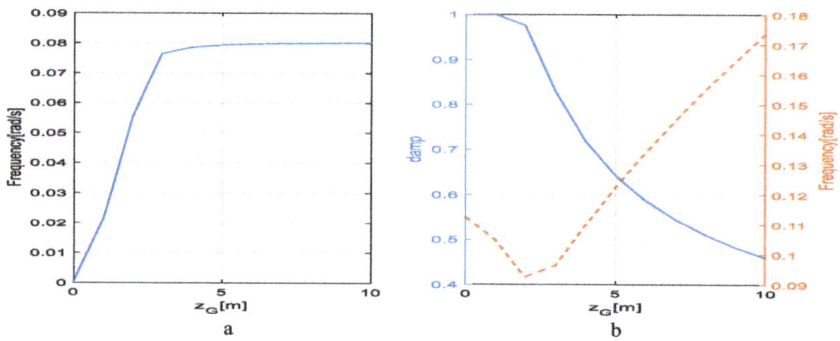

Figure 11. Curve of Influence of Vertical Mass Center Perturbation on Motion Mode of Pitching Channel: (**a**) frequency characteristic curve of the first-order inertia damping mode of the pitching channel affected by the vertical perturbation of the mass center, (**b**) damping and frequency characteristic curves of the second-order oscillation convergence mode of the pitching channel affected by the vertical perturbation of the mass center.

In conclusion, the change in the vertical location of the mass center has a large effect on the motion modes of the pitching channel and a negligible effect on the motion modes of the yaw channel. The above simulation results show that when the vertical location is set in the range between 3 and 8 m, the second-order oscillation convergence mode has better damping characteristics and moderate frequency characteristics, while the first-order inertia damping mode has a small frequency change.

5.3. Location of Buoyant Center

In the aerodynamic modeling process of airship, it is generally assumed that the center of buoyancy is located at the body center; in this way, the buoyancy force does not produce any moment on the airship, and the change of buoyancy force will only affect the motion characteristics of the airship's center of mass but has no effect on the attitude motion characteristics. Therefore, the buoyancy force has no effect on the short-term motion modes. In the process of engineering implementation, it is difficult to configure the airship's center of the buoyancy at its body center, and therefore, it is necessary to study the influence of the center of the buoyancy on the motion mode of the stratospheric airship when it does not coincide with the body center.

According to the linear model of the stratospheric airship, the axial location perturbation of the center of the buoyancy has little effect on the airship's motion modes; this is because the buoyancy has the same characteristics as gravity, and the vertical location perturbation of the center of the buoyancy has a great influence on the pitch disturbed motion mode of the airship but little influence on the yaw disturbed motion mode. Therefore, we only need to consider the influence of the vertical location perturbation of the center of the buoyancy on the pitching channel of the airship when analyzing the influence of the buoyant center perturbation.

According to the linearized model, the associated coupling effect of the buoyant center is shown in the following equation.

$$M^{\theta} = z_B B - z_G G \quad (19)$$

Similar to the vertical center of mass, the vertical perturbation of the buoyant center affects the frequency of the pendulum motion mode of the airship. The above equation shows that when the vertical location of the airship's buoyant center is above the body center, the vertical location perturbation of the buoyant center will not only increase the frequency of the pendulum motion but also increase the stability of the pendulum motion; otherwise, the frequency and stability of pendulum motion will be reduced. When the vertical location of the buoyant center is below the vertical location of the mass center, the pendulum motion is unstable.

According to the requirements of the overall scheme, the vertical perturbation boundary of the buoyant center was selected to simulate and analyze the influence of the perturbation on the disturbed motion mode of the pitching channel; the corresponding curves are shown in Figure 12.

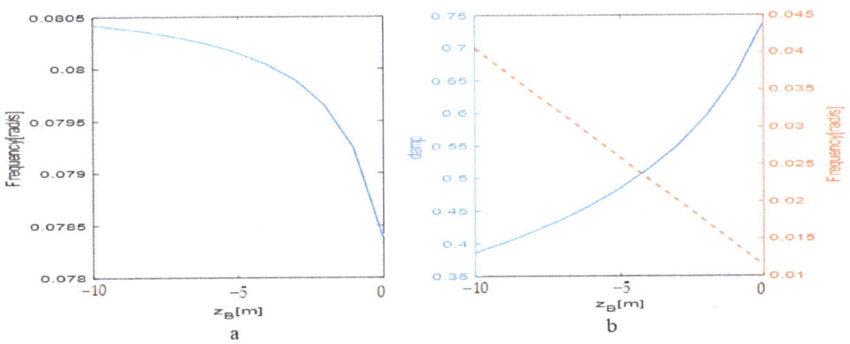

Figure 12. Curve of Influence of Vertical Buoyant Center Perturbation on Motion Mode of Pitching Channel: (**a**) frequency characteristic curve of the first-order inertia damping mode of the pitching channel affected by the vertical perturbation of the buoyant center, (**b**) damping and frequency characteristic curves of the pendulum motion mode of the pitching channel affected by the vertical perturbation of the buoyant center.

Figure 12a depicts the frequency characteristic curve of the first-order inertia damping mode of the pitching channel affected by the vertical perturbation of the buoyant center; Figure 12b depicts the damping and frequency characteristic curves of the pendulum motion mode of the pitching channel affected by the vertical perturbation of the buoyant center. As can be seen from the figures above, the frequency of the first-order inertia damping mode increases with an increase in the distance of the buoyant center deviating from the body center, but the frequency will increase with the decrease of the damping in the pendulum motion mode. As can be seen from the data change amplitude, the buoyant center perturbation mainly affects the characteristics of the pendulum motion mode and has little effect on the first-order inertia damping mode. Moreover, when the buoyant center perturbation is controlled within 3 m, the pendulum motion has better damping characteristics and moderate frequency characteristics.

6. Conclusions

The motion mode describes the characteristics of the free motion of the stratospheric airship. The analysis of motion mode is to obtain the eigenvalues of the stratospheric airship's motion with relevant calculation and analysis methods so as to reveal the kinetic characteristics of the stratospheric airship under disturbance, which can provide a theoretical basis for the optimization of the overall scheme for the airship and the design of the flight control system. In this paper, the changing trends of the stratospheric airship's motion mode under the perturbation of different key parameters are analyzed. According to the simulation and analysis results, unlike high-speed and high-dynamic aircrafts such as airplanes, the pitch and yaw motion modes of the stratospheric airship are hardly affected by the perturbation of aerodynamic stability moment, and therefore, its influence can be ignored in the calculation and mode analysis process, which is also one of the typical characteristics of the differences between stratospheric airships and aircrafts. The vertical center of mass and the vertical center of buoyancy determine the frequency and stability of the pitch pendulum motion mode, and their parameter perturbation greatly affects the pitch motion mode of the stratospheric airship, which is another typical characteristic of the differences between stratospheric airships and aircrafts. The axial location perturbation of the center of mass not only affects the damping of the pitch pendulum motion but also affects the frequency of one yaw motion mode.

Author Contributions: Conceptualization, J.T. and H.J.; methodology, J.T.; software, S.B.; validation, W.X. and J.W.; formal analysis, S.B.; investigation, W.X.; resources, J.W.; data curation, S.B.; writing—original draft preparation, S.B.; writing—review and editing, Y.S.; visualization, H.J.; supervision, J.T.; project administration, J.T.; funding acquisition, J.T. All authors have read and agreed to the published version of the manuscript.

Funding: This work is sponsored by the National Natural Science Foundation of China under Grant No. 51906141 and No. 62073216.

Data Availability Statement: Not applicable.

Conflicts of Interest: The authors declare no conflict of interest.

References

1. *Stratospheric Airship Technology*; Science Press: Beijing, China, 2019.
2. Tang, J.; Xie, W.; Wang, X.; Chen, C. Simulation and Analysis of Fluid-Solid-Thermal Unidirectional Coupling of Near-Space Airship. *Aerospace* **2022**, *9*, 139. [CrossRef]
3. Tang, J.; Xie, W.; Wang, X.; Chen, Y.; Wu, J. Study of the Mechanical Properties of Near-Space Airship Envelope Material based on an Optimization Method. *Aerospace* **2022**, *9*, 655. [CrossRef]
4. Tang, J.; Pu, S.; Yu, P.; Xie, W.; Li, Y.; Hu, B. Research on Trajectory Prediction of a High-Altitude Zero-Pressure Balloon System to Assist Rapid Recovery. *Aerospace* **2022**, *9*, 622. [CrossRef]
5. Xu, Z. *Stratospheric Airship for Early Warning and Detection*; National Defense Industry Press: Arlington, VA, USA, 2017.
6. Fang, Z. *Aircraft Flight Dynamics*; Beihang University Press: Beijing, China, 2005.
7. He, Z.; Gao, H. *The Advanced Flight Dynamics*; Northwestern Polytechnical University Press: Xi'an, China, 1990.
8. Khoury, G.; Gillet, J. *Airship Technology*; Cambridge University Press: London, UK, 2004.
9. Ouyang, J. Research on Modeling and Control of an Unmanned Airship. Ph.D. Thesis, Shanghai Jiaotong University, Shanghai, China, 2003.
10. Li, Y.; Nahon, M. Modeling and Simulation of Airship Dynamics. *J. Guid. Control Dyn.* **2007**, *30*, 1691–1700. [CrossRef]
11. Liu, Y.; Wu, Y.; Wu, Y. Stability and Control Analysis based on Airship Dynamic Modeling. In Proceedings of the IEEE International Conference on Automation and Logistics, Jinan, China, 18–21 August 2007.
12. Yang, Y.; Zheng, W.; Wei, H.; Ye, Y.; Shao, H.B. Flight mode analysis for stratospheric airships. *J. Natl. Univ. Def. Technol.* **2015**, *37*, 57–64.
13. Wang, R. *Mathematics Modeling and Stability Augmentation Control System Design of Stratospheric Airship*; Northwestern Polytechnical University: Xi'an, China, 2006.
14. Liu, Y. *Stratospheric Autonomous Airship: Model, Dynamics and Control*; South China University of Technology: Guangzhou, China, 2009.
15. Mao, J. *Dynamics Analysis & Motion Control of Airship*; Graduate University of Chinese Academy of Sciences: Beijing, China, 2008.
16. Wang, F. *Dynamic Characteristics and Control Method of a New Concept Stratospheric Airship*; Beihang University Press: Beijing, China, 2012.
17. Wang, R.; An, J.; Wu, M.; Chen, L. Stratospheric Airship Longitudinal Intelligence Stability Augment System Design based on EA, LMI and ANN. *J. Proj. Rockets Missiles Guid.* **2007**, *27*, 21–28.

8. Tiwari, A.; Vora, A.; Sinha, N. Airship Trim and Stability Analysis Using Bifurcation Techniques. In Proceedings of the 2016 7th International Conference on Mechanical and Aerospace Engineering, London, UK, 18–20 July 2016.
9. Wu, Y.; Liu, Y. Modelling and Controller-Designing of a Novel Driven Airship. In Proceedings of the 31st Chinese Control Conference, Hefei, China, 25–27 July 2012.
10. Zhang, Y. *Flight Stability Research for Stratospheric Airship*; Beihang University Press: Beijing, China, 2015.
11. Liu, L. *Aerodynamic Model and Linear Motion Analysis of Airship*; Xiamen University: Xiamen, China, 2015.
12. Yuan, J.; Zhu, M.; Guo, X.; Lou, W. Trajectory Tracking Control for a Stratospheric Airship Subject to Constraints and Unknown Disturbances. *IEEE Access* **2020**, *8*, 31453–31470. [CrossRef]
13. Huang, L.; Shen, S. Parameter Range in Hovering Control of Airship. In Proceedings of the 14th International Conference on Computer Science & Education (ICCSE 2019), Toronto, ON, Canada, 19–21 August 2019.
14. Liu, L. On the aerodynamic parameter model of the stratospheric airship. In Proceedings of the 33rd Chinese Control Conference, Nanjing, China, 28–30 July 2014.
15. Li, H. *Dynamic Modeling and Longitudinal Stability Analysis of a Stratospheric Dual-Hull Airship*; Chinese Aeronautical Society: Beijing, China, 2023.
16. Gobiha, D. Numerical Approach to Maneuver Design and Feasibility Evaluation for the Autonomy of Airship, Lighter than Air Systems. In *Lighter than Air Systems*; Springer Nature: Singapore, 2022.
17. Sun, L.; Zhu, M.; Guo, X. Error-Constrained Moving Path-Following Control for a Stratospheric Airship. In Proceedings of the 2021 IEEE International Conference on Mechatronics and Automation, Takamatsu, Japan, 8–11 August 2021.
18. Wang, X.; Ma, Y.; Shan, X. Modeling of Stratosphere Airshi. In Proceedings of the 2010 3rd International Symposium on Systems and Control in Aeronautics and Astronautics, Harbin, China, 8–10 June 2010.
19. Hu, G.; Wu, M. Motion analysis and simulation of a stratospheric airship. *J. Harbin Eng. Univ.* **2011**, *32*, 1501–1508.
20. Gao, M. *Research on Modelling and Control of Near-Space Airship's Flight Movements*; Nanjing University of Aeronautics and Astronautics: Nanjing, China, 2008.
21. Zhang, J. *Trajectory Control Method for Stratospheric Airship in Wind Field*; National University of Defense Technology: Changsha, China, 2017.
22. Zheng, W.; Yang, Y. *Airship Flight Dynamics and Control*; Science Press: Beijing, China, 2016.

aerospace

MDPI

Article

Thrust-Based Stabilization and Guidance for Airships without Thrust-Vectoring

Carlo E.D. Riboldi [1,*] and Alberto Rolando [2]

1 Dipartimento di Scienze e Tecnologie Aerospaziali, Politecnico di Milano, 20156 Milan, Italy
2 PACE Aerospace & IT GmbH, 14059 Berlin, Germany
* Correspondence: carlo.riboldi@polimi.it

Abstract: The concept of thrust-based control without the employment of thrust-vectoring (TVC), already introduced in a previous work by the authors, is further developed in conjunction with an appropriate control suite, tasked with both artificial stabilization and beam-tracking navigation functions. In the paper, the fully non-linear mathematical model employed for testing the controllers in a virtual environment is outlined. Then a comparative approach is adopted in the analysis, where a standard tail-back airship with deflectable aerodynamic surfaces is employed as a baseline, and the performance of a four-thrusters layout with a thrust-based control and no TVC is assessed with respect to it. Featured test cases in forward flight include short climbs, abrupt turns, and multi-checkpoint navigation. The research supports the feasibility and adequate performance of the proposed thrust-based airship layout and control, and presents a critical analysis of the pros and cons with respect to the considered baseline airship configuration featuring standard aerodynamic control.

Keywords: flight dynamics; airship; high-altituda airship (HAA); unmanned aerial vehicle (UAV); autonomous flight; lighter-than-air (LTA); stability augmentation system (SAS); thrust-based stabilization; guidance; navigation; thrust-based guidance; control surfaces; airship design; control design; thrust vector control (TVC)

1. Introduction

The employment of lighter-than-air (LTA) platforms for autonomous flight missions is attracting the interest of the scientific and industrial communities, thanks primarily to the increased flight endurance of such flying vehicles, starkly above that of current rotary-wing UAVs, and the ability to fly in near-hover or hover conditions, which is not achievable in fixed-wing machines [1–6]. However, despite the relative simplicity of the principle, resulting in such interesting endurance and hover performance, i.e., the availability of a significant buoyancy force, several typical features in the layout of LTAs—especially considering the traditional airship layout—it is generally responsible for reduced stability and controllability levels.

These features are primarily bound to the sizeable envelope of the airship, typically designed as a slender body with poor lifting characteristics (at least considering airships that are not specifically designed to obtain high aerodynamic efficiency, and based on a low buoyancy ratio [7]). This general shape is associated with significant mass and aerodynamic inertia, especially around the pitch and yaw axes [8,9], and a large wet area which tends to amplify the dynamic reaction to disturbances in the airstream [10]. Furthermore, the lack of a wing has an effect, primarily felt on the eigendynamics, like a limited damping-in-roll, or in the execution of maneuvers, such as turns. Finally, the positioning of the thrusters and control surfaces, both typically placed close to the stern of the airship, and the lack of ailerons significantly far from the roll axis, tend to produce a peculiar dynamic response to controls, different from that featured by either rotary- or fixed-wing aircraft configurations [9,11].

All such specificity tends to produce a control problem that appears to be rather different with respect to that of standard-winged or rotary-wing aircraft, and a significant amount of research has been developed correspondingly over the years. A common way to deal with this departure from a standard aircraft layout and from the control degrees of freedom found on aircraft is to adopt alternative propulsive configurations, typically optimizing the flying platform for some part of the mission profile (such as hover, or constant-airspeed cruise) [12]. Effective ways of controlling airships may be easily associated with the employment of thrust vector control (TVC), adopted on larger manned designs as well as on smaller, tail-less layouts [13,14]. Control for airships has been analyzed both in terms of stabilization and disturbance rejection [15,16], and considering hovering and forward flight guidance [17,18]. Control for precise positioning in 3D space has been of special interest for HAPS (High-Altitude Pseudo-Satellite) applications [19,20].

However, in the quest for the containment of design and manufacturing costs, construction simplicity and platform versatility often represent points of merit in the design. The former may mitigate time between overhauls (TBO), when the number of actuators and mechanisms is reduced, where the latter allows exploiting the very same LTA for diverse missions. This simplicity-oriented approach to the design of LTAs is also an enabler of their employment in the low-weight, low-size, limited-budget UAV market, since the reduction in the overall structural weight, which includes the weight of actuators and systems, would allow reducing the volume of the LTA without much alteration in power (hence flight performance) and payload weight. Furthermore, the adaptability of the same platform to a generic flight profile would also foster its market desirability.

In pursuing this simplicity-oriented philosophy, the deletion of flight control surfaces and the de-localization of thrusters, which shall be now employed for simultaneously achieving a desired flight speed and controlling the airship, is clearly an interesting concept. A previous study by the authors has investigated a feasible layout and the chance to artificially increase the stability of a solely thrust-based airship design, defining a procedure for the optimal arrangement of the thrusters onboard, and showing that a basic stabilization control can effectively condition the free dynamics of the LTA, making it pitch- and yaw-stable without either TVC or controllable aerodynamic surfaces [10].

In the present work, this concept is further extended. Concentrating on control, a more sophisticated stability-conditioning control suite is deployed [21], comparing the results of its adoption on a baseline airship featuring standard aerodynamic surfaces to the performance of a conceptually similar controller obtained on a four-thrusters, thrust-controlled airship layout emerging from a previous study [10]. Then the problem of guidance is analyzed, and realistic beam-tracking directional and vertical controllers are introduced, on both the baseline and thrust-controlled machines [19,21]. The results, obtained in a virtual environment on a fully non-linear simulator of the LTA flight dynamics, cover several maneuvers in forward flight, with small changes in altitude. They show that migrating from a standard control configuration based on aerodynamic surfaces to a purely thrust-based control with no thrust-vectoring, where the thrusters act simultaneously for stabilization, navigation, and speed-tracking, does not require substantial differences in the control philosophy. Furthermore, it is found that comparable or better maneuverability can be obtained in the thrust-based configuration. The latter may be therefore considered to be a viable design alternative to standard aerodynamics-based control in terms of flight dynamics performance, bringing in also the mentioned advantage of greater simplicity, and the ability to exploit the reduced size and prompter response of de-localized small electric motors. These features in turn may constitute primers for the down-scaling of airship size, and the ensuing wider adoption of LTAs also in the low-weight category of the UAV market.

2. Flight Dynamics Model of the Airship

In the present study, the nonlinear dynamics of the airship were modeled according to the same approach introduced in [10], which will be briefly recalled and slightly expanded in this section.

Considering Figure 1, the dynamic equations of the system represent the dynamic equilibrium of translation and rotation around the center of buoyancy of the body, with respect to an inertial reference $\mathcal{I}$ on the ground. We introduce the velocity vector v_{CB} of the center of buoyancy CB, expressing the rate of change in the position vector x_{CB} pointing CB from the origin of $\mathcal{I}$, and the rotational rate $\omega_{\mathcal{B}/\mathcal{I}}$ of the body frame $\mathcal{B}$ attached to the airship with respect to $\mathcal{I}$.

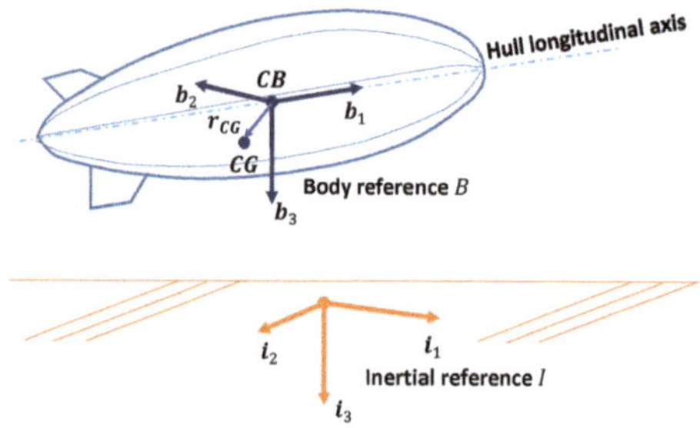

Figure 1. Ground and body references.

Vectors v_{CB} and $\omega_{\mathcal{B}/\mathcal{I}}$ are collected in the generalized velocity vector w_{CB}, as

$$w_{CB} = \{v_{CB}, \omega_{\mathcal{B}/\mathcal{I}}\}^T. \tag{1}$$

It is possible to express dynamic equilibrium as

$$M_{CB}\dot{w}_{CB} + w_{CB_{\times}} M_{CB} w_{CB} = s_{CB}. \tag{2}$$

where in Equation (2), M_{CB} represents the generalized mass matrix, wrapping the static moment tensor S_{CB} and inertia tensor J_{CB} with respect to point CB, thus yielding by definition

$$M_{CB} = \begin{bmatrix} mI & S_{CB}^T \\ S_{CB} & J_{CB} \end{bmatrix}. \tag{3}$$

Expressing Equation (2) by components in the body frame $\mathcal{B}$, assumed centered in CB, a set of six scalar first-order differential equations is obtained, as typical in flight mechanics [22]. In the following expressions, superscript $(\cdot)^{\mathcal{B}}$ applied to a vector or tensor indicates its representation by scalar components in the body frame $\mathcal{B}$. In particular, the generalized velocity in body components is defined as $w_{CB}^{\mathcal{B}} = \{U, V, W, p, q, r\}^T$, where the first three are components of velocity $v_{CB}^{\mathcal{B}}$, and the last three are components of the rotational rate vector $\omega_{\mathcal{B}/\mathcal{I}}^{\mathcal{B}}$.

The generalized forcing term on the right hand side of Equation (2) is composed of four elements, representing as many sources of force and moments, namely aerodynamics, buoyancy, gravity, and thrust. Formally, this can be written as in Equation (4) (with forcing terms appearing in the cited order on the right hand side of the equation),

$$s_{CB} = s_{CB}^a + s_{CB}^b + s_{CB}^g + s_{CB}^t. \tag{4}$$

Each term in Equation (4) takes the form $s_{CB}^{(\cdot)} = \left\{f^{(\cdot)}, m_{CB}^{(\cdot)}\right\}^T$, wherein $f^{(\cdot)}$ is force and $m_{CB}^{(\cdot)}$ moment with respect to point CB.

The terms appearing in Equation (4) are treated making explicit their dependency on either the current state or control setting of the airship, and are expressed in the body frame as required to complete the expression of the scalar equations of motion [10].

In particular, considering aerodynamic force, it can be split into active and reaction terms, yielding

$$\boldsymbol{s}^{a}_{CB} = \boldsymbol{s}^{a,b}_{CB} + \boldsymbol{s}^{a,m}_{CB}, \tag{5}$$

where in Equation (5), $\boldsymbol{s}^{a,b}_{CB}$ is the active aerodynamic term, and $\boldsymbol{s}^{a,m}_{CB}$ the reaction aerodynamic term. Similar to winged aircraft, the active term can be expressed in body components through a linear decomposition with respect to the generalized state $\boldsymbol{w}^{\mathcal{B}}_{CB}$ and, in the case they are present on board, to aerodynamic controls. Considering as a baseline an airship with standard aerodynamic controls, in the form of deflectable surfaces on the vertical and horizontal tail, the very same array of controls to be found on most winged aircraft will be featured, i.e., containing elevator (δ_e), aileron (δ_a), and rudder (δ_r) deflections, together with thrust settings $\delta_{T_1}, \ldots, \delta_{T_{N_t}}$ for each of the N_t thrusters on board. On the other hand, only thrust-setting inputs will be there in the case of a thrust-controlled airship. To provide a general representation of $\boldsymbol{s}^{a,b}_{CB}$ to cover both cases of airship layout at a modeling level, we introduce an array of controls $\boldsymbol{u}$ which is composed of two sub-parts, as

$$u = \left\{ \begin{array}{c} u^a \\ u^t \end{array} \right\} = \left\{ \begin{array}{c} \left\{ \begin{array}{c} \delta_e \\ \delta_a \\ \delta_r \end{array} \right\} \\ \left\{ \begin{array}{c} \delta_{T_1} \\ \vdots \\ \delta_{T_{N_t}} \end{array} \right\} \end{array} \right\}. \tag{6}$$

With the definition in Equation (6), the active force term in Equation (5) can be explicitly written in body components as

$$\boldsymbol{s}^{a,b^{\mathcal{B}}}_{CB} = \left\{ \begin{array}{c} X^{a,b} \\ Y^{a,b} \\ Z^{a,b} \\ L^{a,b}_{CB} \\ M^{a,b}_{CB} \\ N^{a,b}_{CB} \end{array} \right\} = \boldsymbol{U}^{\mathcal{B}}_{CB}\boldsymbol{w}^{\mathcal{B}}_{CB} + \boldsymbol{V}^{\mathcal{B}}_{CB}\boldsymbol{u}, \tag{7}$$

where matrix $\boldsymbol{U}^{\mathcal{B}}_{CB}$ in Equation (7) contains stability derivatives, functions of the shape and of the current aerodynamic state of the airship (i.e., typically depending on the current airspeed, density of air, Reynolds number, etc.), whereas matrix $\boldsymbol{V}^{\mathcal{B}}_{CB}$ defines the sensitivity of active aerodynamic force to controls. The latter can be written for a generic airship layout (i.e., with or without aerodynamic control surfaces) as

$$\boldsymbol{V}^{\mathcal{B}}_{CB} = \left[\begin{array}{cc} \boldsymbol{V}^{a,\mathcal{B}}_{CB} & \boldsymbol{V}^{t,\mathcal{B}}_{CB} \end{array} \right], \tag{8}$$

splitting it explicitly in the components multiplying $\boldsymbol{u}^a$ and $\boldsymbol{u}^t$, respectively, (defined in Equation (6)). On account of the selected layout of the airships considered in this paper, we will neglect the effect of the thrusters on aerodynamics, i.e., in particular, any interaction effects between the stream downwind of the thrusters and the tail surfaces. Therefore, we will consider matrix $\boldsymbol{V}^{t,\mathcal{B}}_{CB} = \boldsymbol{0}$. As for matrix $\boldsymbol{V}^{a,\mathcal{B}}_{CB}$, for a baseline airship featuring aerodynamic controls, this matrix will host control derivatives with respect to $\boldsymbol{u}^a$ components, functions of the current aerodynamic state, whereas when thrust-based control is considered $\boldsymbol{V}^{a,\mathcal{B}}_{CB} = \boldsymbol{0}$ [10].

The reaction aerodynamic term $s_{CB}^{a,m^{\mathcal{B}}}$ in Equation (5), modeled according to Munk-Jones-DeLaurier theory [8,23], can be written in body components as

$$s_{CB}^{a,m^{\mathcal{B}}} = N_{CB}^{\mathcal{B}} \dot{w}_{CB}^{\mathcal{B}}, \tag{9}$$

where matrix $N_{CB}^{\mathcal{B}}$ is diagonal for an axis-symmetric body [24], and is typically merged to generalized mass $M_{CB}^{\mathcal{B}}$ on the left hand side of Equation (2).

The buoyancy and gravity forcing terms s_{CB}^{b} and s_{CB}^{g} in Equation (4) can be expressed as explicit functions of the volume $\mathcal{V}$ of the lifting gas, the density of air ρ and the mass of the airship m, as well as of the three attitude angles $e_{321}^{\mathcal{B}} = \{\varphi, \vartheta, \psi\}^T$. The latter appear nonlinearly in the rotation matrix linking the ground and body frames $\mathcal{I}$ and $\mathcal{B}$. These relationships yield the implicit definitions in body components [10]

$$\begin{aligned} s_{CB}^{b^{\mathcal{B}}} &= s_{CB}^{b^{\mathcal{B}}}(\mathcal{V}, \rho, e_{321}), \\ s_{CB}^{g^{\mathcal{B}}} &= s_{CB}^{g^{\mathcal{B}}}\left(m, e_{321}, r_{CG}^{\mathcal{B}}\right). \end{aligned} \tag{10}$$

Notably, the gravitational forcing term in Equation (10) is also a function of the position r_{CG} of the center of gravity CG with respect to CB (see Figure 1).

Finally, the thrust forcing term $s_{CB}^{t^{\mathcal{B}}}$ can be defined in body reference components as follows [10]:

$$s_{CB}^{t^{\mathcal{B}}} = s_{CB}^{t^{\mathcal{B}}}\left(u^t, \sigma_i, \lambda_i, r_{P_{T_i}}^{\mathcal{B}}\right) = \sum_{i=1}^{N_t} T_i \left\{ \begin{array}{c} \left\{ \begin{array}{c} \cos\sigma_i \cos\lambda_i \\ \sin\sigma_i \\ \cos\sigma_i \sin\lambda_i \end{array} \right\} \\ r_{P_{T_i}}^{\mathcal{B}} \times \left\{ \begin{array}{c} \cos\sigma_i \cos\lambda_i \\ \sin\sigma_i \\ \cos\sigma_i \sin\lambda_i \end{array} \right\} \end{array} \right\}, \tag{11}$$

where $r_{P_{T_i}}^{\mathcal{B}}$ is the representation in body components of the position vector pointing from CB to the point of application (named P_{T_i}) of the thrust force T_i of the i-th thruster, angle σ_i represents the misalignment between the vertical plane of symmetry of the airship and the thrust line of the i-th thruster (positive to starboard), and angle λ_i is defined between the projection of the i-th thrust line on the vertical plane of symmetry and the longitudinal body axis (positive downwards), as accurately described in [10].

Thrust intensity T_i is expressed as a function of the control variable δ_{T_i} as follows:

$$T_i = \tilde{T}_i \tilde{K}_i(\delta_{T_i}) \delta_{T_i}, \tag{12}$$

where $\tilde{T}_i$ is the nominal value of thrust from the i-th thruster. The control δ_{T_i} modulates the nominal value through the shape function $\tilde{K}_i(\delta_{T_i})$. The latter allows the reproduction of possible regime-dependent efficiency effects or non-linear features related to the specific thruster technology. Furthermore, according to the technology implemented in the thruster, a different range of δ_{T_i} values can be considered. For instance, for a piston engine, this control would be limited to a positive value, whereas for electric motors it may be also negative, thus producing an inversion of the thrust force, exploiting a well-known advantage of these types of thrusters with respect to more standard piston-powered ones.

It is noteworthy that the buoyancy and gravity terms in Equation (10) bring the term $e_{321}^{\mathcal{B}}$ in the equilibrium Equation (2), increasing the number of scalar unknowns to 9. A set of three nonlinear differential relations between $\dot{e}_{321}^{\mathcal{B}}$ and $\omega_{\mathcal{B}/\mathcal{I}}^{\mathcal{B}}$ can be written based on kinematic identities [22], yielding

$$\dot{e}_{321}^{\mathcal{B}} = S_{321}^{\mathcal{B}^{-1}} \omega_{\mathcal{B}/\mathcal{I}}^{\mathcal{B}}, \tag{13}$$

with $S_{321}^{\mathcal{B}} = S_{321}^{\mathcal{B}}\left(e_{321}^{\mathcal{B}}\right)$ in Equation (13).

For dynamics simulation purposes, and for illustrating control logics, it is worth presenting a set of non-linear first-order differential equations resulting from the modeling just introduced, which synthetically define the time rate of the state of the airship system, here defined as $x^{AC} = \left\{w^{\mathcal{B}}_{CB}, e^{\mathcal{B}}_{321}\right\}^T$, and yielding

$$\begin{aligned} \dot{x}^{AC} &= \tilde{g}\left(x^{AC}, u\right) = \dots \\ &= \left\{ \begin{array}{c} \left(M^{\mathcal{B}}_{CB} - N^{\mathcal{B}}_{CB}\right)^{-1} \left(\begin{array}{c} \left(-w_{CB_\times} M_{CB} + U^{\mathcal{B}}_{CB}\right) w^{\mathcal{B}}_{CB} + V^{\mathcal{B}}_{CB} u + \dots \\ s^{b^{\mathcal{B}}}_{CB}(\mathcal{V}, \rho, e_{321}) + s^{g^{\mathcal{B}}}_{CB}\left(m, \rho, e_{321}, r^{\mathcal{B}}_{CG}\right) + \dots \\ s^{t^{\mathcal{B}}}_{CB}\left(u^t, \sigma_i, \lambda_i, r^{\mathcal{B}}_{P_{T_i}}\right) \end{array} \right) \\ \\ S^{\mathcal{B}^{-1}}_{321} \omega^{\mathcal{B}}_{\mathcal{B}/\mathcal{I}} \end{array} \right\}. \end{aligned} \quad (14)$$

The partly explicit expression in Equation (14) bends itself to numerical integration, constituting a useful tool for flight simulation purposes (not only for the case of LTAs, but more in general for any craft subject to the set of forces defined in Equation (4)). The integration of Equation (14) is carried out when launching the control schemes described in the following, where it represents the content of the airship dynamics blocks.

3. Airship Stabilization

From the analysis of the linearized dynamics of an airship in forward flight, it is typical to observe a degree of instability or marginal stability over the entire span of operative airspeeds [9]. In particular, the longitudinal pendulum mode is generally associated with little damping and low frequency, whereas the lateral-directional sideslip and roll subsidence modes are, respectively, an unstable real mode and a little-damped oscillatory one, featuring a frequency that increases with speed. Therefore, especially in scenarios where autonomous flight is of interest, deploying an inner stabilization layer is recommendable to make the airship easily steerable by means of an external guidance system.

In this work, three logics are implemented for stabilization, namely a pitch-rate damper, a yaw-rate damper, and a roll stabilization loop (roll-rate damper). The corresponding control laws will be introduced at first for a standard airship taken as a baseline, featuring control surfaces on the trailing edge of the fins. Then a four-thrusters configuration will be employed to show how to possibly port the same logic on a thrust-controlled airship concept, not featuring movable aerodynamic surfaces, nor thrust vector control.

3.1. Stabilization on an Airship Featuring Movable Aerodynamic Surfaces

The main task for the stabilization loop is that of increasing damping around the three body axes. The logic behind stability augmentation systems around the pitch and yaw axes is similar on an airship featuring movable aerodynamic surfaces and on a standard winged aircraft. Actually, for the layout of such an airship, damping around the pitch and yaw axes is typically obtained by action on the control surfaces on the tail, namely the elevator and rudder, which on aircraft are deployed, aiming at increasing damping of the short-period and Dutch-roll modes, respectively. However, since the airship is lacking a wing, compared to aircraft, significant damping in roll needs to be added, especially at lower airspeed values. In this regard, roll stabilization on airships is a control problem more similar to that found on back-tailed missiles, and inspiration to tackle it is therefore taken from that application [21].

The logic implemented for the three stabilizing loops is presented in Figure 2.

Low-pass filters on feedback signals (i.e., body components of rotational rates) are required for disturbance rejection, whereas washout filters are needed on pitch and yaw-rate signals to allow sufficient bandwidth for the guidance controller to operate.

Concerning roll, a washout was found to be unnecessary following experimentation in a virtual environment when piloting the airship in turns. The airship behavior in turn is significantly different from that of winged aircraft, first due to the lack of a wing and the ensuing low contribution of lift to counter centrifugal inertia, and second to the fact that buoyancy, which tends to take the role that lift has on aircraft, is always pointing in a direction opposite to gravity. As will be shown in Section 4, satisfactory almost-flat turns can be performed to keep the airship on track during navigation.

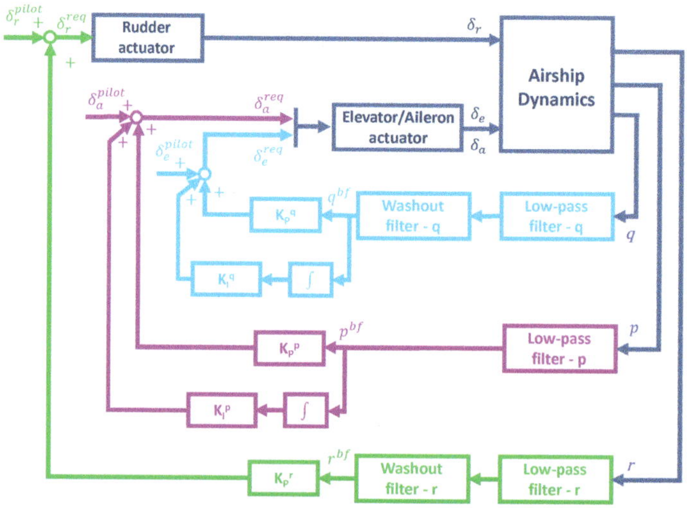

Figure 2. Proposed stabilization control for an airship with deflectable tail control surfaces. Cyan: pitch-rate damper. Purple: roll-rate damper. Green: yaw-rate damper.

According to Figure 2, the elevator, rudder, and aileron control inputs—the latter obtained as a differential use of the two elevator surfaces, located to the left and right of the vertical stabilizer—are the composition of the stabilizing control signal and of an input coming from the pilot or autopilot, responsible for trimming and maneuvering the airship. In analytical terms, the input to the actuators (superscript $(\cdot)^{req}$) can be expressed as follows

$$\begin{aligned}\delta_e^{req} &= \delta_e^{pilot} + k_P^q q^{bf} + k_I^q \int q^{bf} \mathrm{d}t,\\ \delta_a^{req} &= \delta_a^{pilot} + k_P^p p^{bf} + k_I^p \int p^{bf} \mathrm{d}t,\\ \delta_r^{req} &= \delta_r^{pilot} + k_P^r r^{bf}.\end{aligned} \tag{15}$$

In Figure 2, and Equation (15), the superscript $(\cdot)^{bf}$ stands for bandwith-filtered, and the corresponding signals have undergone low-pass and (for pitch- and yaw-rate) washout filtering.

In Equation (15), according to the logics in Figure 2, a purely proportional controller has been implemented on the yaw axis stabilization system, whereas proportional-integral controllers have been envisaged for the pitch and roll axis. Despite being in contrast with the idea of producing only an increase in damping by means of stability augmentation loops, the adoption of small integral components in the feedback loop, based on $\int q^{bf}$ and $\int p^{bf}$ signals, has been found empirically to produce better stabilization results than the proportional control alone. In particular, a value of k_I^p two orders of magnitude below the corresponding proportional gain k_P^p has been considered for roll.

In the baseline configuration featuring aerodynamic surfaces, no thrust control has been considered for stabilization.

3.2. Thrust-Based Stabilization

To achieve a sufficient degree of controllability for stabilization on an airship with thrust control only and no thrust vectoring (i.e., thrust lines not changing attitude with respect to the body reference), a preliminary design of the layout of the thrusters on board needs to be carried out with this goal in mind. As explained in a dedicated publication [10], several configurations can produce a level of controllability such as to allow airship stabilization. A four-thrusters configuration, emerging as the most advantageous one, based on a minimum-energy measure of performance, has been considered in this work as a conceptual test bed, and employed at first for studying the portability of the stabilization laws introduced in Section 3.1 for the baseline airship featuring aerodynamic controls.

It should be stressed that improper placement of the thrusters (e.g., a configuration where all thrusters feature parallel thrust lines, or are all located on the plane of symmetry of the airship [10]) may hamper controllability from the start. Therefore, the definition of the thruster configuration is an essential preliminary step towards a feasible thrust-based control.

The considered conceptual layout of a four-thrusters airship is shown in Figure 3. The thrusters are numbered from #1 to #4, looking at the airship from astern, starting from #1 at the bottom-starboard, and proceeding in a clockwise direction.

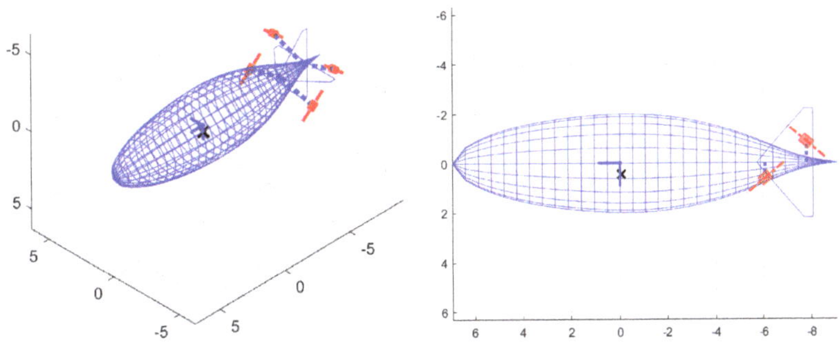

Figure 3. Conceptual layout of a four-thrusters airship with no movable surfaces and no thrust vector control. (**Left**) three-quarters view. (**Right**) view from the port side of the airship.

Considering the modeling approach in Equation (11), the bottom thrusters (#1 and #2) feature a positive tilt $\lambda_1 = \lambda_2 > 0$ (i.e., they are tilted downwards with respect to the longitudinal body axis of the airship), whereas the top thrusters, #3 and #4, feature a negative tilt, $\lambda_3 = \lambda_4 < 0$. The lateral misalignment angle σ_i is null for all thrusters.

In the layout considered in Figure 3, a positive thrust would suffice to obtain control around all three axes. However, hypothesizing the ability of the system to produce both positive and negative thrust, albeit the latter with a limited efficiency as outlined in Equation (12), better stabilization results can be obtained [10], generally requiring less energy to produce an increasingly quickly converging dynamic response. Despite being hardly achievable with piston engines, reversible thrust is more technically viable when considering electric motors driving purpose-designed propellers, or ion thrusters.

The process of porting a control law on an airship with aerodynamic control surfaces to a thrust-controlled one is based on the use of thruster combinations to obtain a response similar to that of the baseline airship. The definition of these specific combinations is intimately related to the adopted thruster layout (i.e., the number of thrusters, their positioning on board, and the direction of the thrust lines).

To show this, considering the concept in Figure 3, an effect around the pitch axis can be obtained by changing the thrust of bottom thrusters (#1 and #2) and top thrusters

(#3 and #4) in an opposing way, i.e., bottom thrusters pushing forward and top thrusters pulling back for a positive pitch-rate, and vice versa. Similarly, an effect around the yaw axis can be obtained with the starboard thrusters (#1 and #4) and port thrusters (#2 and #3) operating in an opposing way, i.e., port thrusters pushing forward and starboard thrusters pulling back for a positive yaw-rate, and vice versa. Finally, control around the roll axis can be obtained by having thrusters #1 and #3 pushing and #2 and #4 pulling for a positive roll rate and vice versa.

These coordinated control actions should be seen as components, contributing to the overall control input from each thruster. The latter is also able to produce the thrust needed for static equilibrium in forward flight, or for carrying out specific maneuvers as required by the guidance system (see Section 4).

The scheme of the proposed controller is presented in Figure 4. All three stabilization loops for pitch, roll, and yaw axes have been migrated from the baseline case with aerodynamic surfaces (Figure 2).

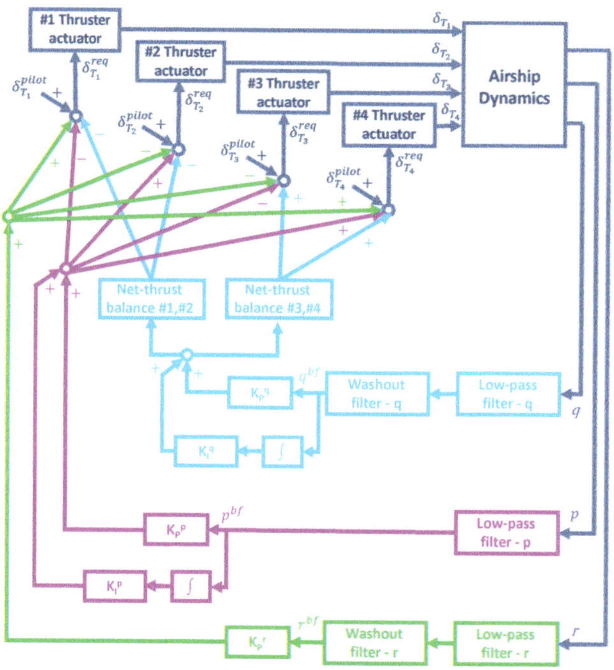

Figure 4. Stability augmentation scheme proposed for the four-thrusters layout concept introduced in Figure 3. Same color scheme as for Figure 2.

The actual signs of the stabilizing contributions reflect the way of reasoning mentioned above. To ease comparison with Figure 2, the same color code has been adopted for the corresponding stabilization loops in Figure 4.

A remark on the scheme in Figure 4 is related to the net thrust balance blocks. These are implemented for a better-balanced control around the pitch axis, and correspond to a normalization of the block input signal by the cosine of the tilt angle of the corresponding thrusters (i.e., multiplication by $1/\cos\lambda_i$). This feature has been implemented to avoid putting a non-null net thrust contribution in the direction of the longitudinal axis of the airship. Due to asymmetry in positive and negative thrust (modeled through the scaling of thrust by $\tilde{K}_i(\delta_{T_i})$ in Equation (12)), a net longitudinal thrust may also result when governing the thrusters to provoke a roll motion (i.e., when trying to reproduce the behavior of the ailerons). However, this has been found in practice to produce a mild effect at least for the proposed four-thrusters layout and control combination.

In analytic terms, the input to each thruster actuator in the considered conceptual four-thrusters layout with no thrust vectoring can be written as follows:

$$
\begin{aligned}
\delta_{T_1}^{req} &= \delta_{T1}^{pilot} - \left(k_P^q q^{bf} + k_I^q \int q^{bf}\right)\frac{1}{\cos\lambda_1} - \left(k_P^p p^{bf} + k_I^p \int p^{bf}\right) + k_P^r r^{bf},\\
\delta_{T_2}^{req} &= \delta_{T2}^{pilot} - \left(k_P^q q^{bf} + k_I^q \int q^{bf}\right)\frac{1}{\cos\lambda_2} + \left(k_P^p p^{bf} + k_I^p \int p^{bf}\right) - k_P^r r^{bf},\\
\delta_{T_3}^{req} &= \delta_{T3}^{pilot} + \left(k_P^q q^{bf} + k_I^q \int q^{bf}\right)\frac{1}{\cos\lambda_3} - \left(k_P^p p^{bf} + k_I^p \int p^{bf}\right) - k_P^r r^{bf},\\
\delta_{T_4}^{req} &= \delta_{T4}^{pilot} + \left(k_P^q q^{bf} + k_I^q \int q^{bf}\right)\frac{1}{\cos\lambda_4} + \left(k_P^p p^{bf} + k_I^p \int p^{bf}\right) + k_P^r r^{bf},
\end{aligned}
\tag{16}
$$

which reflects the logics portrayed in Figure 4.

4. Airship Guidance

For automatic guidance, the problem of following a pre-assigned set of checkpoints in 3D space, connected by straight legs, has been considered to be the scenario of interest. In that setting, both lateral and longitudinal guidance need to implement a beam-tracking logic. This compares well to the scenario of aircraft in approach making use of an ILS, with lateral and glide-slope guidance, which has provided some inspiration for this work [21]. However, coping with the specific dynamics of an airship, different from those of fixed-wing aircraft, especially in terms of turn maneuvers, climbs, and descents in the longitudinal plane, some features of the proposed guidance controller differ from those most typically found on winged aircraft for automatic approach.

The logic implemented will be introduced at first for the more intuitive case of the baseline airship with movable aerodynamic surfaces. Then, the porting of the guidance laws on the four-thrusters conceptual layout will be described.

It should be remarked that the dynamics of the airship to which guidance is applied are artificially stabilized by means of the stability augmentation system described in Section 3. The input to the airship system required by guidance will enter the schemes in Figures 2 and 4 through the $\delta_{(\cdot)}^{pilot}$ signals. Due to the unstable or marginally stable free response of most airships, guidance may not be successfully pursued without employing such an inner stabilization layer.

4.1. Guidance in the Longitudinal Plane

Longitudinal guidance on the baseline airship is carried out by the simultaneous employment of elevator and thrust. Similar to winged aircraft, beam-tracking in the longitudinal plane is not achievable without simultaneously controlling airspeed along the trajectory, i.e., the modulus $v_{CB} = |\boldsymbol{v}_{CB}|$, and the rate of climb. The latter is analytically defined as $RC = -\boldsymbol{v}_{CB} \cdot \boldsymbol{i}_3$, where $\boldsymbol{i}_3$ is the third unit vector in the inertial frame $\mathcal{I}$, which according to the sketches in Figure 5 is pointing to the ground.

A first loop has been closed on $|\boldsymbol{v}_{CB}|$, where the thrust setting is changed according to a proportional law trying to keep this variable close to an assigned set-point v_{CB}^*. An integrator is put on the forward branch, so as to allow the guidance controller to trim the airship to a steady state condition. Introducing the error $e_S = |\boldsymbol{v}_{CB}| - v_{CB}^*$, the guidance law for thrust writes

$$
\delta_T^{pilot} = k_P^{e_S} e_S + \delta_T^{pilot,last}, \tag{17}
$$

where superscript $(\cdot)^{pilot}$ identifies the guidance control requirement from the automatic controller.

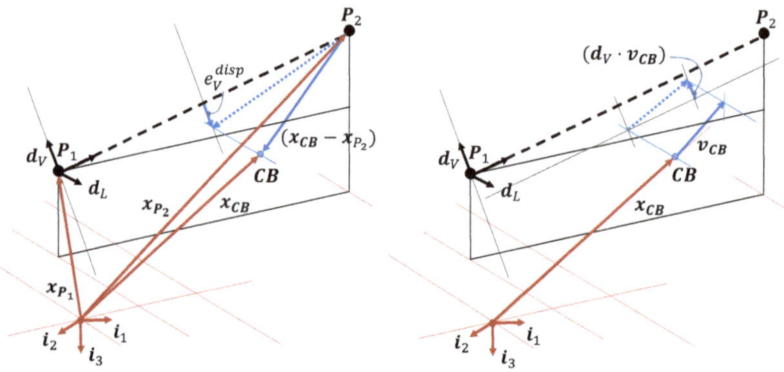

Figure 5. Sketch of beam-tracking measurements in the longitudinal plane. (**Left**) definition of e_V^{disp}. (**Right**) computation of $d_V \cdot v_{CB}$.

Elevator control is employed for beam-tracking, and its value is computed proportionally to two error components.

1. a distance between the airship CB and the beam, computed in the vertical plane containing the beam, and in a perpendicular direction with respect to the beam. Analytically, considering the unit vector d_V in Figure 5, which is defined as normal to the rectilinear beam going from P_1 to P_2 and contained in the vertical plane where the target beam is, this distance is easily measured as $e_V^{disp} = d_V \cdot (x_{CB} - x_{P_2})$, where clearly the reference for this distance is taken as null. By definition, e_V^{disp} will be positive when the airship is above the track, and negative vice versa.
2. the error e_V^{vel} between the velocity component of the reference point v_{CB} along the same cross-beam vertical unit vector d_V just introduced and a reference value v_v^* of the cross-beam speed. This reference value is obtained as a bounded linear function of the cross-beam displacement error e_V^{disp}, such that

$$v_v^* = \begin{cases} v_V^{*,top}, & e_V^{disp} < e_V^{disp,bot} \\ \frac{e_V^{disp}}{e_V^{disp,bot}} \cdot v_V^{*,top}, & e_V^{disp,bot} \leq e_V^{disp} < 0 \\ \frac{e_V^{disp}}{e_V^{disp,top}} \cdot v_V^{*,bot}, & 0 \leq e_V^{disp} < e_V^{disp,top} \\ v_V^{*,bot}, & e_V^{disp} \geq e_V^{disp,top} \end{cases} . \tag{18}$$

In Equation (18) the boundary values $v_V^{*,top}$, $v_V^{*,bot}$, $e_V^{disp,top}$ and $e_V^{disp,bot}$ are tunable parameters, similar to control gains. In particular, top values will be positive in sign, whereas bottom values will be negative. Based on Equation (18), the value of the error on cross-beam vertical speed can be defined as $e_V^{vel} = d_V \cdot v_{CB} - v_v^*$.

In Figure 5, two sketches help to intuitively capture the measures just introduced for longitudinal beam-tracking.

Based on the two errors just introduced, the longitudinal beam-tracking control law for the elevator can be assigned as in Equation (19), which yields

$$\delta_e^{pilot} = k_P^{e_V^{disp}} e_V^{disp} + k_P^{e_V^{vel}} e_V^{vel} + \delta_e^{pilot,last}. \tag{19}$$

Similar to Equation (17), an integrator is put on the forward branch of the scheme in Equation (19).

The proposed scheme for the guidance system for the baseline airship is shown in Figure 6 in all its components, including airspeed and beam-tracking in the longitudinal plane (red and brown, respectively). Low-pass filters are included as customary when dealing with potentially noisy measures (albeit not treated in this paper, turbulence may produce a polluting effect on measure signals).

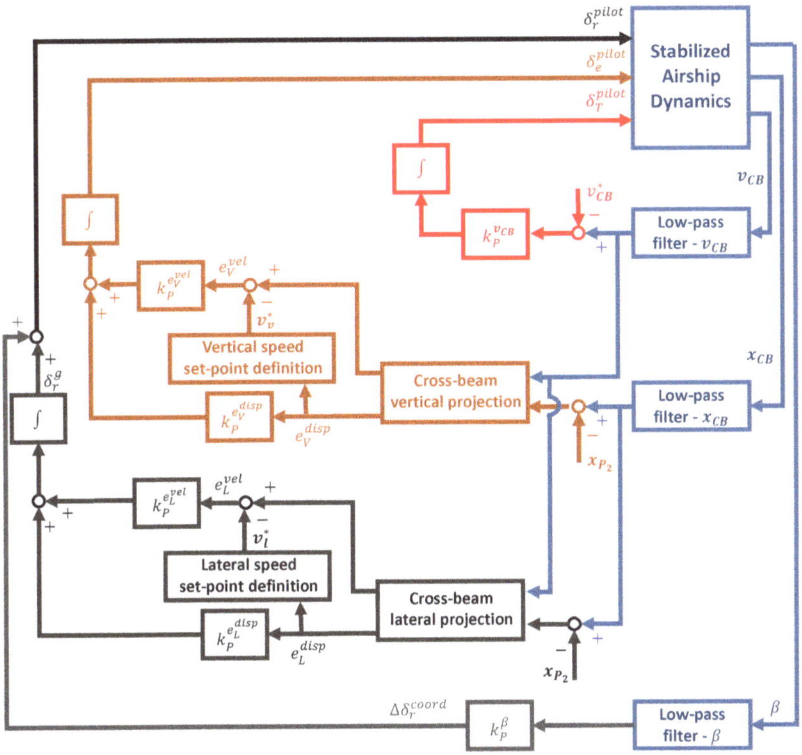

Figure 6. Proposed guidance control for the baseline airship with deflectable aerodynamic surfaces and a single thruster. Red: airspeed tracking. Brown: longitudinal beam-tracking. Grey: lateral beam-tracking with turn coordination.

4.2. Lateral-Directional Guidance

Guidance in the horizontal plane has been designed based on two cooperating loops, namely for turn coordination and for beam-tracking.

Turn coordination has been implemented targeting the sideslip angle in the body frame, which is pushed toward zero by an action of the rudder. In analytical terms, the control law is, correspondingly,

$$\Delta\delta_r^{coord} = k_P^{\beta}\beta, \tag{20}$$

where β is the sideslip angle.

From a conceptual standpoint, lateral beam-tracking is performed in a similar way with respect to longitudinal beam-tracking (see Figure 7). A lateral displacement $e_L^{disp} = \boldsymbol{d}_L \cdot (\boldsymbol{x}_{CB} - \boldsymbol{x}_{P_2})$ is introduced, based on the unit vector $\boldsymbol{d}_L$. The latter is normal to the vertical plane where the beam to be tracked is, and positive to the right of the trajectory leading to $\boldsymbol{P}_2$. A lateral speed error is defined as $e_L^{vel} = \boldsymbol{d}_L \cdot \boldsymbol{v}_{CB} - v_l^*$, where v_l^* is a target lateral speed value, defined in a totally similar fashion to Equation (18), based on the current value of e_L^{disp} and a set of four assigned parameters $v_L^{*,top}$, $v_L^{*,bot}$, $e_L^{disp,top}$ and $e_L^{disp,bot}$. For clarity, a positive value of e_L^{disp} is obtained when flying to the right of the beam, therefore producing a negative v_l^* setting, which would push the airship towards the desired trajectory. The opposite happens for a negative e_L^{disp}.

The measures required for lateral beam-tracking are graphically explained in the sketches in Figure 7.

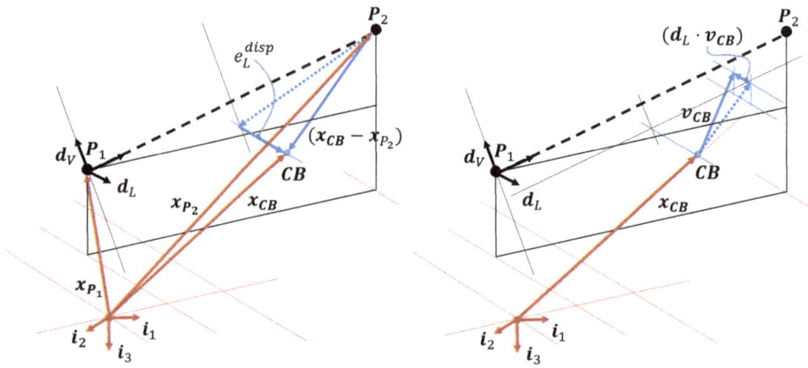

Figure 7. Sketch of beam-tracking measurements for lateral guidance. (**Left**) definition of e_L^{disp}. (**Right**) computation of $d_L \cdot v_{CB}$.

Only rudder control is directly employed to steer the airship for lateral guidance. The corresponding guidance law for lateral beam-tracking writes

$$\delta_r^g = k_P^{e_L^{disp}} e_L^{disp} + k_P^{e_L^{vel}} e_L^{vel} + \delta_r^{g,last}. \tag{21}$$

Similar to Equations (12) and (19), the control law in Equation (21) includes an integrator, which allows obtaining a non-null control input when the airship is on track, thus providing the controller with flight trimming capability.

The overall automatic guidance control on the rudder finally writes

$$\delta_r^{pilot} = \delta_r^g + \Delta\delta_r^{coord}. \tag{22}$$

The scheme in Figure 6 synthetically portrays the complete proposed guidance system for the baseline airship.

4.3. *Thrust-Based Guidance: Navigation of a Four-Thrusters Airship with No Aerodynamic and Thrust Vector Control*

A process similar to that considered for porting the stabilization system laws designed for the baseline airship with aerodynamic surfaces to the four-thrusters concept has also been envisaged for the guidance system (see Section 3.2).

Considering airspeed tracking, all thrusters contribute towards this goal, and their contribution is weighed on account of their respective tilt angles, so as to produce a resultant change in thrust only in the direction of the longitudinal body axis of the airship. This implies dividing the control input bound to speed, as shown in Equation (17), by $\sin \lambda_i$.

Concerning beam-tracking in the longitudinal plane, a straightforward possibility is offered by adopting the same philosophy considered for porting the pitch-rate stabilization algorithm from the baseline airship to the thrust-controlled concept. This requires reproducing the effect of the elevator with the differential use of the top and bottom thrusters (see Section 3.2). In the process, weighing the corresponding thrust setting inputs by $\cos \lambda_i$ would allow avoiding a change in the thrust component along the longitudinal axis of the airship.

This solution, however, was found to produce an excess of authority of the guidance law around the pitch axis, potentially resulting in interference between the stabilization system and the longitudinal guidance, making the respective tuning of the pitch-rate damper—including its washout filters and gains—and of the longitudinal beam-tracking system delicate and difficult.

An alternative solution has been found in employing longitudinal beam-tracking, with either the bottom thrusters (#1 and #2) or the top ones (#3 and #4), employing positive thrust, respectively, to obtain a positive or negative pitch-rate. Clearly, in this way, a contribution of thrust in the direction of the longitudinal axis is unavoidable and will tend to alter the velocity of the airship. However, it was found that the additional burden on the speed-tracking system can be managed by the collective use of all thrusters as explained above in this section, whereas interference between the pitch-rate stabilizer and the longitudinal guidance system is somewhat reduced, thus allowing for an easier tuning of both control layers.

Lateral beam-tracking is more straightforwardly ported from the baseline case to the thrust-based case, employing the thrusters differentially to obtain an effect similar to that of the rudder (see Section 3.2).

Similarly, the correction to rudder control to achieve turn coordination has been ported in exactly the same way as for lateral beam-tracking guidance.

In analytical terms, this produces the following guidance law for each thruster

$$\begin{aligned}
\delta^g_{T_1} &= +\left(k_P^{e_S} e_S \frac{1}{\sin\lambda_1}\right) - \left(k_P^{e_V^{disp}} e_V^{disp} h_{1,2}^{disp} + k_P^{e_V^{vel}} e_V^{vel} h_{1,2}^{vel}\right) + \left(k_P^{e_L^{disp}} e_L^{disp} + k_P^{e_L^{vel}} e_L^{vel}\right) + \delta^{g,last}_{T_1},\\
\delta^g_{T_2} &= +\left(k_P^{e_S} e_S \frac{1}{\sin\lambda_2}\right) - \left(k_P^{e_V^{disp}} e_V^{disp} h_{1,2}^{disp} + k_P^{e_V^{vel}} e_V^{vel} h_{1,2}^{vel}\right) - \left(k_P^{e_L^{disp}} e_L^{disp} + k_P^{e_L^{vel}} e_L^{vel}\right) + \delta^{g,last}_{T_2},\\
\delta^g_{T_3} &= -\left(k_P^{e_S} e_S \frac{1}{\sin\lambda_3}\right) + \left(k_P^{e_V^{disp}} e_V^{disp} h_{3,4}^{disp} + k_P^{e_V^{vel}} e_V^{vel} h_{3,4}^{vel}\right) - \left(k_P^{e_L^{disp}} e_L^{disp} + k_P^{e_L^{vel}} e_L^{vel}\right) + \delta^{g,last}_{T_3},\\
\delta^g_{T_4} &= -\left(k_P^{e_S} e_S \frac{1}{\sin\lambda_4}\right) + \left(k_P^{e_V^{disp}} e_V^{disp} h_{3,4}^{disp} + k_P^{e_V^{vel}} e_V^{vel} h_{3,4}^{vel}\right) + \left(k_P^{e_L^{disp}} e_L^{disp} + k_P^{e_L^{vel}} e_L^{vel}\right) + \delta^{g,last}_{T_4},
\end{aligned} \tag{23}$$

where an integrator has been included, similar to the baseline case, to add a trimming capability to the guidance system. In Equation (23), functions $h_{1,2}^{(\cdot)}$ and $h_{3,4}^{(\cdot)}$ take a logical value of 0 or 1, based on the following definition

$$\begin{aligned}
h_{1,2}^{(\cdot)} &= \begin{cases} 1, & e_V^{(\cdot)} < 0 \\ 0, & \text{otherwise} \end{cases},\\
h_{3,4}^{(\cdot)} &= \begin{cases} 0, & e_V^{(\cdot)} < 0 \\ 1, & \text{otherwise} \end{cases},
\end{aligned} \tag{24}$$

and are responsible for the alternative activation of top and bottom thrusters for a required negative or positive pitch rate, respectively. The definitions in Equation (24) hold for both errors on displacement and velocity (to be alternatively put in the empty superscript placeholders).

The control inputs in Equation (23) are finally summed to a lateral contribution for turn coordination, so that the final input to the thrusters writes as in Equation (25)

$$\begin{aligned}
\delta^{pilot}_{T_1} &= \delta^g_{T_1} + k_P^{\beta}\beta,\\
\delta^{pilot}_{T_2} &= \delta^g_{T_2} - k_P^{\beta}\beta,\\
\delta^{pilot}_{T_3} &= \delta^g_{T_3} - k_P^{\beta}\beta,\\
\delta^{pilot}_{T_4} &= \delta^g_{T_4} + k_P^{\beta}\beta.
\end{aligned} \tag{25}$$

The corresponding scheme for guidance control for the thrust-based concept is represented in Figure 8.

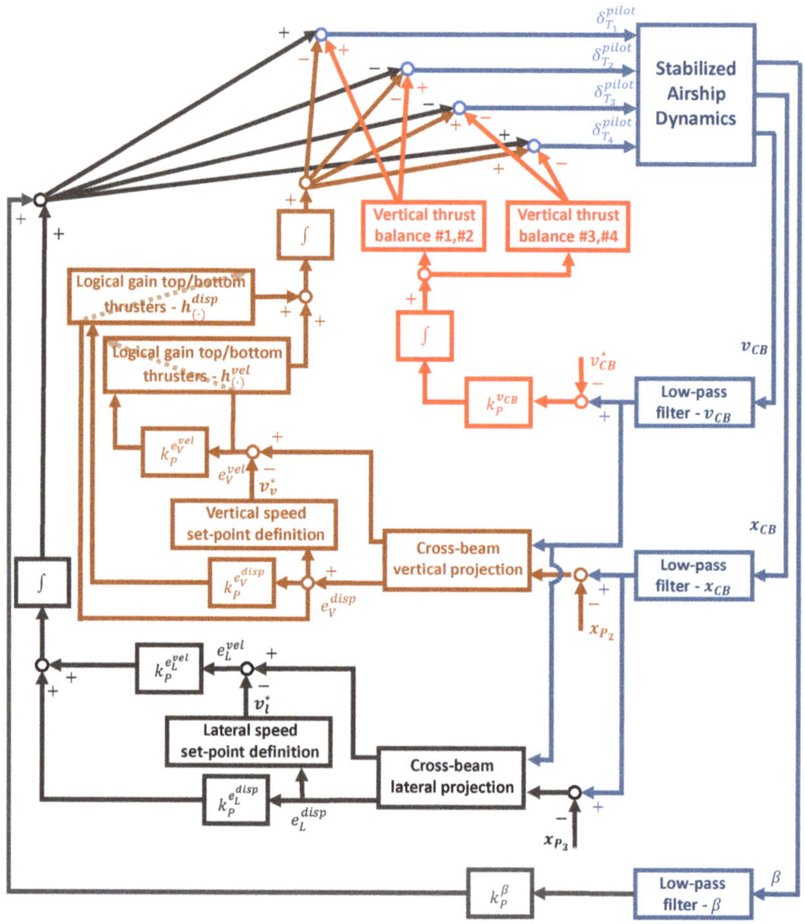

Figure 8. Proposed guidance control for the four-thrusters concept. Same color codes as for Figure 6.

5. Application Results

In this section, some example case studies will be presented, to assess the capabilities of the proposed stabilization and guidance systems, both in the baseline case of an airship with movable aerodynamic control surfaces, and in the case of a four-thrusters concept.

The baseline test bed is that of the Lotte airship [9,11]. Already considered to be a test-case in previous studies [10], a virtual model of the airship was implemented for the present research in `SILCROAD` (*Si*mulation *L*ibrary for *Cr*aft-*O*bject *A*dvanced *D*ynamics), a novel object-oriented library developed in `Matlab (R2019b)`® at the Department of Aerospace Science and Technology (DAER), Politecnico di Milano, to accurately simulate or co-simulate (in the case of interacting objects in the same scenario) the non-linear response of several machine types (generically named craft), subject to aerodynamic, gravity, buoyancy, and thrust forcing terms. Among the sub-classes in the library is the airship class, which was employed for the fully non-linear simulations presented in this work. This highly customizable library lends itself to multiple uses, including the systematic design, implementation, and testing of control laws applied for instance to the dynamic models of winged aircraft and airships (as well as submarines or torpedoes).

Previously considered in the methodological part of this text as a baseline, the Lotte experimental airship features a single engine astern of the envelope and a cruciform tail with vertical stabilizer and rudder, horizontal stabilizer, and elevons [11].

The four-thrusters concept, implemented in this work again employing the SILCROAD library, reproduces the layout obtained from a previous study [10], which emerged from an optimal approach in the arrangement of thrusters on board, based on an energy measure of performance, including control use and response to disturbance. Concerning aerodynamics, the envelope shape and size are the same as the Lotte airship, whereas the entire area of the tail surfaces has been considered fixed, thus removing the control degrees of freedom for elevons or rudder motion. The rear thruster of the baseline airship has been removed as well, and replaced by four thrusters arranged as in Figure 3. The mass of the actuation systems and rear thruster has been estimated and removed, whereas the mass for the four new thrusters has been added. Correspondingly, a slight forward shift of the CG (by 5.9 mm) was observed, as well as an increase in mass of 3 kg [10], with respect to the original Lotte design.

The basic data of the two airships are reported in Table 1.

Table 1. Reference data for the considered airship test beds.

Parameter	Baseline Airship	4-Thrusters Airship Concept
Mass (kg)	134.28	137.28
Envelope volume (m^3)	107.42	107.42
Overall length (m)	16.0	16.0
Hor. disp. CG from CB (mm)	0	5.9
Ver. disp. CG from CB (mm)	455.0	455.0
Number of thrusters	1	4
Nominal thrust (each unit) (N)	500	250
Aerodynamic control	Rudder, elevons	(no aerodynamic control)

The following subsections illustrate the behavior and quality of the response obtained at first from the stabilization system alone, and second from the guidance system working in conjunction with the artificially stabilized airships. Showing the functions of the stabilization system, working alone in a simplified scenario, has been deemed useful to better appreciate the features of this controller, which is pivotal in granting good flying qualities in navigation problems. Test cases for guidance are more elaborated, including climbs, steep turns, and multi-checkpoint navigation along a circuit, in still air, or in the presence of constant wind. According to the schemes in Figures 6 and 8, these navigation test cases are based on the employment of the artificially stabilized system.

5.1. Stability Augmentation System: Response to Perturbation

The goal of the stabilization system for both configurations is that of reducing the amplitude of oscillations of the rotational rates p, q, and r, in a bandwidth where the control actuators can operate and do not interfere with guidance control.

For both the baseline and four-thrusters concept configurations, the bandwidth where the stabilizing controller may operate is selected through a set of low-pass filters and washout filters (see Figures 2 and 4).

Since the trials have been carried out in a turbulence- and signal-noise-free virtual testing environment, the effect (and distortion) due to the low-pass filter, implemented as a first-order system, is minimal, and mostly felt in terms of a mild lag in the filter output signal. The time for the three low-pass filters on p, q, and r has been set to a value of $T^{lp} = 0.005$ s.

The washout filter instead is significantly impacting the synthesis of the control input and the response, in particular allowing residual almost-steady or slowly varying components in the target signals (i.e., rotational rates) to survive. This is a positive effect, needed in order to provide for a sufficient degree of maneuverability for the guidance system to operate. In this sense, the tuning of the washout filters requires some experimentation, to satisfactorily weigh the needs for a fast and effective stabilization (which is obtained by extending the bandwidth where the stabilizing controller is operating, thus reducing the cutoff frequency of the washout filter) and for an effective guidance (which would benefit from a an increased cutoff frequency, for instance in aggressive maneuvers). The results proposed herein have been obtained with a time constant of the single-state washout filter of $T^{wo} = 5$ s.

For the electro-mechanical actuators controlling the motion of the aerodynamic surfaces on the baseline airship, corresponding first-order system actuators have been considered in the simulation environment, with a time constant $T^{act} = 0.03$ s, and boundary deflection values of $+/-25$ degree for both the rudder and elevons.

For thrusters, an assembly of a propeller driven by an electric motor was considered conceptually. A first-order system with a time constant $T^{thr} = 0.001$ s has been considered, on account of a generally fast response of electric motors. Referring to Equation (12), the output thrust of thrusters saturates at the value of thrust reached for $\delta_{T_i} = +/-1$, and is modulated by the input δ_{T_i} through a value $\tilde{K}_i(\delta_{T_i})$ defined as follows [10]

$$\tilde{K}_i(\delta_{T_i}) = \begin{cases} 1.0, & \delta_{T_i} \geq 0 \\ -0.5, & \delta_{T_i} < 0 \end{cases}, \tag{26}$$

where the modulation in Equation (26) was employed to coarsely model the loss in efficiency of a propeller working in the opposite direction with respect to the intended one. The nominal thrust for the sole propeller in the baseline configuration has been set to $\tilde{T}_1 = 500$ N, whereas for the thrusters on the four-thrusters concept $\tilde{T}_{1\rightarrow 4} = 250$ N.

In Figure 9, the time response of the outputs and controls of the baseline airship with stabilization is presented, starting from a trimmed condition in symmetric horizontal flight, at a ground speed of 8 m/s, and considering an initial perturbation of $\Delta W = \Delta V = 0.5$ m/s. The black dash-dotted lines represent the trimmed values of states and controls. The guidance system is switched off in this test.

Observing the evolution of control in Figure 9, it is apparent that the perturbation is contrasted by the controller within the considered time frame, i.e., the control setting returns to the trimmed values by the end of the simulation. Correspondingly, the rotational rates q, p, and r are reduced to zero or a near-zero condition, whereas residual unbalances to the speed components (especially U and W) are still visible.

The integral terms on q and p are interestingly left to non-null values, due to the low value attributed to k_I^q and k_I^p (see Equation (15)). The latter is one order of magnitude below the corresponding proportional gain k_P^q for the pitch rate and two orders of magnitude below k_P^p for the roll rate. However, it was observed that the integral components of stabilization control, albeit residual (not visible to the naked eye from the control plots in Figure 9, where it appears that control is returning to the trimmed condition, as observed), ameliorate the response in terms of stability, especially in guidance problems.

As anticipated, it can be observed that the stabilization system implemented in this work, due to the presence of washout filters on the feedback line, allows the persistence of constant or slowly varying rates, which in turn produce a slow divergence from the original state. In Figure 9, the pitch angle ϑ is not directly controlled, and remains relatively far from the original trimmed value at the end of the considered time frame. It should be underlined that the proposed controller is mostly aimed at increasing damping around the body axes, without any automatic piloting function, taken over by the guidance control layer. As a result, the flight corresponding to the proposed test will not be straight, and an expected drift in the direction of the velocity vector was observed (not shown in the plots).

In Figure 10 the same scenario just analyzed for the baseline airship in Figure 9 is proposed for the four-thrusters concept. Notably, the guidance system is off in this testing scenario.

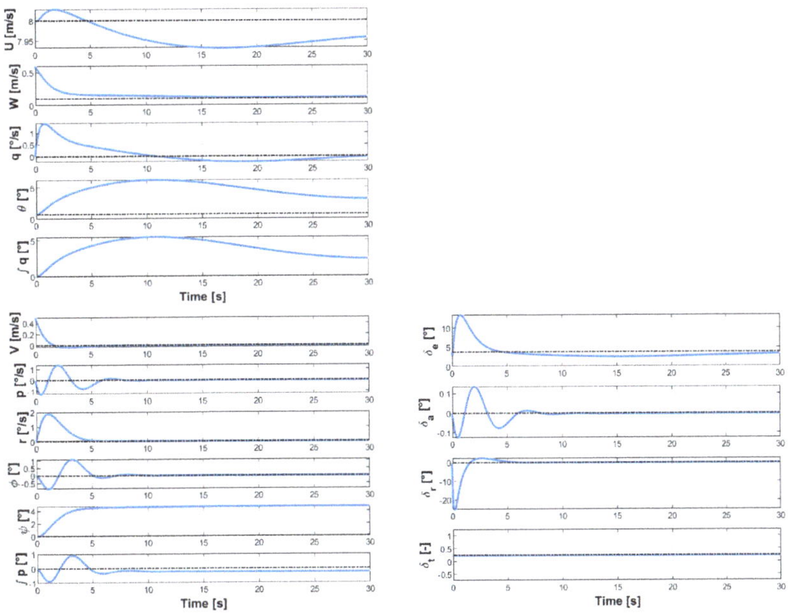

Figure 9. Time response of baseline airship with stabilizing control, in trimmed horizontal flight at a ground speed of 8 m/s, to an initial perturbation of $\Delta W = \Delta V = 0.5$ m/s. (**Top-left**) and (**bottom-left**) longitudinal and lateral-directional states. (**Bottom-right**) controls. Black dash-dotted lines: trimmed value.

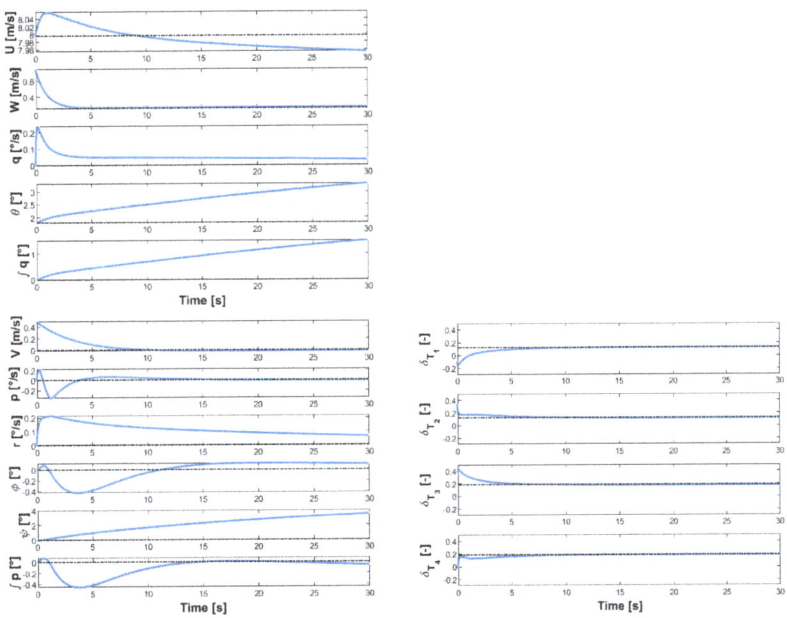

Figure 10. Time response of four-thrusters airship with thrust-based stabilizing control, in trimmed horizontal flight at a ground speed of 8 m/s, to an initial perturbation of $\Delta W = \Delta V = 0.5$ m/s. (**Top-left**) and (**bottom-left**) longitudinal and lateral-directional states. (**Bottom-right**) controls. Black dash-dotted lines: trimmed value.

From the test in Figure 10, as expected [10], the proposed layout of the thrusters and the adopted control law are capable of stabilizing the system in a similar way to aerodynamic surfaces on the baseline airship, and with comparable effectiveness.

Interestingly, except for an initial peak on #1 thruster (bottom-starboard), no negative absolute thrust is employed to react to the perturbation. Actually, the non-negligible positive trimmed value of the thrust setting δ_{T_i}, even summed to the stabilization control contribution, keeps the total to a globally positive figure.

Looking at the pitch-rate signal q, it is apparent that a steady-state value of the latter is left over by the control action, due to the washout filtering effect. Correspondingly, a slow departure from the trimmed reference is observed especially in terms of the pitch attitude ϑ in the response in Figure 10. As observed for the baseline airship and controller, this is expected from the proposed stabilization system, which is tasked with increasing the damping around the body axes of the airship, leaving the task of steering or trimming the airship, as well as the control margin to carry out this task at a lower frequency, to the guidance system. In a similar fashion, the value of the roll angle φ is not returned to its null trimmed value. Instead, a non-null roll-rate p in the initial phase, even though damped over time, still produces a non-null steady-state value of this attitude angle.

5.2. *Performance of the Guidance Control System: Examples of Navigation*

The tuning of the guidance system requires the assignment of the gains in Equations (17) and (19)–(21) for the baseline airship controlled via aerodynamic surfaces, and of those in Equations (23) and (25) for the thrust-based guidance controller on the four-thrusters concept. Additionally, the settings of the vertical and lateral speed set-point selectors, shown in Equation (18) for the vertical rate, namely $v_V^{*,top}$, $v_V^{*,bot}$, $e_V^{disp,top}$ and $e_V^{disp,bot}$ for v_V^*, as well as $v_L^{*,top}$, $v_L^{*,bot}$, $e_L^{disp,top}$ and $e_L^{disp,bot}$ for v_L^*, need to be assigned. In this study, for both the vertical and lateral speed set-point selection, the top and bottom errors have been set to $e_{(\cdot)}^{disp,top} = 10$ m, $e_{(\cdot)}^{disp,bot} = -10$ m, whereas the top and bottom set-point values have been set to $v_{(\cdot)}^{*,top} = 1$ m/s, $v_{(\cdot)}^{*,bot} = -1$ m/s, respectively.

Considering the task for which the automatic guidance system has been designed, i.e., beam-tracking, three scenarios for testing have been envisaged. These are:

1. following a straight climbing track, ascending 20 m over a horizontal length of 200 m
2. following a 200 m long horizontal track, misaligned with respect to the initial velocity of the airship by 60 degree
3. following a six-checkpoint path, with checkpoints forming a regular hexagon viewed from top (60 degree angles, side length of 200 m), and with an altitude of alternatively 0 m or 10 m above ground, starting from 0 m at the beginning of the circuit.

In all considered cases, the airship is at the initial checkpoint at the beginning of the simulation, in a trimmed horizontal flight condition. Checkpoint capture is defined as entering a spherical space around the checkpoint, with a radius of 10 m. In the multi-checkpoint scenario (3.), the guidance system automatically switches to the next checkpoint in the flight plan on reaching a plane normal to the current desired track and including the current target checkpoint. This happens irrespective of the successful capture of the current target checkpoint.

5.2.1. Ascending Track

The trajectory resulting from the application of the proposed guidance law in the case of an ascending track for the baseline airship is presented in Figure 11. The considered starting point is a trimmed horizontal flight at a ground speed of 6 m/s. The same value has been selected as ground speed set-point v_{CB}^* for the speed-tracking loop.

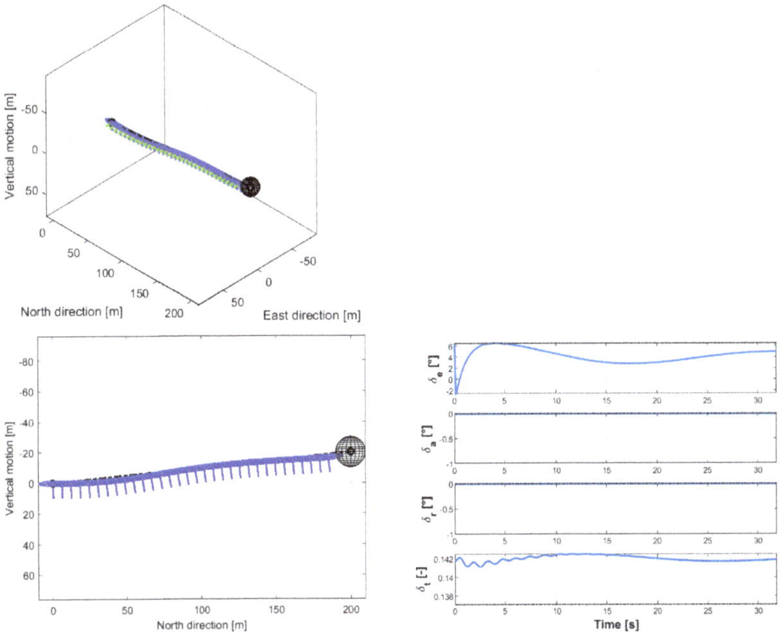

Figure 11. Trajectory and control behavior for an ascending track, as flown by the stabilized baseline airship with the proposed guidance control system. Ground speed setting 6 m/s. (**Top-left**) three-quarter view of the 3D trajectory from the northeast. (**Bottom-left**) view of the trajectory from the east. Black dash-dotted lines: target track. Black spheres: target checkpoint capture volume. (**Bottom-right**) time histories of controls. Black dash-dotted lines: trimmed values.

At the start, the guidance system receives a sudden requirement for a change in altitude, which translates into the action on the elevator control δ_e, visible on the bottom-right plot in Figure 11. This triggers a mild reaction by the velocity-tracking thrust controller. The view from the east of the trajectory shows only mild altitude oscillations along the beam, and a successful capture of the target checkpoint. No event takes place on the lateral-directional plane in this totally symmetrical flight condition.

Looking at the direction of the body unit vectors (left plots in Figure 11), it can be seen that the attitude angles are well under control, showing that the guidance system effectively trims the airship, without an actual divergence of these signals. As observed in the previous section, Section 5.2, this task was excluded from those of the stabilization system, and it is satisfactorily managed by the guidance layer.

In Figure 12, the same scenario is flown by the thrust-controlled four-thrusters concept.

The flight trajectory (left plots in Figure 12) compares well with the one in Figure 11, featured for the baseline airship. This supports the good ability of the proposed thrust-based guidance controller in effectively steering the four-thrusters concept airship, achieving a behavior generally similar to that of a traditionally designed and controlled platform.

It is interesting to analyze the behavior of thrust controls (bottom-right plot in Figure 12) near the start. The black dash-dotted lines refer to the trimmed condition at the start. The sudden change in target altitude at the start of the simulation implies a peak in thrust from the bottom thrusters #1 and #2, whereas no similar action is found in the top thrusters #3 and #4, due to the asymmetry of the vertical guidance law explained in Section 4.3 (see Equations (23) and (24)). The asymmetric vertical law demand here tends to produce an unwanted increase in speed, which is countered by a fast reaction of all thrusters, as per Equation (23). Actually, this contribution can be checked in the top thrusters #3 and #4, which react by suddenly reducing to near-zero thrust at the initial instant, whereas on #1 and #2 the response of the speed-tracking and vertical guidance loops are mixed. This coupling between speed-tracking and beam-tracking control is generally observed with thrust-based control, and is expected to be more marked than for the baseline case with movable aerodynamic surfaces.

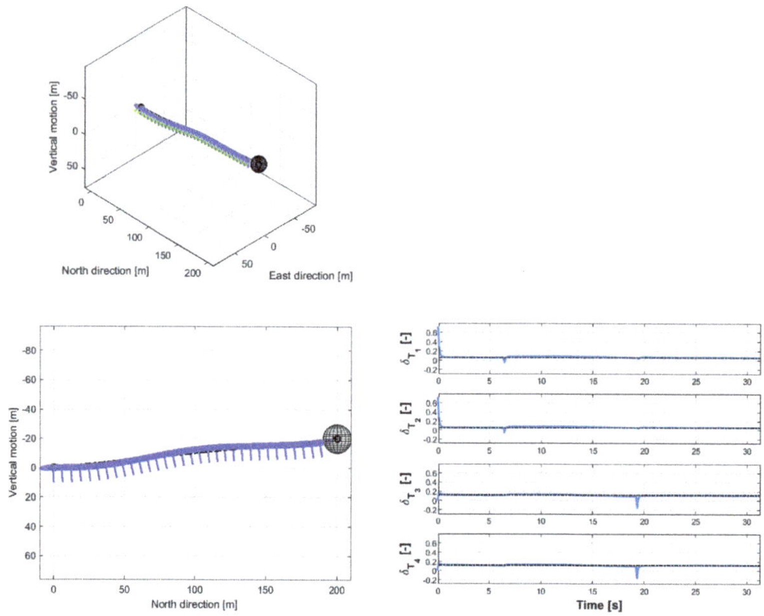

Figure 12. Trajectory and control behavior for an ascending track, as flown by the thrust-controlled, four-thrusters stabilized platform, with the proposed guidance control system. Ground speed setting 6 m/s. (**Top-left**) three-quarter view of the 3D trajectory from the northeast. (**Bottom-left**) view of the trajectory from the east. Black dash-dotted lines: target track. Black spheres: target checkpoint capture volume. (**Bottom-right**) time histories of controls. Black dash-dotted lines: trimmed values.

5.2.2. Misaligned Horizontal Track

In the next scenario ((2.) in the listing in the introduction to Section 5.2), the airship starts from the very same trimmed condition of the previous case (1.), with the only exception of an initial yaw misalignment $\Delta\psi_0 = 60$ degree with respect to the desired track. The track to follow is aligned in a north–south direction, and is horizontal (i.e., no altitude change required).

Figure 13 shows the trajectory and the time histories of controls for the baseline airship, considering an initial trimmed ground speed of 10 m/s, which is also the set-point v^*_{CB}.

As shown in Figure 13, as a result of this initial misalignment, the airship is driven by the guidance controller into a steep turn to port, associated with a brief initial time span where the rudder control is saturated. Following the turn, due to the choice of the gains for directional guidance, which weigh velocity error more than displacement error, the controller reacts to the misplacement with respect to the track by setting the cross-beam target airspeed to its extreme value, but without an aggressive correction of the position error, thus obtaining an almost straight trajectory, and nearing the target track only in close proximity to the target checkpoint.

Furthermore, the reduced aileron action (i.e., cross-action of the elevons) is noteworthy. This is due to the adopted laws, where aileron control is employed for roll stabilization and has no guidance function. The outcome of the test in this sense shows a reduced coupling between the roll and yaw motions on the airship, which is expected in the considered baseline configuration (and makes it rather different from winged aircraft in this respect). As a matter of fact, the turn in the trajectory shown in Figure 13 is basically a flat turn.

The plots in Figure 14 portray the trajectory and controls of the four-thrusters case for the same initial condition and target track as in Figure 13.

The trajectory obtained by the four-thrusters concept in this testing scenario is different from that flown by the baseline airship, and based on a smooth and sustained turn that takes almost the entire simulation, ending with the capture of the checkpoint. A lateral top misplacement of over 70 m (bottom-left plot in Figure 14) is significantly more pronounced than the lateral trajectory elongation of about 50 m found with the baseline airship. Notwithstanding a slightly more intense use of roll in the turn, reaching a steady angle of around $\varphi = -5$ degree for the four-thrusters configuration (not shown), this layout appears to manage this misalignment scenario with more difficulty than the baseline layout. It should be observed that the speed of 10 m/s (which in this still air test is both a ground speed and an airspeed) is significantly close to the design top speed value for the Lotte airship (12 m/s).

Looking at controls in Figure 14, the sudden request for a turn to the left produces a strong push of the starboard thrusters #1 and #4, and a corresponding pull (with intensely negative thrust settings) for port thrusters #2 and #3. Once the turn has been initiated, the mixed requests of the stabilization and guidance controllers evolve with a smooth time history, with very limited values of the controls (roughly similar to a trimmed forward flight) and limited control use (i.e., little oscillation).

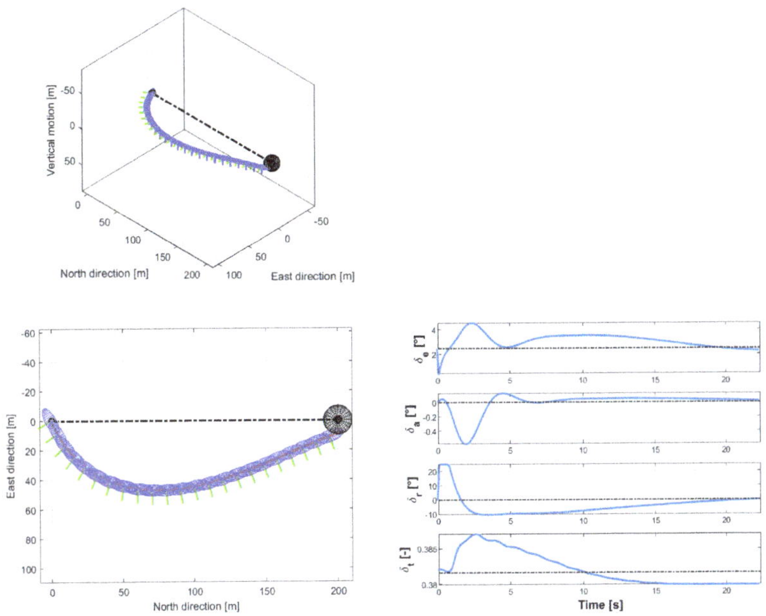

Figure 13. Trajectory and control behavior for an initial misalignment $\Delta\psi_0 = 60$ degree with respect to a horizontal track, as flown by the stabilized baseline airship with the proposed guidance control system, at 10 m/s of ground speed. (**Top-left**) three-quarter view of the 3D trajectory from the northeast. (**Bottom-left**) view of the trajectory from the top. Black dash-dotted lines: target track. Black spheres: target checkpoint capture volume. (**Bottom-right**) time histories of controls. Black dash-dotted lines: trimmed values.

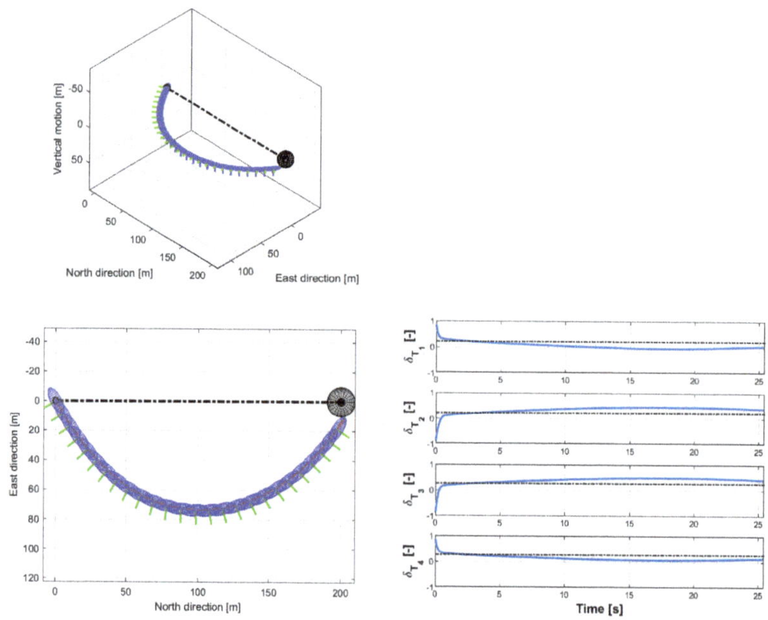

Figure 14. Trajectory and control behavior for an initial misalignment $\Delta\psi_0 = 60$ degree with respect to a horizontal track, as flown by the four-thrusters stabilized airship, with the thrust-based guidance control system, at 10 m/s of ground speed. (**Top-left**) three-quarter view of the 3D trajectory from the northeast. (**Bottom-left**) view of the trajectory from the top. Black dash-dotted lines: target track. Black spheres: target checkpoint capture volume. (**Bottom-right**) time histories of controls. Black dash-dotted lines: trimmed values.

5.2.3. Multi-Checkpoint Track with Variable Altitude

In the last scenario ((3.) in the listing reported on top of Section 5.2), the airship starts being aligned with the north (and at zero coordinates in the following trajectory plots), and the first segment requires a climb to 10 m and a turn to port of 60 degree. Each segment is then at an angle of 60 degree with respect to the previous one, forming a regular hexagon from the top. The checkpoints are at alternate altitudes of 0 m or 10 m, starting from the origin at 0 m. Similar to the previous scenarios, the track length on the ground is 200 m.

Figure 15 shows three views of the trajectory flown by the baseline airship, considering a ground speed setting of 8 m/s, and the time history of controls.

From Figure 15 it can be observed that the baseline airship, autonomously piloted by the proposed controller, successfully captures all checkpoints, with a rather regular action on the controls. In particular, at the start of the circuit (leftmost checkpoint, zero coordinates), the airship is forced into a climb and a steep turn to port, being misaligned with respect to the track by 60 degree. As a result of the trajectory flown along the first track, the airship enters the second track with a misalignment of nearly 90 degree (i.e., on a western course, where a northern course is required). Even this requiring condition does not lead to a miss of the next checkpoint, showing adequate accuracy of the control system in capturing the checkpoint position, at the price of a certain elongation of the trajectory with respect to the expected track.

In this regard, in analyzing this and other similar cases for the present research, it was found that due to the limited authority of the controls (as required for a realistic simulation), constrained between hard limits of the actuator dead ends, a trade-off scenario is configured between the accuracy along the track and in the capture of checkpoints. In other words, when trying to increase control gains to achieve better accuracy along the tracks, control saturation is more often encountered, and this in turns produces a detrimental effect on both track and checkpoint capture accuracy. By reducing control gains, thus limiting the reactivity of control to feedback variables (errors with respect to set-points), significant elongations with respect to the trajectory are encountered (as in the test in Figure 15), but a smoother overall trajectory and control time history are obtained, which produce as a welcome side-effect a better ability to capture checkpoints. Better track-keeping performance is generally encountered for lower speeds or longer tracks.

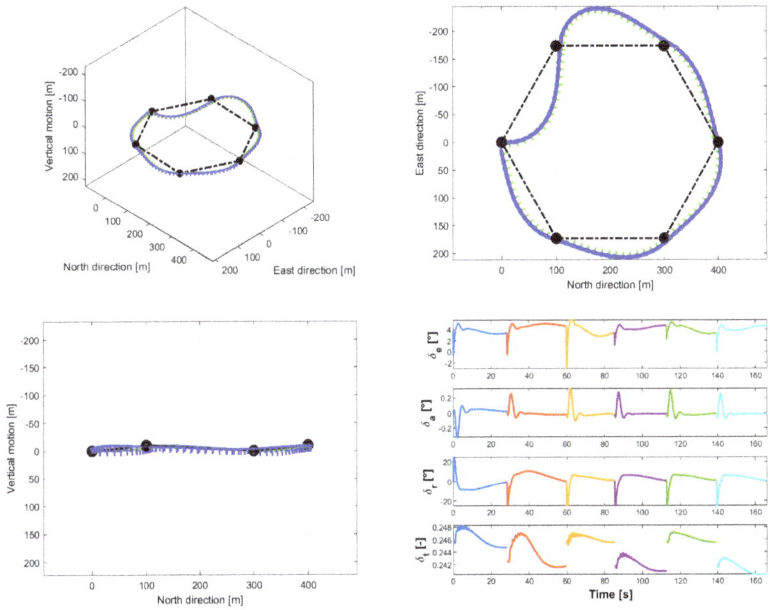

Figure 15. Trajectory and control behavior for a six-checkpoints circuit, flown by the stabilized baseline airship, at 8 m/s of ground speed. (**Top-left**) three-quarter view of the 3D trajectory from the northeast. (**Bottom-left**) view of the trajectory from the east. (**Top-right**) view of the trajectory from the top. Black dash-dotted lines: target track. Black spheres: target checkpoint capture volume. (**Bottom-right**) time histories of controls (a different color for each segment).

From the time history of control, it can be observed that, with the proposed control gains, only the rudder is operated near the actuator limit values (+/−25 degree), thus avoiding (even if only by a short measure) any saturation.

Similar to previous scenarios, it can be observed that an aileron control is only employed for stabilization in conjunction with steep changes in the rudder request at the start of each new track, and is in general only keeping the airship at a close-to-null roll angle. Therefore, turns are basically flat. Thrust control is on average higher for ascending tracks and lower for descending ones, as expected.

The plots in Figure 16 correspond to the flight of the four-thrusters, thrust-controlled airship in the same scenario as in Figure 15.

A more aggressive set of gains has been employed here (for the vertical and lateral beam-tracking only, whereas stabilization, speed-tracking, and turn coordination are unchanged) with respect to the previous two scenarios ((1.) and (2.) in the listing on top of Section 5.2), to better show the qualities of this airship layout and the corresponding guidance control. The gains have been tuned according to the same consideration just outlined for the baseline case, i.e., properly balancing accuracy in the track-following problem and use of controls, which ensues the need for limited control saturation.

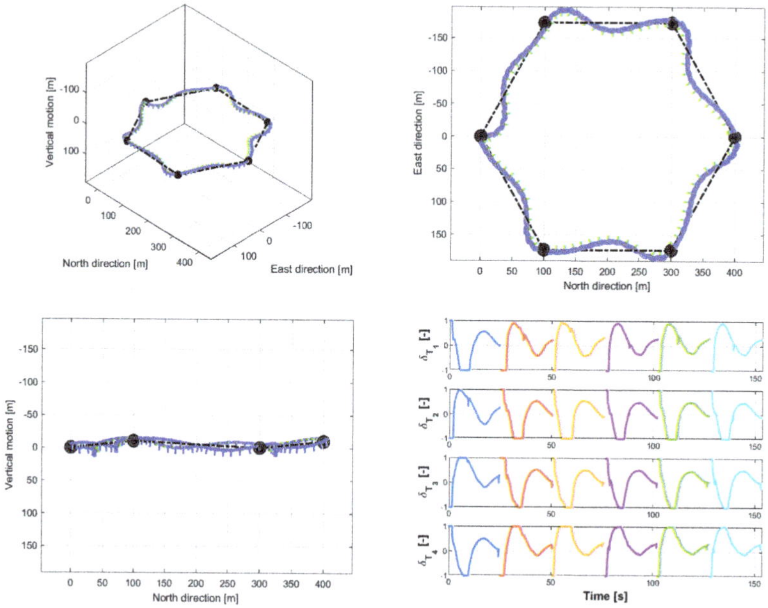

Figure 16. Trajectory and control behavior for a 6-checkpoints circuit, flown by the thrust controller on the stabilized four-thrusters concept airship, at 8 m/s of ground speed. (**Top-left**) three-quarter view of the 3D trajectory from the northeast. (**Bottom-left**) view of the trajectory from the east. (**Top-right**) view of the trajectory from the top. Black dash-dotted lines: target track. Black spheres: target checkpoint capture volume. (**Bottom-right**) time histories of controls (a color for each segment).

Generally speaking, comparing the trajectory in Figure 16 both to the baseline in Figure 15 and to the same four-thrusters airship flying in the misaligned track test (2.) in Figure 14, it is apparent that the airship is making more intense use of controls, which in several instances produce saturation. This is especially true for thrusters #2 and #4, which are put to full negative and positive thrust, respectively, to obtain a turn to port following the initial steep starboard turn in all tracks except the first.

The accuracy in both track-following and checkpoint capture is generally greater than for the baseline airship (Figure 15), with smaller top elongation at least in terms of lateral guidance. However, a more intense higher-frequency oscillation is encountered in the flight trajectory for the four-thrusters airship with this gain setting, which is especially visible in the side-view, showing some difficulty in steadily keeping the mildly ascending or descending tracks. The proposed hexagonal circuit is generally more requiring in terms of lateral navigation than in vertical navigation (i.e., the changes in altitude are relatively small, whereas horizontal changes in course are very significant and abrupt). A spillover of oscillation from the lateral to the vertical flight trajectory is a result of the higher degree of control coupling, expected to be encountered with thrust-based control compared to standard aerodynamics-based control. It should be recalled that the same four thrust inputs are responsible for beam-tracking and speed-tracking, whereas the thrust control on the baseline airship is basically responsible for speed-tracking only, leaving attitude control and navigation to the aerodynamic surfaces. In combination with the increased gains, this is also reflected in the more intense rotation around the roll axis, with bank angles φ between $+/-20$ degree (not shown), much greater than for the baseline airship (as observed, turning basically flat in every instance).

5.2.4. Flight with Constant Wind

Extensive testing has been carried out considering a windy condition, easy to treat within the SILCROAD library with arbitrary (including stochastic) wind. Sample results with a constant wind of 3 m/s from the west will be shown here.

Figure 17 shows the top and side view of the ascending trajectory considered in Section 5.2.1, for both airship test beds. The ground speed modulus set-point is here 6 m/s. The composition of this value and the western wind of 3 m/s will produce the actual value of the airspeed, responsible for the aerodynamic forces on the airship. The initial trimmed condition is such that the ground speed vector be aligned with the north, with a wind of 3 m/s, and a roll angle of $\varphi = 0$ degree. Therefore, an adjustment following the start of the simulation is required to cope with the requested altitude increase.

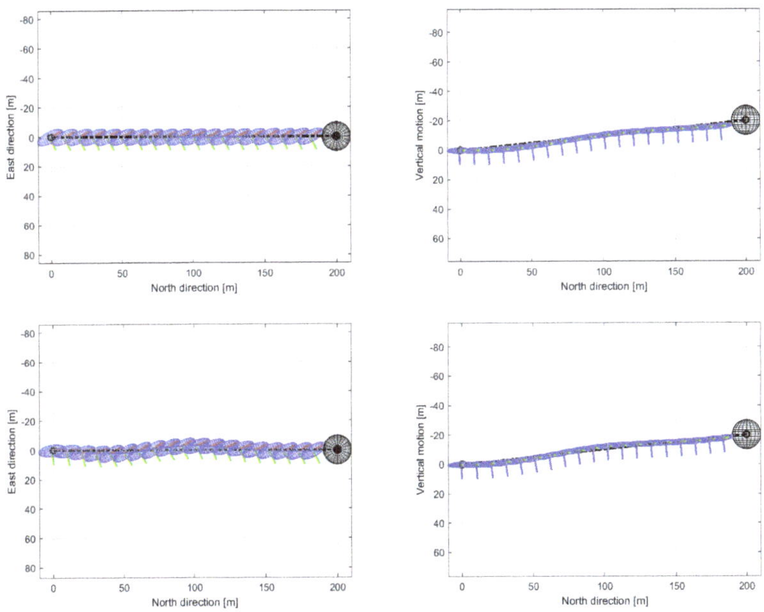

Figure 17. Trajectory views in the presence of a western wind of 3 m/s, considering the same target track and checkpoint of Figures 11 and 12. Ground speed setting of 6 m/s. (**Top**) baseline airship. (**Bottom**) four-thrusters thrust-controlled airship. (**Left**) top view. (**Right**) view from the east. Black dash-dotted lines: target track. Black spheres: target checkpoint capture volume.

Both airships and guidance systems perform satisfactorily in terms of track and checkpoint capture accuracy, both laterally and vertically. Significantly more action is required for the four-thrusters layout and thrust-based control to adjust to a change in the trimmed condition (bottom plots in Figure 17). As observed, this is again the result of control-induced coupling, which is higher than for the aerodynamics-based control on the baseline airship.

Figure 18 shows the flight trajectories in the scenario considered in Section 5.2.3 flown by both airship test beds, with a 3 m/s western wind, and a ground speed setting of 6 m/s.

As a general comment on Figure 18, the baseline airships and the four-thrusters concept behave in a qualitatively similar way as in the scenarios previously analyzed. In particular, the baseline airship features a less oscillating, smooth trajectory with some elongation in terms of beam-tracking, whereas the four-thrusters concept displays a more oscillating trajectory, but with a lower top elongation. Oscillation for the four-thrusters airship is especially visible at altitude, and, as observed, roll motion is more intense for this airship, due to the higher degree of lateral-directional vs. vertical coupling induced by control.

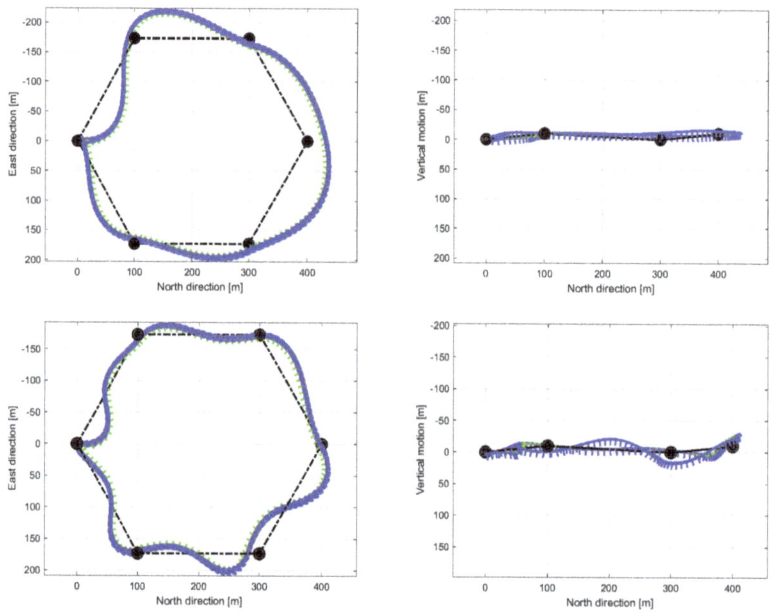

Figure 18. Trajectory views in the presence of a western wind of 3 m/s, considering the same 6-checkpoints circuit of Figures 15 and 16. Ground speed setting of 6 m/s. (**Top**) baseline airship. (**Bottom**) four-thrusters thrust-controlled airship. (**Left**) top view. (**Right**) view from the east. Black dash-dotted lines: target track. Black spheres: target checkpoint capture volume.

Considering the complexity of the scenario in terms of circuit geometry and wind intensity, as well as the large sail area of the airship, the result of the test is promising. However, interestingly, both missions feature checkpoint misses. These are all localized in the downwind component of the circuit.

The baseline airship misses the third checkpoint (at the end of the third track) due to lateral elongation (top plots in Figure 18). Actually, following the capture of the second checkpoint, the target track–going northeast–puts the airship in an almost tailwind condition, which tends to increase its ground speed. As a result of that, the engine control δ_T is reduced to almost zero, reducing the airspeed to cope with the requirement on ground speed (constant at 6 m/s for all tracks), in turn reducing the effectiveness of the control surfaces and producing a sluggish response to the controls.

On the other hand, the four-thrusters airship misses the fourth checkpoint (bottom plots in Figure 18). After reaching the third checkpoint, the fourth target track implies a sharp turn to starboard, where the wind passes from tailwind-port to tailwind-starboard. The reaction of the controller to this perturbation is very intense, despite remaining out of saturation (not shown), and markedly around the rolling axis. The ensuing motion also produces a spillover of oscillation in the vertical plane, which triggers an additional control component from the vertical guidance. The complex response of the thrust-controlled layout results in an excessively oscillating maneuver, which despite not missing excessively the track, causes a miss on the checkpoint.

A closing note concerns gain tuning. For all testing on the baseline airship and control, no specific tuning was required on the gains (clearly following the definition of a satisfactory set of values in a preliminary phase), for the entire range of trials carried out, included those exemplified here. Clearly, for further decreasing airspeeds below the minimum of 6 m/s considered in this research, a worsening authority and performance of aerodynamics-based control can be expected (in hover, to the limit, the baseline airship will feature a completely uncontrollable behavior). On the other hand, the tuning of the gains for the four-thrusters layout and thrust-based control required more effort, and the remarked-on coupling between vertical and lateral beam-tracking control, as well as the underlying coupling with speed-tracking, required some tweaking of controller parameters when approaching a higher speed (10 m/s). In particular, a reduction in navigation gains and an increase in turn coordination gain was found to help in governing the airship with less oscillation at the higher speed, especially in the presence of wind. That said, a single set of gains can be employed also for thrust-based control on the four-thrusters layout for the vast majority of the considered test cases, and whenever operating in still air. Furthermore, hover has been left out from the present research. The thrust-based layout with a corresponding thrust-based control should retain good controllability features close to and in hovering flight conditions.

6. Conclusions and Outlook

In the quest to increase the acceptability of airships as useful and reliable platforms, especially for the UAV market, the major shortcomings of LTAs—namely limited controllability, excessive sensibility to disturbances, and, broadly speaking, unusual dynamic response compared to winged aircraft and rotorcraft—tend to overshadow the obvious advantages of a longer endurance and ability to manage both hover and forward flight.

In this paper, the problem of control and navigation in forward flight is treated, showing that a control law taken from the realm of winged aircraft and missiles is capable of stabilizing the airship, in particular increasing damping around the body axis, leaving sufficient authority to a guidance control layer to autonomously pilot the airship along a pre-defined track.

Satisfactory controllability characteristics have been found on a realistic airship test bed, stabilized and piloted with controllers based on the deflection of standard tail control surfaces, similar to those found on aircraft and on most standard airships for low atmospheric flight. A three-axis damping stabilization layer is implemented, which is tested alone, and as a base layer for a guidance controller, implementing speed- and beam-tracking logics.

Trying to achieve a greater design and construction simplicity, a four-thrusters airship layout has been considered, and an exclusively thrust-based control suite with the same tasks as those developed for the baseline case was designed, implemented, and tested. In particular, no thrust-vectoring, task separation between thrusters for propulsion, or attitude control or maneuvering were considered.

A critical comparison from testing carried out in a fully non-linear simulation environment (employing the novel `SILCROAD` library and simulation tool) showed that both airships with the respective controllers perform realistically in terms of control use, and with good accuracy in the required navigation tasks.

Critical points emerge in terms of control-induced coupling on the four-thrusters configuration, which tends to produce unnecessary oscillations in the flight trajectory, due to the spillover of the control action from the three branches of the navigation system (i.e., speed-tracking, and longitudinal- and lateral-beam-tracking). Furthermore, the tuning of the gains especially for higher forward speed, or in the presence of constant wind, may suggest the need for a gain-scheduling to ensure better precision, even though generally good results can be obtained for a broad range of speeds with a single gain set.

Major points in favor of the thrust-based layout and control with respect to the reference airship include generally better proximity to the target track, even in complex navigation scenarios, and the preserved authority of controls in windy conditions, which conversely, as expected tends to decrease for aerodynamic surfaces in tailwind conditions corresponding to lower airspeed values.

The present analysis supports that a control suite not substantially dissimilar from that found on standard winged aircraft produces adequate stabilization and navigation results for aerodynamically governed airships. Furthermore, it suggests that a thrust-based and thrust-controlled layout with no thrust vectoring is an achievable and interesting possibility, with advantages and disadvantages with respect to the aerodynamics-based baseline, but generally reliable and not unbearably difficult to design.

Following suit from the latest observations, future research will explore the transition to hovering flight, and the management of hover and near-hover maneuvers, of great relevance besides forward flight, on a thrust-controlled airship without thrust-vectoring, thus extending the analysis of this platform to a broad range of maneuvers and flight conditions of practical interest. In particular, both automatic and semi-automatic maneuvers will be studied, benefitting from the employment of a guidance suite already developed by the authors (currently applied to rotorcraft), and suitably adapted for obstacle avoidance in terminal maneuvers [25,26].

Of further relevance to airships is the tackling of longer climbs and transitions to higher altitudes, where the machine is expected to express modified dynamics. In particular, this may require recurring more heavily to control scheduling.

Author Contributions: The Authors contributed equally to the production of the research and the preparation of the present article. All authors have read and agreed to the published version of the manuscript.

Funding: This research received no external funding.

Data Availability Statement: The data presented in this study are partially available on request from the corresponding author.

Conflicts of Interest: The authors declare no conflict of interest.

Abbreviations

The following abbreviations are used in this manuscript:

$\mathcal{B}$	Body reference (supposed centered in ***CB*** in this paper)
$\mathcal{I}$	Inertia reference (ground in this paper)
$\mathcal{V}$	Volume of lifting gas on airship
$\boldsymbol{CB}$	Center of buoyancy
$\boldsymbol{CG}$	Center of gravity
J_{CB}	Inertia tensor in ***CB***
M_{CB}	Generalized mass matrix in ***CB***
N_{CB}	Matrix of aerodynamic reaction forcing term
S_{CB}	Static moment in ***CB***
$S_{321}^{\mathcal{B}}$	Kinematic matrix for rotational rate
U_{CB}	Sensitivity matrix of active aerodynamic force in ***CB*** wrt. generalized velocity
V_{CB}	Sensitivity matrix of active aerodynamic force in ***CB*** wrt. controls
V_{CB}^{a}	Sensitivity matrix of active aerodynamic force in ***CB*** wrt. aerodynamic controls
V_{CB}^{t}	Sensitivity matrix of active aerodynamic force in ***CB*** wrt. thrust controls
d_L	Lateral unit vector (for guidance control computations)
d_V	Vertical unit vector (for guidance control computations)
$e_{321}^{\mathcal{B}}$	Array of attitude angles
f	Force vector
m_{CB}	Moment vector in ***CB***
r_{CG}	Position of center of gravity from ***CB***
$r_{P_{Ti}}$	Position of point of application of *i*-th thrust force from ***CB***
s_{CB}	Generalized forcing term vector in ***CB***
s_{CB}^{a}	Aerodynamic forcing term vector in ***CB***
$s_{CB}^{a,b}$	Active component of aerodynamic forcing term vector in ***CB***

$s^{a,m}_{CB}$	Reactive component of aerodynamic forcing term vector in CB
s^{b}_{CB}	Buoyancy forcing term vector in CB
s^{g}_{CB}	Gravity forcing term vector in CB
s^{t}_{CB}	Thrust forcing term vector in CB
u	Array of controls
u^a	Array of aerodynamic controls
u^t	Array of thrust controls
v_{CB}	Velocity vector of CB
w_{CB}	Generalized velocity vector of CB
x^{AC}	State array of flying craft
x_{CB}	Position vector of CB from the origin of reference $\mathcal{I}$
x_P	Position vector of navigation checkpoint from the origin of reference $\mathcal{I}$
$\omega_{\mathcal{B}/\mathcal{I}}$	Rotational speed of body reference wrt. inertial reference
$\tilde{K}_i$	Modulating function of thrust vs. thrust setting for i-th thruster
$L^{a,b}_{CB}$	First (roll) component of aerodynamic active moment in CB, in body ref.
$M^{a,b}_{CB}$	Second (pitch) component of aerodynamic active moment in CB, in body ref.
$N^{a,b}_{CB}$	Third (yaw) component of aerodynamic active moment in CB, in body ref.
RC	Rate of climb of airship
$T^{act}, T^{lp}, T^{thr}, T^{wo}$	Time constants (actuator, low-pass, thruster, washout)
$\tilde{T}_i$	Nominal thrust intensity for i-th thruster
U	First (longitudinal) component of v_{CB} in body ref.
V	Second (lateral) component of v_{CB} in body ref.
W	Third (vertical) component of v_{CB} in body ref.
$X^{a,b}$	First (longitudinal) component of aerodynamic active force in body ref.
$Y^{a,b}$	Second (lateral) component of aerodynamic active force in body ref.
$Z^{a,b}$	Third (vertical) component of aerodynamic active force in body ref.
e^{disp}_L	Lateral error on position (for guidance)
e^{vel}_L	Lateral error on velocity (for guidance)
e^{disp}_V	Vertical error on position (for guidance)
e^{vel}_V	Vertical error on velocity (for guidance)
$h_{1,2}, h_{3,4}$	Logical functions on four-thrusters configuration
k_I	Integral gain
k_P	Proportional gain
m	Mass of airship
p	First component of $\omega_{\mathcal{B}/\mathcal{I}}$ in body ref.
q	Second component of $\omega_{\mathcal{B}/\mathcal{I}}$ in body ref.
r	Third component of $\omega_{\mathcal{B}/\mathcal{I}}$ in body ref.
v^*_{CB}	Velocity set-point (scalar)
v^*_l	Lateral velocity set-point
v^*_v	Vertical velocity set-point
β	Airship sideslip angle
δ_e	Elevator deflection
δ_a	Aileron deflection
δ_r	Rudder deflection
δ_{T_i}	Thrust setting of i-th thruster
ϑ	Pitch attitude angle
λ_i	Tilt of i-th thrust line
ρ	Density of air
σ_i	Lateral misalignment of i-th thrust line
φ	Roll attitude angle
ψ	Yaw attitude angle
$(\cdot)^{\mathcal{B}}$	Representation in body components
$(\cdot)^{\mathcal{I}}$	Representation in inertial components
$(\cdot)^{bf}$	Bandwidth-filtered signal
$(\cdot)^{coord}$	Turn coordination input signal
$(\cdot)^{g}$	Guidance input signal
$(\cdot)^{pilot}$	Pilot/Auto-pilot input signal
$(\cdot)^{req}$	Required command value set as input to actuator

References

1. Chu, A.; Blackmore, M.; Oholendt, R.G.; Welch, J.V.; Baird, G.; Cadogan, D.P.; Scarborough, S.E. A Novel Concept for Stratospheric Communications and Surveillance: Star Light. In Proceedings of the AIAA Balloon Systems Conference, Williamsburg, VA, USA, 21–24 May 2007. [CrossRef]
2. Miller, S.H.; Fesen, R.; Hillenbrand, L.; Rhodes, J.; Baird, G.; Blake, G.; Booth, J.; Carlile, D.E.; Duren, R.; Edworthy, F.G.; et al. Airships: A New Horizon for Science. *arXiv* **2014**, arXiv:1402.6706.
3. Elfes, A.; Bueno, S.; Bergerman, M.; Ramos, J. A semi-autonomous robotic airship for environmental monitoring missions. In Proceedings of the 1998 IEEE International Conference on Robotics and Automation (Cat. No. 98CH36146), Leuven, Belgium, 16–20 May 1998; Volume 4, pp. 3449–3455.
4. Jon, J.; Koska, B.; Pospíšil, J. Autonomous airship equipped with multi-sensor mapping platform. *ISPRS-Int. Arch. Photogramm. Remote Sens. Spat. Inf. Sci. XL-5 W* **2013**, *1*, 119–124. [CrossRef]
5. Fedorenko, R.; Krukhmalev, V. Indoor autonomous airship control and navigation system. In Proceedings of the 3rd International Conference on Control, Mechatronics and Automation (ICCMA 2015), Barcelona, Spain, 21–22 December 2015; Volume 42, p. 01006.
6. Riboldi, C.E.D.; Rolando, A.; Regazzoni, G. On the feasibility of a launcher-deployable high-altitude airship: Effects of design constraints in an optimal sizing framework. *Aerospace* **2022**, *9*, 210. [CrossRef]
7. Manikandan, M.; Pant, R.S. Research and advancements in hybrid airships—A review. *Prog. Aerosp. Sci.* **2021**, *127*, 100741.
8. Jones, S.P.; DeLaurier, J.D. Aerodynamic estimation techniques for aerostats and airships. *J. Aircr.* **1982**, *20*, 120–126. [CrossRef]
9. Kämpf, B. Flugmechanik und Flugregelung von Luftschiffen. Ph.D. Thesis, University of Stuttgart, Stuttgart, Germany, 2004. [CrossRef]
10. Riboldi, C.E.D.; Rolando, A. Layout Analysis and Optimization of Airships with Thrust-Based Stability Augmentation. *Aerospace* **2022**, *9*, 393. [CrossRef]
11. Kornienko, A. System Identification Approach for Determining Flight Dynamical Characteristics of an Airship from Flight Data. Ph.D. Thesis, University of Stuttgart, Stuttgart, Germany, 2006. [CrossRef]
12. Nagabhushan, B.L.; Tomlinson, N.P. Thrust-vectored takeoff, landing, and ground handling of an airship. *J. Aircr.* **1986**, *23*, 250–256. [CrossRef]
13. Liesk, T.; Nahon, M.; Boulet, B. Design and experimental validation of a nonlinear low-level controller for an unmanned fin-less airship. *IEEE Trans. Control Syst. Technol.* **2013**, *21*, 149–161. [CrossRef]
14. Chen, L.; Zhang, H.; Duan, D.P. Control system design of a multivectored thrust stratospheric airship. *Proc. Inst. Mech. Eng. Part G J. Aerosp. Eng.* **2014**, *228*, 2045–2054. [CrossRef]
15. Nagabhushan, B.L.; Tomlinson, N.P. Dynamics and control of a heavy lift airship hovering in a turbulent cross wind. *J. Aircr.* **1982**, *19*, 826–830. [CrossRef]
16. Cai, Z.; Qu, W.; Xi, Y.; Wang, Y. Stabilization of an underactuated bottom-heavy airship via interconnection and damping assignment. *Int. J. Robust Nonlinear Control* **2007**, *17*, 1690–1715. [CrossRef]
17. Beji, L.; Abichou, A. Tracking control of trim trajectories of a blimp for ascent and descent flight manoeuvres. *Int. J. Control* **2005**, *78*, 706–719. [CrossRef]
18. Paiva, E.; Benjovengo, F.; Bueno, S.; Ferreira, P. Sliding mode control approaches for an autonomous unmanned airship. In Proceedings of the 18th AIAA Lighter-Than-Air Systems Technology Conference, Seattle, WA, USA, 4–7 May 2009.
19. Yang, Y.; Wu, J.; Zheng, W. Station-keeping control for a stratospheric airship platform via fuzzy adaptive backstepping approach. *Adv. Space Res.* **2013**, *51*, 1157–1167. [CrossRef]
20. Yang, Y.; Wu, J.; Zheng, W. Positioning control for an autonomous airship. *J. Aircr.* **2016**, *53*, 1638–1646. [CrossRef]
21. Blakelock, J.H. *Automatic Control of Aircraft and Missiles*; John Wiley & Sons: New York, NY, USA, 1991.
22. Pamadi, B.N. *Performance, Stability, Dynamics, and Control of Airplanes*; AIAA Education Series; American Institute of Aeronautics and Astronautics, Inc.: Reston, VA, USA, 2004.
23. Munk, M.M. *The Aerodynamic Forces on Airship Hulls*; NACA Technical Report; Report No.184; National Advisory Committee for Aeronautics: Washington, DC, USA, 1926.
24. Lamb, H. *Hydrodynamics*; Dover Publications: New York, NY, USA, 1945.
25. Rolando, A.; Rossi, F.; Riboldi, C.E.D.; Trainelli, L.; Grassetti, R.; Leonello, D.; Redaelli, M. The pilot acoustic indicator: A novel cockpit instrument for the greener helicopter pilot. In Proceedings of the 41st European Rotorcraft Forum, Munich, Germany, 1–4 September 2015.
26. Trainelli, L.; Gennaretti, M.; Bernardini, G.; Rolando, A.; Riboldi, C.E.D.; Redaelli, M.; Riviello, L.; Scandroglio, A. Innovative helicopter in-flight noise monitoring systems enabled by rotor-state measurements. *Noise Mapp.* **2016**, *3*, 190–215. [CrossRef]

aerospace

MDPI

Article

Development and Evaluation of an Enhanced Virtual Reality Flight Simulation Tool for Airships

Mohsen Rostami, Jafer Kamoonpuri, Pratik Pradhan and Joon Chung *

Department of Aerospace Engineering, Toronto Metropolitan University, Toronto, ON M5B 2K3, Canada
* Correspondence: j3chung@torontomu.ca

Abstract: A real-time flight simulation tool is proposed using a virtual reality head-mounted display (VR-HMD) for remotely piloted airships operating in beyond-line-of-sight (BLOS) conditions. In particular, the VR-HMD was developed for stratospheric airships flying at low/high altitudes. The proposed flight simulation tool uses the corresponding aerodynamics characteristics of the airship, the buoyancy effect, mass balance, added mass, propulsion contributions and ground reactions in the FlightGear Flight Simulator (FGFS). The VR headset was connected to the FGFS along with the radio controller containing the real-time orientation/state of each button, which is also simulated to provide better situational awareness, and a head-up display (HUD) that was developed to provide the required flight data. In this work, a system was developed to connect the FGFS and the VR-capable graphics engine Unity to a PC and a wireless VR-HMD in real time with minimal lag between data transmission. A balance was found for FGFS to write to a CSV file at a period of 0.01 s. For Unity, the file was read every frame, which translates to around 0.0167 s (60 Hz). A test procedure was also conducted with a similar rating technique based on the NASA TLX questionnaire, which identifies the pilot's available mental capacity when completing an assigned task to assure the comfortability of the proposed VR-HMD. Accordingly, a comparison was made for the aircraft control using the desktop simulator and the VR-HMD tool. The results showed that the current iteration of the system is ideal to train pilots on using similar systems in a safe and immersive environment. Furthermore, such an advanced portable system may even increase the situational awareness of pilots and allow them to complete a sizeable portion of actual flight tests with the same data transmission procedures in simulation. The VR-HMD flight simulator was also conceived to express the ground control station (GCS) concept and transmit flight information as well as the point of view (POV) visuals in real-time using the real environment broadcast using an onboard camera.

Keywords: virtual reality; flight simulation; ground control station; airship; flight test; training

1. Introduction

Maintaining stability and operating autonomously are the main challenges related to the prolonged deployment of airships, and in-depth studies of flight operations are an important step toward deploying stratospheric hybrid airships [1]. A substantial amount of work has been carried out to enhance the design and development of airships, including several research projects and some stratospheric airship technology demonstrations such as HiSentinel by the Southwest Research Institute (SwRI) and Aerostar International Inc. (Sioux Falls, SD, USA) [2], the Sceye Stratospheric Airship by Sceye Inc. (Moriarty, NM, USA) [3], the Integrated Sensor is Structure (ISIS) by DARPA [4], the High-Altitude Long-Endurance Demonstrator (HALE-D) by Lockheed Martin (Bethesda, MD, USA) [5] and Thales Alenia Space Airship by the Thales group (New Castle, DE, USA) [6].

With the help of modelling and simulation, the development cost of an aerial vehicle may be lowered [7,8]. Moreover, the development of a flight simulation structure for a remotely piloted air vehicle using the same techniques and subsystems enables the validation and verification of concepts and systems, optimization of the design and, additionally, the enhancement of flying techniques and performance [9]. Nowadays, several types of flight simulators are available for different purposes, ranging from desktop platforms to more sophisticated training simulators and advanced X-DOF cockpit simulators [10]. Traditional simulators integrate flight dynamics models and simulation subsystems with a hardware cockpit mock-up, an outside visual and motion simulation [11]. Sometimes, these complex tools may exceed the cost of the actual product due to highly demanding quality standards for both hardware and software. Therefore, it is important to find the optimum cost/benefit ratio to reach an acceptable level of realism while considering the final cost [12].

VR provides a promising substitute to traditional simulators [13]. VR technology provides the opportunity to replace the entire reality with virtual objects with no limitations to what this could be. The difference between VR technology and conventional screens is the three-dimensional capability of VR compared to the two-dimensional representation of traditional screens. While most research and commercial applications use VR to put the pilot in a virtual cockpit, some use it to emulate an HUD with the out-the-window (OTW) laid behind the HUD visuals [14,15]. In any case, the following three "pillars" play major roles to ensure the usability of the technology among all targeted users: immersion, presence and interactivity. Immersion is defined as a psychological experience often governed by various technological limitations including the field of view (FOV), resolution, latency and trackability. Presence refers to the state or experience of a user in the virtual environment that may or not be related to the physical environment. Interactivity is the degree to which the user can control the virtual environment using a medium. According to some studies, these three pillars together constitute user satisfaction and comfort while using VR headsets [16,17].

However, there are several concerns that are specifically involved when dealing with VR technology to ensure the comfort of users. Motion sickness or spatial disorientation are usually the first things that are discussed when talking about VR. This is because of the disconnect the user experiences between what they see and what they feel [18]. One way to lessen the chance of motion sickness is to provide a portable infrastructure with a masked simulated environment. The enhanced portable VR-HMD for flight simulation and evaluation is a great training tool for pilots [19,20]. VR-based flight simulators are small and more portable than a full-size cockpit mock-up simulator and are much less expensive. This also makes them the best choice for most operators who are looking to train pilots in remote locations and perform flight operations in remote locations. Utilizing the same digital telemetry radio system in real and simulated flights can provide more consistency between each phase of the design [21]. Another key aspect to consider when working with VR headsets is human perception and behavior when using the application. One of such factors is the FOV [22]. Figure 1 compares the FOV of human eyes with various VR and mixed reality (MR) headsets. As illustrated, human eyes have an average total FOV of about 180° × 120°, while the Oculus Quest only has an FOV of around 94° × 90°. In this work, an Oculus Quest 2 (Meta, Cambridge, MA, USA) has been used to study the simulated flight environment.

Augmented reality (AR) is a display device that enhances the real-world environment using virtual content and shows this on a screen. The usage of AR and VR technologies for futuristic vehicles or cockpit design is prevailing widely. In 2014, Aero Glass developed an application that could populate a large volume of ADS-B and other avionic information in the Osterhout Design Group (ODG) R-7 smart glasses [23,24]. It generated the information with respect to the pilot's head position and orientation in a 3D environment. In 2016, Epson Moverio BT-300 smart glasses were used as a head-mounted display (HMD) to fly DJI Phantom quadcopters [25]. In 2016, Japan Airlines (JAL) demonstrated the idea of using AR to train pilots with a realistic flight simulator [24,25]. In 2017, Bell Helicopters began exploring the idea of an augmented reality futuristic cockpit system for their FCX-001 concept helicopter [25]. In 2021, Bombardier Aviation developed a full-size AR-based cockpit display system (CDS) for next-generation flight deck HMI design practices [25].

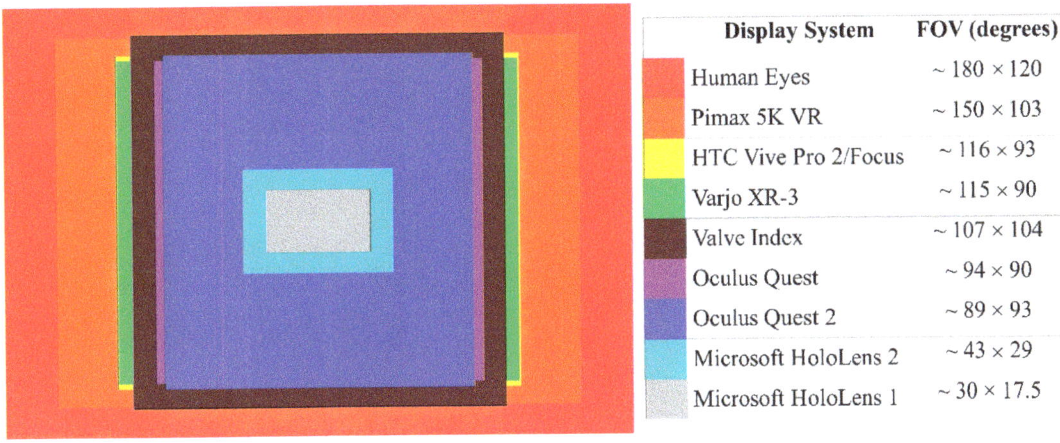

Figure 1. Field of View (FOV) comparison between different visual systems [26,27].

The VR-HMD simulator allows the operator to speed up plans to use VR headsets for real flights and to provide an integrated product that aims to train pilots from flight simulation to real flights using the same data transmission techniques. Utilizing the same digital telemetry radio system in real and simulated flights can provide more consistency between each phase of the design. Furthermore, the new flight simulator equipped with VR can be provided to customers for training and operational flying purposes with the same data transmission techniques in a simulated environment. Moreover, VR setups are small and portable, which makes them the best choice for most operational cases where customers are interested in operating in remote locations, particularly for airship-related missions. Using this new technology, the required time to train pilots will significantly drop compared to the existing technologies as the integrated VR-HMD system will be used from start to finish to train the pilots for actual flight. Consequently, the VR-HMD flight simulator will allow the operator to accomplish the goal of transiting from simulated training to actual flight with a single united system. The proposed portable integrated VR-HMD flight simulator has the potential to open up a new generation of technology that costs just a few thousand dollars compared to existing separate GCS and full-size cockpit mock-up simulators, which are significantly costlier. Such an advanced system is important to complete missions in real time from remote locations, lower pilot workload and increase situational awareness.

2. Methodology

In Section 2, the methodology used for the development of the VR-HMD flight simulator is presented. Accordingly, in Section 2.1, the airship architecture is presented with the details of the geometry parameters. Next, in Section 2.2, the flight simulator architecture including the flight dynamics model and the sub-sections of the flight simulator are presented in detail. In Section 2.3, the VR simulation application developed for the purpose of this study is introduced. Finally, in Section 2.4, the test procedure for the evaluation of the proposed tool is presented along with the research ethics board (REB) questionnaire.

2.1. Airship Architecture

Herein, two models for stratospheric airship design are introduced. The first one is high-altitude architecture, with the mission to carry the desired payload to 70,000 ft and station-keep for a long time (e.g., several months and even years). The precise sizing characteristics are not the purpose of this work. The stratospheric airship uses two electrical engines, both placed at ~30% of the airship length from the nose, as presented in Figures 2 and 3. The engine location allows the airship to use thrust vectoring to generate sufficient pitch and yaw moments. The hybrid architecture is designed to employ the communication system, solar panels, fuel cells and batteries and station-keep for a long time. The hull uses a circular cross-section multi-ellipsoid geometry, and the tail has a cross-shaped four-fin empennage configuration (X layout) with inflated fixed fins and ruddervator configuration to provide stability in both longitudinal and lateral directions.

Figure 2. The airship model equipped with two electrical engines.

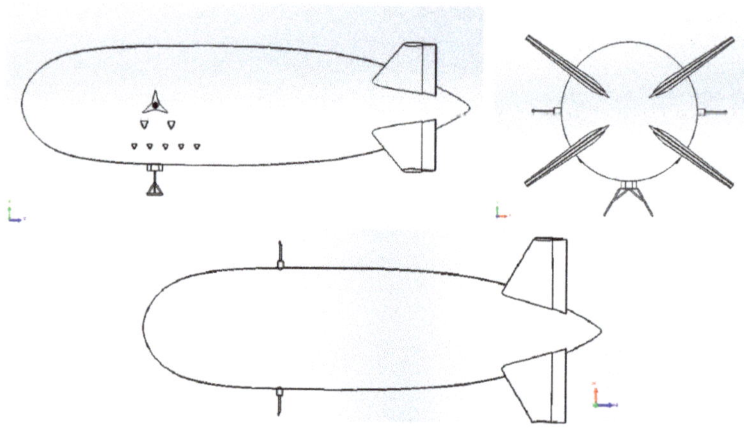

Figure 3. Side, back and top view of the airship geometry.

The second model is a scaled-down version of the base architecture to carry the desired payload to 10,000 ft and corresponds to the low-altitude demonstrator to validate the control and navigation systems. The low-altitude demonstrator model is considered to validate the simulator model in low-altitude scenarios using actual flight test results and then scale up to the stratospheric model. The geometry of the second airship is presented in Figure 2, with different parameters in Table 1.

Table 1. Airship geometry parameters.

Variable	Value	Description
d_{max}	17 ft	Maximum hull diameter
f_r	3.86	Fineness ratio ($l_{airship}/d_{max}$)
X_{CG}	28.9 ft	Center of gravity location from nose
X_{CB}	28.9 ft	Centre of buoyancy location from nose
$Z_{CG\text{-}CB}$	3.28 ft	Vertical distance between CG and CB (CG below CB)
$l_{airship}$	65.62 ft	Total airship length
m_e	375 lb	Empty mass
S_f	63.5 ft^2	Fin reference area
S_h	228.2 ft^2	Hull reference area
Vol	10,594.4 ft^3	Hull volume
T_{max}	45 lbf	Maximum engine thrust (each)

2.2. Flight Simulator Architecture

The desktop flight simulator was developed for the hybrid airship where beyond-line-of-sight operations are of interest. The proposed flight simulator was developed as a modular platform using the FGFS such that it is scalable and low-cost. FGFS offers a free and open-source multi-platform capability with atmospheric and orbital flight characteristics. It has been widely used in aerospace research and industry. The flight simulation uses a simulated atmosphere, power, equations of motion, aerodynamics, buoyant forces, gravity, mass, added mass and inertia, as shown in Figure 4, to study the behavior of the airship in different flight conditions [28]. Hence, different types and scales of the airship model can be simply simulated to study the behavior of the base design at various altitudes and flight conditions. To set up the flight simulator architecture, the geometry of the airship described in Table 1 was modelled in FGFS using the Flight Dynamics Model (FDM) provided by JSBSim [29]. Accordingly, the corresponding aerodynamics of the hybrid airship, along with the buoyant forces, added mass, mass balance, ground reactions, atmosphere and propulsion contributions, were considered in the development of the flight simulator. These details are all gathered in separate extensible markup language (XML) configuration files to enable a modular platform for the flight simulator architecture.

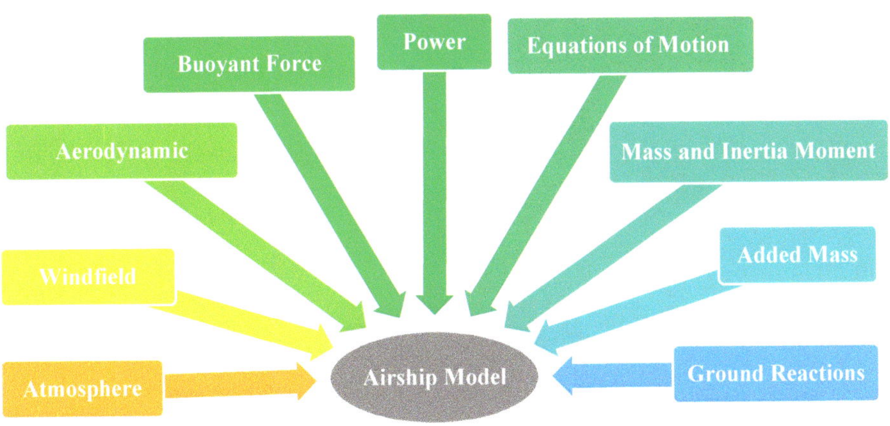

Figure 4. Required sub-sections to construct the flight simulation module.

The matrix form of the equations of motion for an aircraft can be shown using the following equations for the longitudinal and lateral direction motions, respectively [30] (see nomenclature list):

$$\begin{bmatrix} (s - X_u - X_{T_u}) & -X_\alpha & g\cdot\cos\theta_1 \\ -Z_u & (s(U_1 - Z_1) - Z_\alpha) & (-(Z_q + U_1)s + g\sin\theta_1) \\ -(M_u + M_{T_u}) & -(M_{\dot\alpha}s + M_\alpha + M_{T_\alpha}) & (s^2 - M_q s) \end{bmatrix} \begin{Bmatrix} \frac{u(s)}{\delta_e(s)} \\ \frac{\alpha(s)}{\delta_e(s)} \\ \frac{\theta(s)}{\delta_e(s)} \end{Bmatrix} = \begin{Bmatrix} X_{\delta_e} \\ Z_{\delta_e} \\ M_{\delta_e} \end{Bmatrix} \tag{1}$$

$$\begin{bmatrix} (sU_1 - Y_\beta) & -(sY_p + g\cos\theta_1) & s(U_1 - Y_r) \\ -L_\beta & (s^2 - L_p s) & -\left(s^2\bar{A}_1 + sL_r\right) \\ -N_\beta - N_{T_\beta} & -\left(s^2\bar{B}_1 + N_p s\right) & (s^2 - sN_r) \end{bmatrix} \begin{Bmatrix} \frac{\beta(s)}{\delta(s)} \\ \frac{\phi(s)}{\delta(s)} \\ \frac{\psi(s)}{\delta(s)} \end{Bmatrix} = \begin{Bmatrix} Y_\delta \\ L_\delta \\ N_\delta \end{Bmatrix} \tag{2}$$

For airships, in addition to the aerodynamic terms, significant force and moment terms due to the static buoyancy and inertial terms due to the displaced mass of air may be seen. Accordingly, the equations of motion for airships may be written as follows [31] (see nomenclature list):

$$\begin{aligned} &M\begin{bmatrix} \dot{u} \\ \dot{v} \\ \dot{w} \\ \dot{p} \\ \dot{q} \\ \dot{r} \end{bmatrix} = \\ &F_d(u, v, w, p, q, r) + S\left(\dot{u}_g, \dot{v}_g, \dot{w}_g, u_g, v_g, w_g, p_g, q_g, r_g\right) \\ &+A(u + v + w + p + q + r) + G(\lambda_{13}, \lambda_{23}, \lambda_{33}) \\ &+C(\text{control forces and moments}) + P(\text{propulsion forces and moments}) \end{aligned} \tag{3}$$

Considering the equations of motion presented in Equation (3), the decoupled longitudinal equations of motions of the airship could be presented as follows [31] (see nomenclature list):

$$\begin{aligned} &m_x\dot{u} + \left(ma_z - x_{\dot{q}}\right)\dot{q} = \\ &-m_z qW + X_{a_0} + X_u u + X_w w + X_q q - \theta(mg - B)\cos\theta_0 - (mg - B)\sin\theta_0 + T_t + \left(T_{d_S} + T_{d_p}\right)\cos\mu + X_c \end{aligned} \tag{4}$$

$$\begin{aligned} &m_z\dot{w} - \left(ma_x + z_{\dot{q}}\right)\dot{q} = \\ &m_x qU + Z_{a_0} + Z_u u + Z_w w + Z_q q + (mg - B)\cos\theta_0 - \theta(mg - B)\sin\theta_0 - \left(T_{d_s} + T_{d_p}\right)\sin\mu + Z_c \end{aligned} \tag{5}$$

$$\begin{aligned} &\left(m_{a_z} - M_{\dot{u}}\right)\dot{u} - (ma_x + M_{\dot{w}})\dot{w} + I_{yy}\dot{q} = \\ &-mg(a_x U + a_z w) + M_{a_0} + M_u u + M_w w + M_q q + \theta\{mg(a_x\sin\theta_0 - a_z\cos\theta_0) + B(b_x\sin\theta_0 - b_z\cos\theta_0)\} \\ &-mg(a_x\cos\theta_0 + a_z\sin\theta_0) - B(b_x\cos\theta_0 + b_z\sin\theta_0) + T_t c_z \\ &+\left(T_{d_s} + T_{d_p}\right)(d_z\cos\mu - d_x\sin\mu) + M_c \end{aligned} \tag{6}$$

Additionally, the lateral equations are expressed as follows [31] (see nomenclature list):

$$\begin{aligned} &m_y\dot{v} - \left(m_{a_z} - Y_{\dot{p}}\right)\dot{p} + (m_{a_x} - Y_{\dot{r}})\dot{r} = \\ &-m_x rU + m_z pW + Ya_0 + Y_v v + Y_p p + Y_r r + \phi(mg - B) + Y_c \end{aligned} \tag{7}$$

$$\begin{aligned} &-(ma_z + L_{\dot{v}})\dot{v} + I_{xx}\dot{p} - I_{xz}\dot{r} = \\ &ma_z(rU - pW) + La_0 + L_v v + L_p p + L_r r - \phi(a_z mg + b_z B) + \left(T_{d_p} - T_{d_s}\right)d_y\sin\mu + L_c \end{aligned} \tag{8}$$

$$
\begin{aligned}
&(ma_x - V_{\dot{v}})\dot{v} - I_{xz}\dot{p} + I_{zz}\dot{r} = \\
&-ma_x(rU - pW) + Na_0 + N_v v + N_p p + N_r r + \phi\cos\theta_0(a_x mg + b_x B) + \left(T_{d_p} - T_{d_s}\right)d_y \cos\mu + N_c
\end{aligned}
\tag{9}
$$

To define the equations of motion, two different functions are usually used. The first one is based on the Euler angle estimations providing roll, pitch and yaw angles. A direction–cosine matrix (DCM) function is usually defined to estimate the Earth-to-body relationship using the roll, pitch and yaw angles in radians. The second function is responsible for estimating the equations of motion in accordance with the quaternion definition. Likewise, a DCM function is defined to consider the Earth-to-body axis rotation working with the four components of the quaternion. The definition of the state vectors is explained in Table 2 as follows:

Table 2. The state vectors used in the Euler angle and quaternion simulations [32].

Vector	Euler Angle	Quaternion
X (1)	Body axis x inertial velocity, m/s	Body axis x inertial velocity, m/s
X (2)	Body axis y inertial velocity, m/s	Body axis y inertial velocity, m/s
X (3)	Body axis z inertial velocity, m/s	Body axis z inertial velocity, m/s
X (4)	North position of center of mass WRT Earth, m	North position of center of mass WRT Earth, m
X (5)	East position of center of mass WRT Earth, m	East position of center of mass WRT Earth, m
X (6)	Negative of CG altitude WRT Earth, m	Negative of CG altitude WRT Earth, m
X (7)	Body axis roll rate, rad/s	Body axis roll rate, rad/s
X (8)	Body axis pitch rate, rad/s	Body axis pitch rate, rad/s
X (9)	Body axis yaw rate, rad/s	Body axis yaw rate, rad/s
X (10)	Roll angle of body WRT Earth, rad	q1, x component of quaternion
X (11)	Pitch angle of body WRT Earth, rad	q2, x component of quaternion
X (12)	Yaw angle of body WRT Earth, rad	q3, x component of quaternion
X (13)	NA	q4, cos (Euler) component of quaternion

The control surfaces are bound with the FrSky Taranis X9 radio transmitter (FrSky Electronic Co., Jiangsu, China) through a wireless USB dongle to transfer the pilot's commands and control the airship during flight. Additionally, an HUD was designed for the desktop simulator to provide aircraft performance data and environment information on the screen, as shown in Figure 5, to increase pilots' situational awareness. Finally, as the airship flight simulator is designed to enhance the flight test and training procedures, the autopilot features consist of basic modes such as pitch hold and altitude hold developed with the help of PID controllers. Features for data logging and real-time plotting are also available with a ".CSV" output file working in real time, and these can be connected to the real-time plotting tools. The CSV file is used by the VR-HMD to update the HUD elements in real time.

Figure 5. HUD elements were presented on the screen of the FGFS to increase pilots' situational awareness.

2.3. VR Simulation

To accurately simulate the final goal of deploying a VR-based solution to monitor and pilot an airship in BVLOS scenarios, a VR application was developed using Unity software (Unity Technologies, San Francisco, CA, USA). Unity was chosen as the main VR engine due to its ease of use and multi-platform compatibility capabilities. Within the VR application, the user/pilot is able to view a POV screen showing the simulated view from a POV camera located on the airship. Overlaid on top of the POV screen and locked to the pilot's view will be the HUD, which will display all the necessary flight information the pilot will need to safely operate the airship. To give the pilot better awareness of the controls, a digital twin of the radio flight controller is also included within the VR environment. Not only is the position of the virtual radio flight controller tracked, but the orientation of the individual buttons, switches and control sticks is also tracked and displayed, thus reducing the disconnect the pilot might experience between their physical environment and the virtual environment while wearing the VR-HMD.

Figure 6 shows a simplified process flow of how the flight simulator graphics and data are handled and displayed on the wireless VR-HMD. The airship simulator receives flight input to control the simulated airship, and at the same time, an extra script records the flight data and controller data and exports them in a single CSV file in real time. The VR application reads the CSV file and updates the HUD elements as well as the controller elements. The desktop with the POV of the simulator is also captured and displayed in the VR environment. Using the flight data, controller data and desktop capture, the VR environment is rendered within the application. VR pixel streaming applications such as Virtual Desktop, ALVR and Air Link are used to wirelessly stream the rendered view of the VR application onto the VR-HMD. Each frame on the VR application is rendered based on the orientation of the VR-HMD in its physical space, providing more immersion to the pilot within the VR environment.

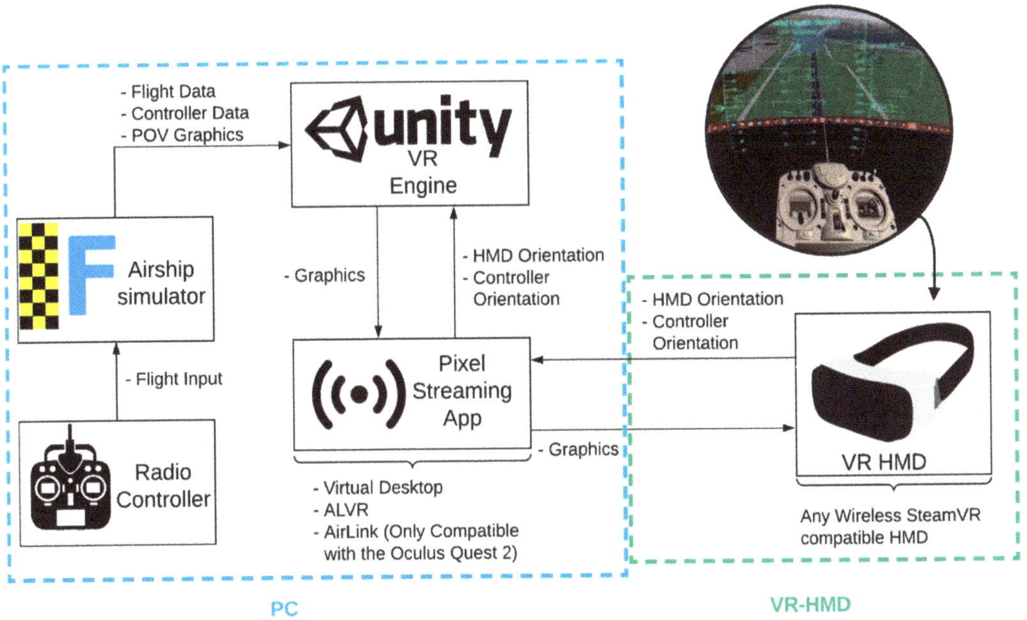

Figure 6. Simplified flowchart describing the process of displaying real-time flight information from a PC-based flight simulator on a wireless VR-HMD.

The resulting flight simulation described in Figure 6 is presented in Figure 7 with a view from the proposed VR-HMD. Using a live data export script written in Nasal within FlightGear (FlightGear developers & contributors, Burnsville, MI, USA), the live flight data, as well as the radio controller input values, were saved in a CSV file. A custom ReadCSV script within Unity was created to locate and open the data file exported by FlightGear. The ReadCSV script queries the last line within the CSV file on every frame and stores the values in an array variable simply named 'DataArray'. The order in which the values being read from FlightGear are placed within DataArray is determined by the order used to save the data.

Many of the variables within the ReadCSV script were made public to create the GUI seen in Figure 8. This GUI is used to set up information such as the file 'Path' to the data CSV file. Another important setup through the GUI in Figure 8 is pairing the data received from FlightGear to the different elements within Unity. For example, the 'Heading Text' variable seen in Figure 8 refers to a text renderer object in Figure 9 that displays the heading information in the HUD, which is locked to the users' view. The 'Heading Text' value changes based on the values within the DataArray variable. To let the application know which index within DataArray refers the heading of the aircraft, the 'DataName' and 'DataIndexInt' variables are set. In this example, the DataName is set to 'heading', and the DataIndexInt is set to 8, which is the index within the DataArray variable that corresponds to the heading of the aircraft. A similar setup can be seen for pitch, which pairs pitch to the index value of 2.

Figure 7. View from within the Oculus Quest running the simulator.

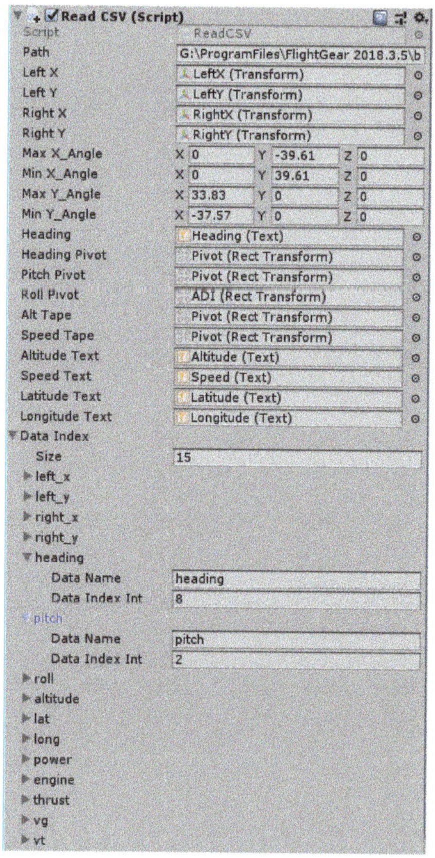

Figure 8. Data import GUI used to set up data from FlightGear in the simulation.

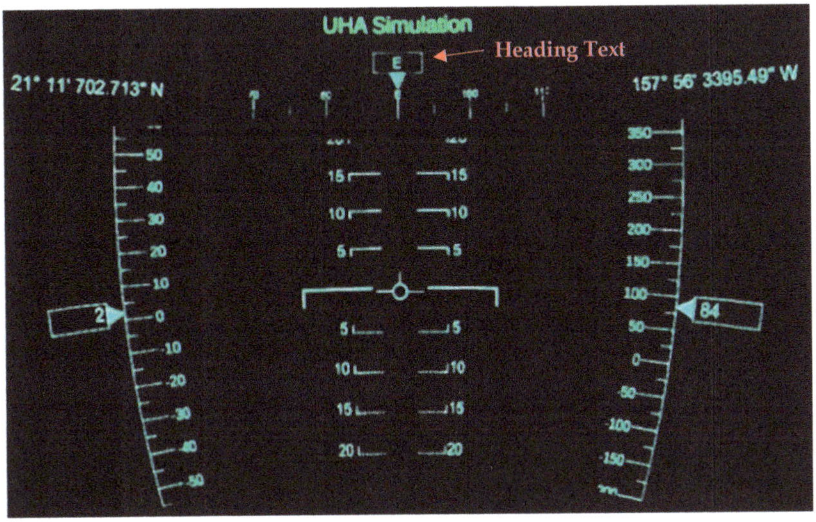

Figure 9. Close-up of the heading tape on the head-locked HUD.

Using this single GUI, all the data received from FlightGear are paired to the objects within the virtual environment. This pairing is what allows the animations on the controller and the values on the HUD to be updated.

The screen displaying the POV of the aircraft is created using a desktop duplication plugin for Unity called uDesktopDuplication. The plugin provides the ability to use the built-in capability of the Windows OS to capture the desktop and display it within the Unity environment. Using the plugin to display the desktop screen on a curved surface within the virtual environment provides more immersion to the user.

As described in Figure 5, a secondary third-party pixel streaming application is used to transmit the visuals from the PC to the wireless VR-HMD. In the case of the test shown in Figure 6, ALVR was the streaming application. Although any VR pixel streaming application can be used depending on personal preference, ALVR was used for this paper as it was free to use. VR pixel streamers such as ALVR stream visual data from the PC to the wireless VR-HMD, while streaming head and controller orientation as well as any inputs from the HMD to the PC. A common downside of using streaming applications such as ALVR and Virtual Desktop is that they have an inherent lag in streaming over Wi-Fi. The lag in streaming can be made worse with a poor internet connection and/or a resource-hungry application.

2.4. *VR Flight Simulator Test and Analysis*

After the successful completion of the proposed symbology design, the VR-HMD flight simulator presented in Figure 6 was utilized for human flight tests. To understand the effect of the primary flight display (PFD) symbology on pilots, a complete flight envelope from takeoff to the desired station altitude was simulated. For all test cases, Billy Bishop Toronto City Airport (CYTZ), situated in the city of Toronto, was used as the local terrain, and a suitable flight path was developed to test users on specific operations that affect the airship safety (CFIT, loss of control and collision with object).

The test procedure was defined as follows:

(1) Start the flight and take-off at CYTZ airport (Runway 08) on reaching ~10 knots and maintain a constant pitch of within 5 degrees.

(2) Reaching an altitude of 2000 ft, turn to the right and try to reach a roll rate of 3 deg/s and then turn left and try to reach a zero roll rate by maintaining a constant pitch of always less than 5 deg.

(3) Adjust the ballonet and helium volume and station for 10 s at a cruise altitude of ~5000 feet with airspeed below 15 knots.
(4) On crossing the cruise altitude of 7000 ft, turn to the left and try to reach a roll rate of 3 deg/s and then turn right and try to reach a zero roll rate by maintaining a constant pitch of always less than 5 deg.
(5) After crossing the altitude of 9000 ft, adjust the ballonet and helium volume and reach a climb rate of less than 20 ft per s to prepare for station-keeping at an altitude of 10,000 ft by aligning the airship.
(6) Reduce the speed and maintain the 10,000 ft altitude.
(7) Align the flight path angle with the other airships in the network and station-keep.

This flight plan tests the user's ability to avoid all unsafe conditions while maintaining proper flight parameters. Towards the end of the test, the user must follow the proper flight path angle and other parameters to keep the airship at the desired station-keeping altitude at the proper waypoint.

Table 3 presents different variables considered while recording each user's flight and the reason for choosing them. Each one of these parameters helps to study the impact of the proposed VR-HMD tools compared to the existing desktop flight simulators.

Table 3. Flight variables under consideration.

Variable (Unit)	Description
Time (s)	Helps in calculating the time taken by the user to accomplish the test phase.
Ground Airspeed (knots)	As speed plays an important factor throughout the flight phase, it is helpful to determine any unsafe conditions (loss of control due to a reduction in speed) that could lead the airship into an accident.
Altitude (ft)	Gives a realistic idea about the flight regime and the user's ability to fly the airship at proper altitude to avoid any unsafe conditions (collision with terrain).
Rate of Climb (ft/s)	It is important to understand the rate of change of height while performing each step of the proposed test procedure.
Roll Rate (deg/s)	Helps to evaluate the difference between the flight plan and the user's performance to calculate any significant difference that may affect the overall flight safety.
Pitch Angle (deg)	As an essential parameter that aids in providing stall information to the pilot, this is useful to evaluate pilot performance.

REB Questionnaire

(a) Mental Demand (Low/High): How much mental activity was required for thinking, deciding, calculating, and remembering or searching for certain information that is being presented?
(b) Physical Demand (Low/High): How much physical activity was required for pushing, pulling, turning, controlling or activating the controllers?
(c) Temporal Demand (Low/High): How much pressure does the operator feel due to the rate or pace at which the tasks or task elements occurred?
(d) Performance (Good/poor): How successful was the operator in accomplishing the goals of a given task set by the experimenter?
(e) Effort (Low/High): How hard did the operator have to work both mentally and physically to accomplish the level of performance?
(f) Frustration level (Low/High): How insecure, discouraged, irritated, stressed and annoyed versus secure, gratified, content, relaxed and complacent did the operator feel with the information presented?

The users were supposed to rate each individual component on a scale of 1 to 10, with 1 being the lowest and 10 being the highest.

3. Results and Discussion

The resulting flight simulation view of the proposed simulator described in Figure 4 is presented in Figure 10 for different POVs. As the VR-HMD flight simulator was developed in the environment of the FGFS, the user could choose, organize or modify the graphical modes using the available setup menus and camera views, such as from the control tower, from the cockpit and from different positions on the airship, as shown in Figure 10.

Figure 10. Different camera views within the flight simulator developed using the FGFS.

The technical goal of this study was to develop a system that could connect two different software packages and two different pieces of hardware in real time with minimal lag between data transmission, the two different software packages being a flight simulator with a VR-capable graphics engine. In this case, the flight simulator that was chosen was FlightGear and the VR-capable graphics engine was Unity. The two pieces of hardware that were connected were a PC and a wireless VR-HMD, in this case, the Oculus Quest. The connection point between FlightGear and Unity was a single CSV file that both applications used to share information. FlightGear was the source for stored data and Unity was used to read the stored data. The connection point for the PC and Oculus Quest was the pixel streaming application ALVR. ALVR transmitted graphical data to the Oculus Quest and transmitted positional and input data to Unity on the PC.

The connection points between the software and hardware are where the biggest issue in lag needs to be addressed. In the case of the CSV file, lag occurs during the writing and reading process. To successfully read or write to a file, it must first be opened, then written to or read from, and finally closed and repeated individually by each application. The speed at which the reading and writing processes occur depends on the language being used. Another limiting factor is the frequency at which FlightGear logs data in the CSV file. If the frequency is too high, it can place a significant load on the PC, which will affect the performance of any application running at the time. If the frequency is too low, then the data received by the user will be too outdated to use and the transmission cannot be considered real-time. A balance was found for FlightGear to write to a CSV at a period of

0.01 s. For Unity, the file was read every frame, which translates to around 0.0167 s (60 Hz). Since the speed of the aircraft is not very high, 0.01 s was considered an acceptable margin to work within at this stage.

For the connection point between the PC and Oculus Quest, the bottleneck occurs at the pixel streaming application and Wi-Fi connection. On average, the acceptable latency of data transmission between the PC and VR for any pixel streaming application is between 22 and 32 ms in ideal conditions. Ideal conditions are where a strong Wi-Fi connection for the VR-HMD, a LAN connection for the PC and a 5 GHz connection are available. Internet speeds do not affect the latency. In situations of high latency, the resolution of the graphics is affected, until no transmission occurs. Another effect of high latency is that the user will receive outdated visual information. This means the user will see information that has already occurred in the flight simulator and it will be too late to react to any situation. Another issue will be input lag. Input lag stands for the time it takes between when the user inputs a command and when the resulting action occurs. This would cause less than ideal flying conditions for the user.

A final global aspect affecting lag and latency is the hardware specifications of the PC and VR-HMD. The tests referred to in this paper were conducted on a desktop PC and a laptop. The desktop PC ran on an AMD Ryzen 9, 3900X CPU, dual Nvidia GeForce RTX 2080 GPU, 64 GB RAM and 1 Tb SSD. The laptop ran on a Core i7 (7th Gen) CPU, Nvidia GeForce GTX 1050 GPU, 32 GB RAM, and 1 TB HDD. Both devices were able to run FlightGear, Unity and ALVR/Virtual Desktop with no latency issues. Performance on differently configured PCs will give a better understanding of which configurations would be best for the proposed system.

The test subjects were tested on three different scenarios with 10 different users while following the same flight path. All participants in the test were amateur users with limited experience using VR headsets and flight simulation tools. In the first test case, the test subjects flew the airship with the desktop flight simulator. Secondly, the test subjects flew the airship via the VR-HMD flight simulator equipped with an internal HUD. Finally, the test subjects flew the airship via the VR-HMD flight simulator equipped with an internal HUD that tracked the remote controller motions along with test procedure projection on the screen of the virtual FrSky controller. As presented in Figure 11, the test procedure was also shown on the screen of the virtual FrSky controller and the VR screen to let users know about each step of the test well ahead of time. For all cases, the operator's feedback was recorded.

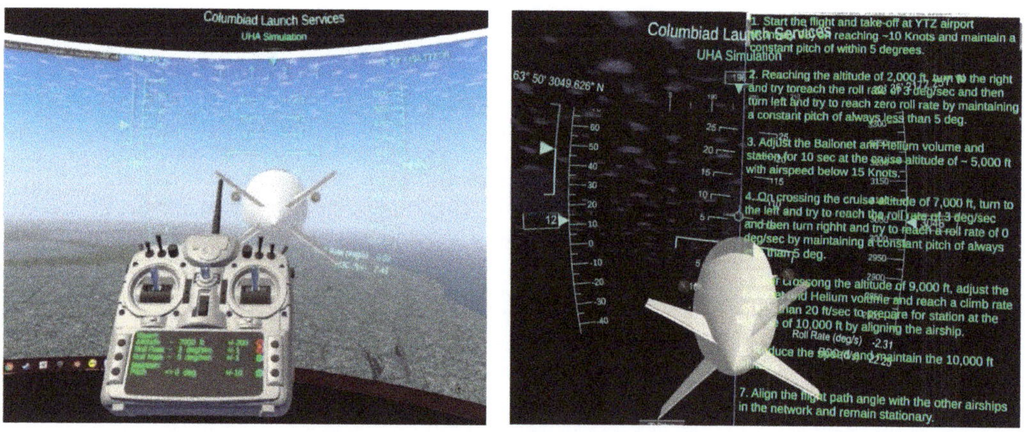

Figure 11. Test procedure that was shown on the virtual RC controller screen and VR screen to let users know about each step.

Upon completing the flight tests, participants' flight test results were recorded, and they were requested to fill out the NASA TLX rating scale forms along with their feedback. In the following, the results are tabulated based on their respective flight tests for further analysis. It should be noted that in the following, the results from the VR-HMD flight simulator equipped with an internal HUD are presented as VR-HMD 1 and the VR-HMD flight simulator equipped with the internal HUD that tracked the remote controller motions along with test procedure projection on the screen of the virtual FrSky controller is described as VR-HMD 2.

Figure 12 presents the altitude versus flight time for each participant. As can be seen, all participants were able to fly the airship at proper altitude and maintain flight plan requirements. However, the required time to accomplish the flight plan for the VR-HMD 1 simulator was higher than the desktop simulator and the VR-HMD 2 simulator. The end time of each flight scenario for pilots is distinguished in Figure 12 using a circle marker.

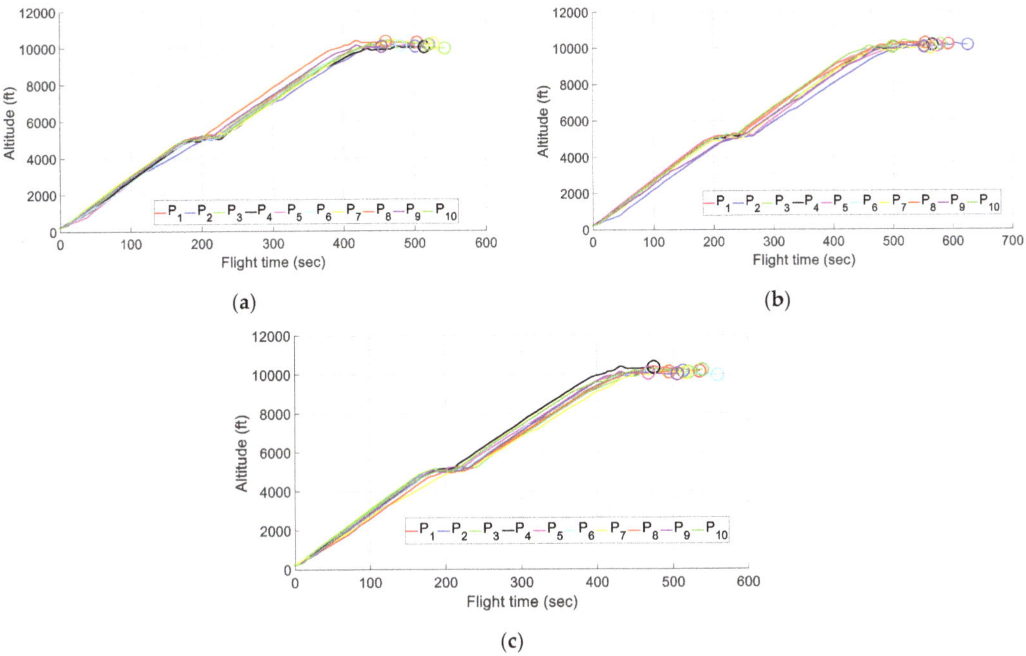

Figure 12. Recorded data for the altitude against the flight time for different pilots. (**a**) Desktop flight simulator; (**b**) VR-HMD 1 simulator; (**c**) VR-HMD 2 simulator.

Figure 13 presents the pitch angle versus flight time for each participant. As can be seen, users had better performance while operating the VR-HMD 2 simulator and the desktop simulator compared to the VR-HMD 1 simulator. For the case of the VR-HMD_1 simulator, the resulting pitch reached almost 30 degrees in some cases, which was much higher than the cases reported for the desktop flight simulator and the VR-HMD_2. Users were expected to maintain a value below 5 degrees for the pitch angle. However, during the second phase of the flight test plan and during the fourth phase, where the users were supposed to maintain the required roll rate, most of them struggled with maintaining the pitch requirements up to 5 degrees. The main reason for defining this scenario was to study the user behavior in multi-tasking scenarios and challenge their capabilities and mental/physical demands, which are presented in the next step.

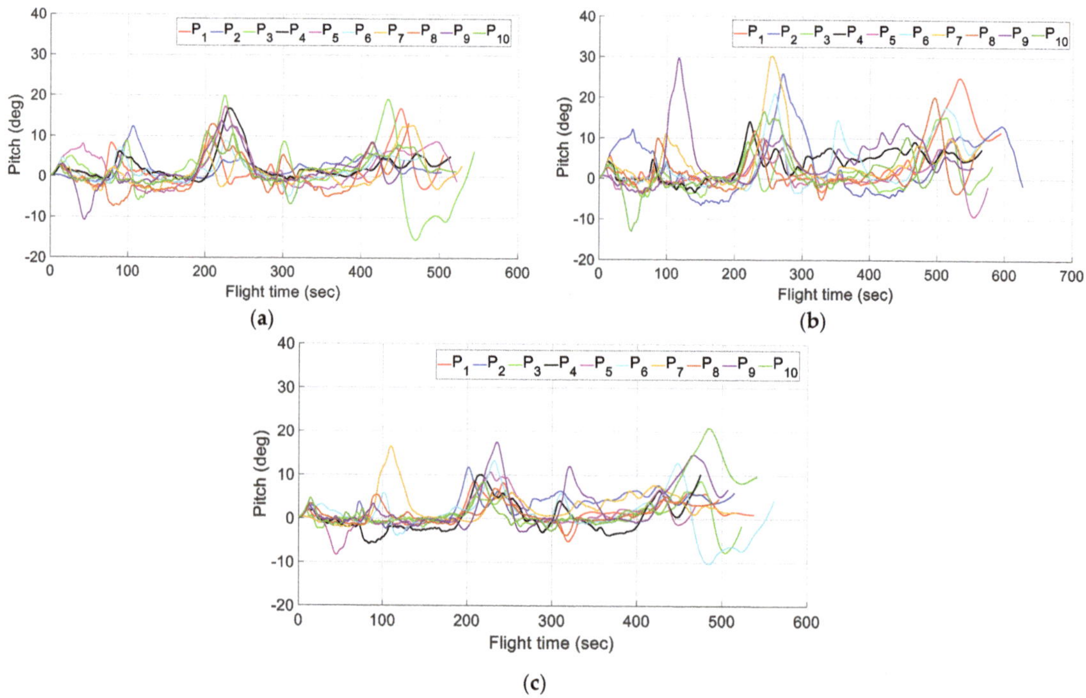

Figure 13. Recorded data for the pitch angle against the flight time for different pilots. (**a**) Desktop flight simulator; (**b**) VR-HMD 1 simulator; (**c**) VR-HMD 2 simulator.

Figure 14 presents the roll rate versus flight time for each participant. The users were supposed to reach a roll rate of 3 deg/sec at certain altitudes to calculate any significant difference that may affect the overall flight safety. As can be seen, almost all participants were able to maintain the flight plan requirements. However, as can be seen in Figure 13, not all of them were able to maintain the required pitch below 5 degrees at all times.

Figure 15 presents the ground airspeed versus flight time for each user. The participants were supposed to maintain different ground speed values at certain altitudes to complete the flight plan. The participants were supposed to remain below 15 knots, reaching an altitude of 15,000 ft, and, as can be seen in Figure 15, all pilots were able to maintain the required ground speed limit.

Finally, Figure 16 presents the rate of climb versus flight time for each user. The participants were supposed to maintain the desired altitudes at each phase of the flight plan while controlling the rate of height change as well. As can be seen, some pilots were struggling with maintaining the required rate of climb using VR-HMD_1, as a higher demand was required to complete the flight plan using this method. This is supported by the results reported in the following for the NASA TLX questionnaire, which identifies the pilots' available mental capacity.

The mental workload experienced by all the participants during all three flight tests is shown in Figure 17. The mental workload seems to be high among participants using the VR-HMD 1 simulator compared to the desktop simulator and the VR-HMD 2 simulator. Additionally, the HOD decreases the mental workload in most of the participants. The main reason for the reduction in mental workload could be due to the test procedure projection on the screen of the virtual FrSky controller, which helps the user have a step-by-step guideline. The dotted lines show the average values for each method.

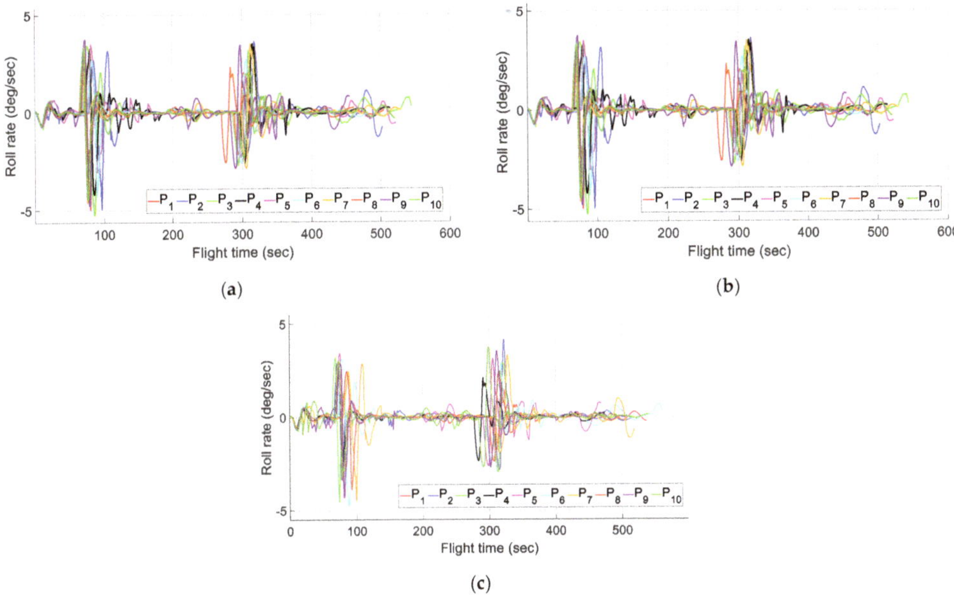

Figure 14. Recorded data for the roll rate against the flight time for different pilots. (**a**) Desktop flight simulator; (**b**) VR-HMD 1 simulator; (**c**) VR-HMD 2 simulator.

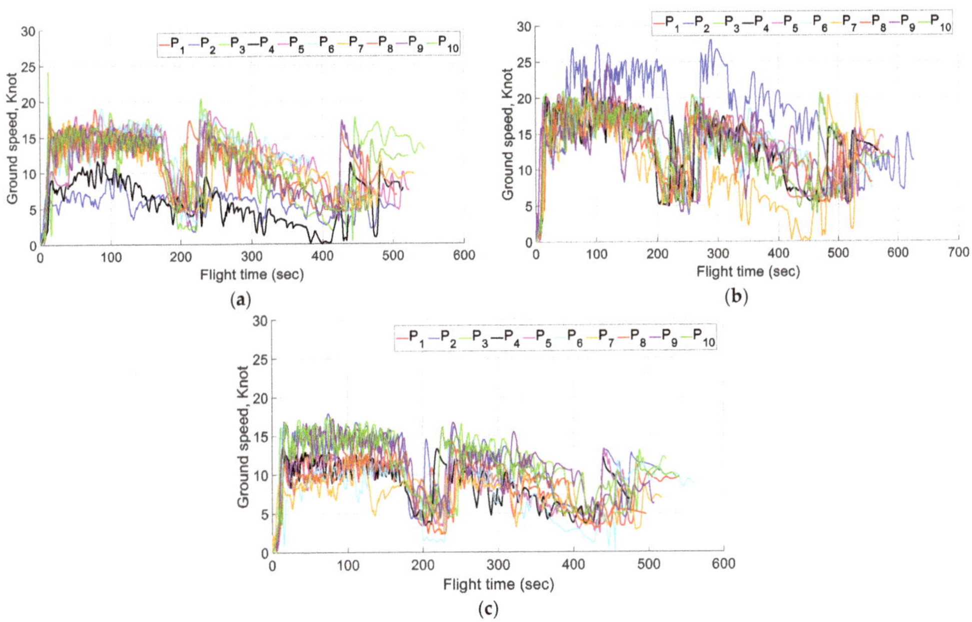

Figure 15. Recorded data for the ground airspeed against the flight time for different pilots. (**a**) Desktop flight simulator; (**b**) VR-HMD 1 simulator; (**c**) VR-HMD 2 simulator.

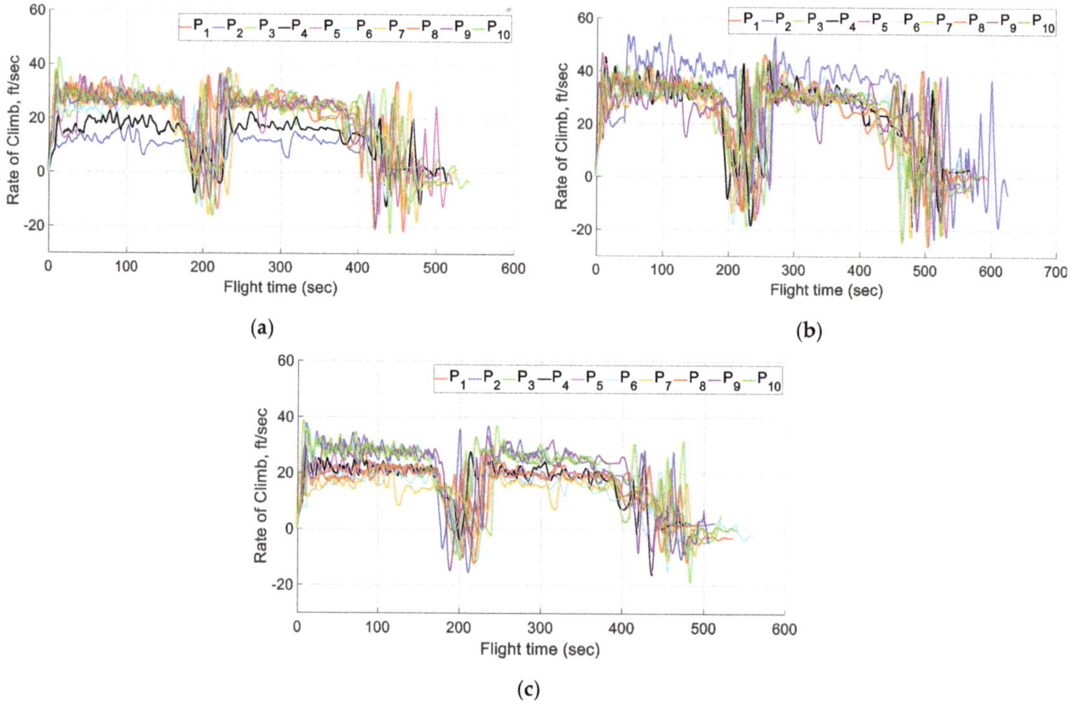

Figure 16. Recorded data for the rate of climb against the flight time for different pilots. (**a**) Desktop flight simulator; (**b**) VR-HMD 1 simulator; (**c**) VR-HMD 2 simulator.

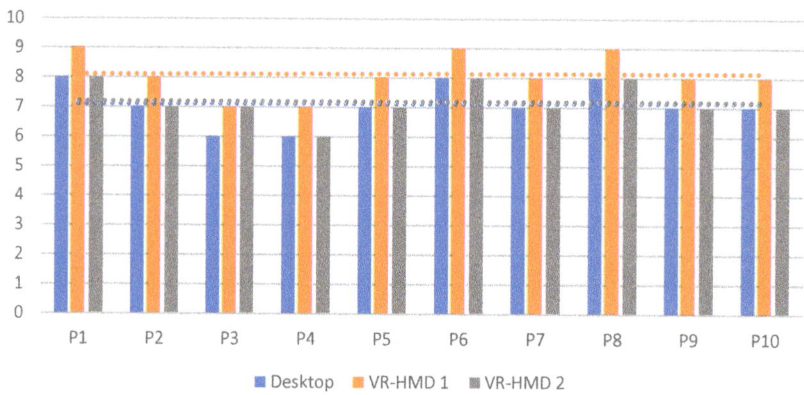

Figure 17. Mental demand experienced by all the participants using the three test procedures.

Figure 18 presents the physical demand experienced by all the participants during the three flight tests. Similar to mental workload, the physical demand seems to be highest among participants using the VR-HMD 1 simulator compared to the desktop and VR-HMD 2 simulators. The dotted lines present the average values for each method.

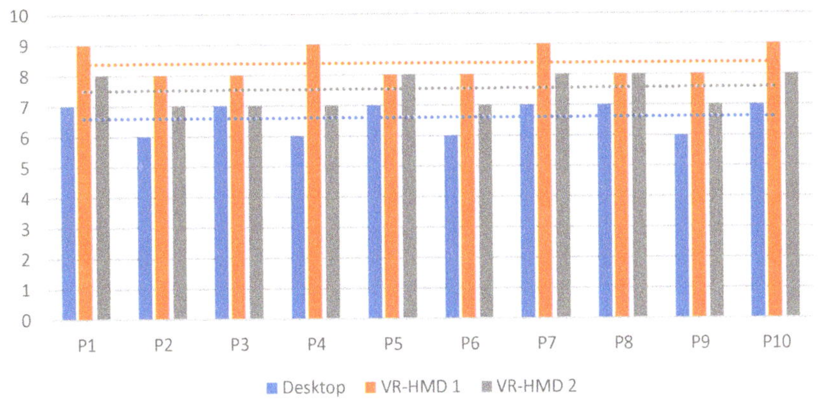

Figure 18. Physical demand experienced by all the participants using the three test procedures.

Figure 19 depicts the overall temporal demand experienced by the participants during the three flight tests. The flight plan was carefully designed not to exert any sense of rush on the participants. Additionally, they were reminded to complete the tasks in their own time during the pre-flight briefing. The average amount of rush experienced by the participants using the desktop and VR-HMD 1 simulators was higher than that for the VR-HMD 2 simulator. The dotted lines show the average values for each method.

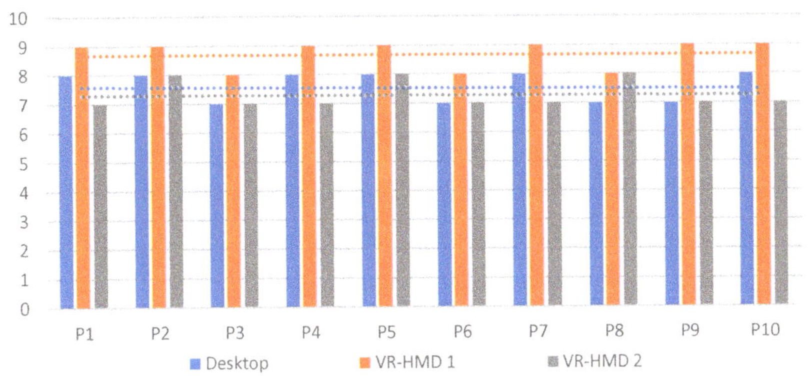

Figure 19. Temporal demand experienced by all the participants using the three test procedures.

In the next step, the participants were asked to rate their performance on a scale of 1 to 10. Their performance rating could depend on their ability to maneuver the airship properly, follow the flight path and station-keep. As can be seen in Figure 20, the participants performed well when using the VR-HMD 2 and desktop simulators compared to the VR-HMD 1 simulator.

Figure 21 presents the amount of effort required to complete the flight test using the different methods. As expected, the effort needed to accomplish the flight test procedure using the desktop and VR-HMD 2 simulators was less than that for the VR-HMD 1 simulator. The dotted lines present the average values for each method.

Finally, the participants were asked to rank the amount of frustration experienced while using all three different methods, and the results are graphed in Figure 22. The participants reported the lowest frustration values for VR-HMD 2 compared to the other two methods. The dotted lines show the average values for each method.

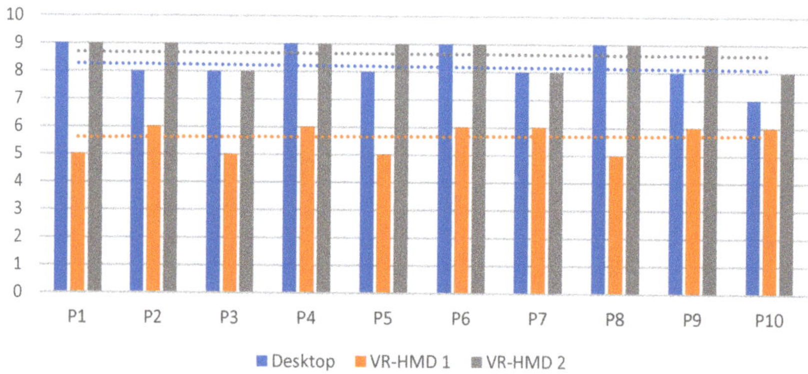

Figure 20. Performance of all the participants using the three test procedures.

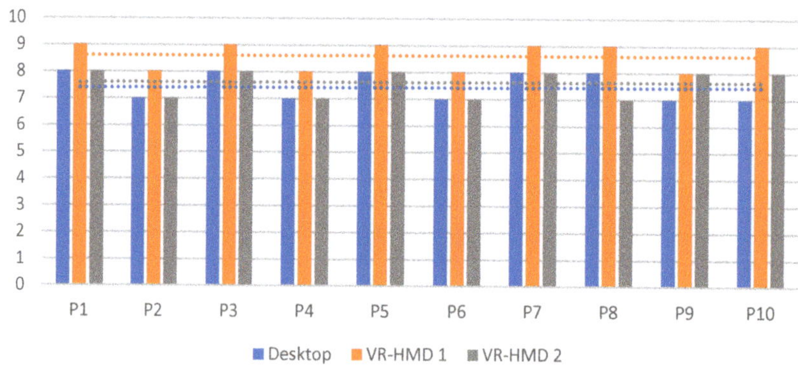

Figure 21. Effort needed by all the participants using the three test procedures.

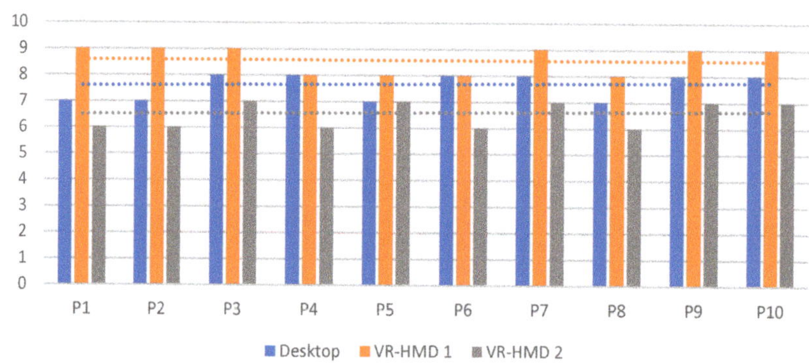

Figure 22. Frustration experienced by all the participants using the three test procedures.

Further Developments

The proposed VR-HMD flight simulator can also be seen as a GCS system, and as shown in Figure 23, it is conceived to fully define the GCS concept and transmit flight information as well as POV visuals in real time via the actual images broadcast by the onboard video cameras. Accordingly, the same concept developed for the VR-HMD simulator may be used for the VR-HMD GCS. A 360-degree camera (Insta360, Irvine, CA, USA) of some sort, as shown in Figure 23a, to allow a VR-oriented view for the GCS may be integrated with the Raspberry PI (Raspberry Pi Foundation, Cambridge, England,

UK) and Pixhawk (Pixhawk by Auterion, Zürich, Switzerland) shown in Figure 23e using the available 4G/5G network. The VR-HMD simulator allows pilots to accomplish a part of real flight tests with the same data transmission techniques used in the simulated environment. In addition, the development of the VR-HMD GCS may be enhanced by using an AR headset to visualize augmented flight using a first-person POV or a third-person POV such as the one shown in Figure 7 with the use of a graphical model of the actual airship. Furthermore, haptics and gloves may be integrated with the VR/AR-HMD GCS to increase situational awareness of the pilot in BLOS scenarios. Our current focus is to use the latest technology in the field to enhance the command-and-control station systems and move towards the goal of transition from simulated training to actual flight with a single unified system.

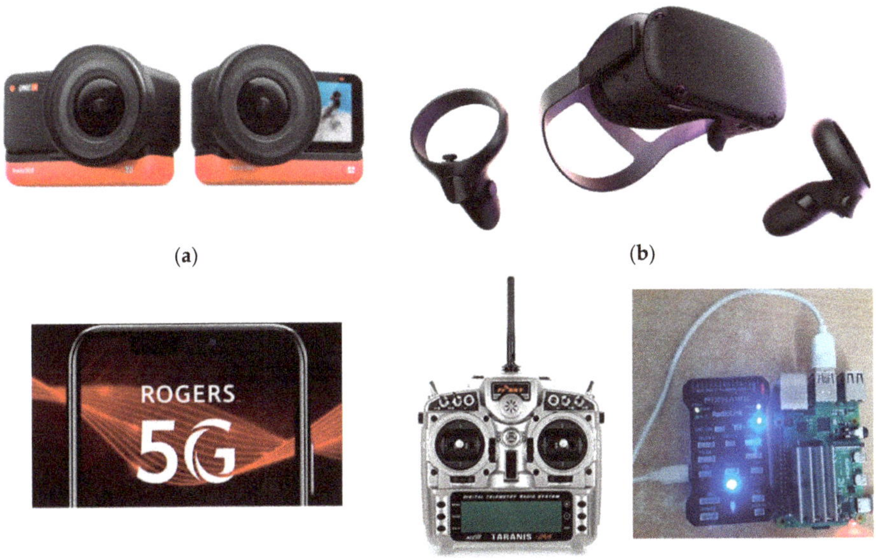

(a) (b) (c) (d) (e)

Figure 23. Connecting the VR headset (**b**) and radio controller (**d**) with an on-board 360-degree camera (**a**) and Raspberry PI (**e**) to obtain the flight data and display them on the VR screen in real time using the 4G/5G network (**c**).

Utilizing haptic sense in the proposed flight simulation tool can enhance the current work. The term "haptics" was proposed by Revesz in 1950, after observing blind performance and referring to an unconventional experience rather than traditional methods of touch and kinesthesis. More specifically, this term means "active touch" rather than passive touching [33,34]. For the next step, we want to incorporate haptic gloves to control the aerial vehicle with the use of a virtual controller defined within the virtual environment. The stand-alone VR headset will be connected to the haptics and the flight simulation tool. The VR headset will be used to visualize basic flight simulation while the vehicle could be controlled via the haptic gloves and a virtual controller defined within the virtual environment. Figure 24 shows the initial haptic-based VR flight simulator where a pair of SenseGlove Nova Gloves (SenseGlove, Delft, The Netherlands) were connected with a VR Oculus Quest and its controller to control the airship via a virtual roll/pitch/yaw controller and a throttle. Initial results showed a significant decrease in the amount of physical and mental demand required by the users during flight.

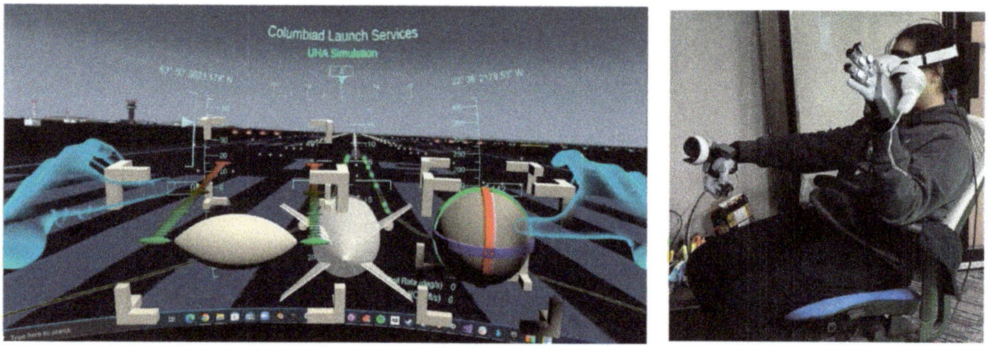

Figure 24. View from within the Oculus Quest VR running the flight simulator and the actual pilot wearing the SenseGlove Nova set for command and control.

4. Conclusions

This paper describes a system that can connect a flight simulator to a wireless VR-HMD and transmit simulated flight information as well as POV visuals in real time. The purpose of such a system was to test the flight of stratospheric airships in BLOS scenarios. In this work, a desktop flight simulator was developed using the scaled-down geometry of an airship designed for stratospheric applications along with the corresponding aerodynamics and other characteristics of the aircraft in the FlightGear flight simulator. Moreover, the control surfaces were connected with a FrSky Taranis X9 radio transmitter. The VR-HMD streamed the flight environment and enhanced the design procedure of the stratospheric airship via the simulation tool. Our results showed that the VR-HMD flight simulator equipped with the internal HUD that tracked the remote controller motions along with the test procedure projection on the screen of the digital twin of the FrSky controller showed better performance, less effort, less temporal demand and less frustration compared to the desktop flight simulator. The current iteration of the system is ideal to train pilots using similar systems in a safe and immersive environment. The proposed VR-HMD flight simulator may also be seen as a GCS prototype, and it was conceived to define the GCS concept and transmit flight information as well as POV visuals in real time via the real images broadcast using an onboard camera. Additionally, incorporating haptic sense into the proposed flight simulation tool may enhance the current study to lower the physical and mental demand during flight.

Author Contributions: Conceptualization, M.R. and J.C.; Methodology, M.R., J.K., P.P. and J.C.; Software, M.R. and J.K.; Formal analysis, M.R. and J.K.; Writing—original draft, M.R.; Preparation, M.R.; Data curation, J.K.; Writing—review and editing, J.K., P.P. and J.C. All authors have read and agreed to the published version of the manuscript.

Funding: This work was funded by Mitacs through the Mitacs Accelerate Program (reference number: IT16648) and Columbiad Launch Services Inc.

Data Availability Statement: Not applicable.

Conflicts of Interest: The authors declare no conflict of interest.

Nomenclature

Roman Symbols

$\bar{A}_1$	Coefficient in denominator of the transfer function
a	Distance from aerodynamic center to the center of gravity
$\bar{B}_1$	Coefficient in denominator of the transfer function
b	Distance from the aerodynamic center to the center of buoyancy

33. Robles-De-La-Torre, G. The importance of the sense of touch in virtual and real environments. *IEEE Multimed.* **2016**, *13*, 24–34. [CrossRef]
34. Srinivasan, M.A. Haptic Interfaces, In Virtual Reality: Scientific and Technical Challenges. In *Report of the Committee on Virtual Reality Research and Development*; The National Academies Press, Computer Science and Telecommunications Board: Washington, DC, USA, 1995.

MDPI AG
Grosspeteranlage 5
4052 Basel
Switzerland
Tel.: +41 61 683 77 34

Aerospace Editorial Office
E-mail: aerospace@mdpi.com
www.mdpi.com/journal/aerospace

www.ingramcontent.com/pod-product-compliance
Lightning Source LLC
Chambersburg PA
CBHW061812210926
51850CB00006B/681
9783725846757